ANALYTICAL ATOMIC SPECTROSCOPY

MODERN ANALYTICAL CHEMISTRY

Series Editor: **David Hercules**
University of Georgia

ANALYTICAL ATOMIC SPECTROSCOPY
By William G. Schrenk • 1975

PHOTOELECTRON AND AUGER SPECTROSCOPY
By Thomas A. Carlson • 1975

In preparation

MODERN FLUORESCENCE SPECTROSCOPY
Edited by Earl L. Wehry

A Continuation Order Plan is available for this series. A continuation order will bring delivery of each new volume immediately upon publication. Volumes are billed only upon actual shipment. For further information please contact the publisher.

ANALYTICAL ATOMIC SPECTROSCOPY

William G. Schrenk
Kansas State University
Manhattan, Kansas

PLENUM PRESS • NEW YORK AND LONDON

Library of Congress Cataloging in Publication Data

Schrenk, William G 1910-
  Analytical atomic spectroscopy.

  (Modern analytical chemistry)
  Includes bibliographies and index.
  1. Atomic spectra. I. Title.
  QD96.A8S37   543'.085   75-22102
  ISBN 0-306-33902-1

Lanchester Polytechnic Library

© 1975 Plenum Press, New York
A Division of Plenum Publishing Corporation
227 West 17th Street, New York, N.Y. 10011

United Kingdom edition published by Plenum Press, London
A Division of Plenum Publishing Company, Ltd.
Davis House (4th Floor), 8 Scrubs Lane, Harlesden, London, NW10 6SE, England

All rights reserved

No part of this book may be reproduced, stored in a retrieval system, or transmitted,
in any form or by any means, electronic, mechanical, photocopying, microfilming,
recording, or otherwise, without written permission from the Publisher

Printed in the United States of America

*To
the women in my life
Bertha, Ellen,
Sara, and Sue*

Preface

This textbook is an outgrowth of the author's experience in teaching a course, primarily to graduate students in chemistry, that included the subject matter presented in this book. The increasing use and importance of atomic spectroscopy as an analytical tool are quite evident to anyone involved in elemental analysis. A number of books are available that may be considered treatises in the various fields that use atomic spectra for analytical purposes. These include areas such as arc–spark emission spectroscopy, flame emission spectroscopy, and atomic absorption spectroscopy. Other books are available that can be catalogued as "methods" books. Most of these books serve well the purpose for which they were written but are not well adapted to serve as basic textbooks in their fields.

This book is intended to fill the aforementioned gap and to present the basic principles and instrumentation involved in analytical atomic spectroscopy. To meet this objective, the book includes an elementary treatment of the origin of atomic spectra, the instrumentation and accessory equipment used in atomic spectroscopy, and the principles involved in arc–spark emission, flame emission, atomic absorption, and atomic fluorescence.

The chapters in the book that deal with the methods of atomic spectroscopy discuss such things as the basic principles involved in the method, the instrumentation requirements, variations of instrumentation, advantages and disadvantages of the method, problems of interferences, detection limits, the collection and processing of the data, and possible applications. Since the book is intended to serve as a textbook, principles are stressed. Detailed methods of analysis for specific elements are not included. It is the hope of the author, however, that the presentation of basic information is sufficiently detailed so the students can develop their own methods of analysis as needed.

Included in the textbook are several appendixes that should be valuable to the student as well as to atomic spectroscopists. These include a compilation of frequently used spectroscopic terms and units, tables of sensitive spectral lines of 70 elements abridged from National Bureau of Standards publications, absorbance values and the Seidel function calculated from percentage transmittances, and a table of elemental detection limits for flame emission and atomic absorption spectroscopy.

The author wishes to acknowledge the assistance of a number of people in the preparation of this book. Dr. W. G. Fateley, Head, Department of Chemistry, Kansas State University, was most encouraging, cooperative, and helpful. Dr. Clyde Frank, The University of Iowa, was of great assistance in planning the text and in reading the manuscript. Dr. Lorin Neufield, Tabor College, Hillsboro, Kansas, Dr. S. E. Valente, Regis College, Denver, Colorado, Dr. Delbert Marshall, Ft. Hays Kansas State College, and Dr. Ron Popham, Chairman, Department of Chemistry, Southeast Missouri State College, Cape Giraudeau, Missouri, all have read and aided in the revision of certain chapters. All those mentioned above, including Dr. Frank, are former graduate students of the author.

Mr. Allan Childs, a graduate student in chemistry, aided in preparing a number of the spectral plates used as illustrations in the text and Mr. Truman Waugh of the Kansas Geological Survey, Lawrence, Kansas, was very helpful with the preparation of Appendix IV, which shows wavelength positions of some 50 elements together with a reference iron spectrum.

Ms. Diane Janke and Ms. Jane Adams were the most cooperative of typists and Steven Wallace made the line drawings. Their help is most appreciated.

Contents

Chapter 1
Historical Introduction 1

   1. Early Developments 1
   2. The Newtonian Era 2
   3. The Early 1800's (to Kirchhoff and Bunsen) 3
   4. The Later 1800's 4
   5. Arc and Spark Excitation 6
   6. Flame Emission Spectroscopy 6
   7. Atomic Absorption Spectroscopy 7
   8. Atomic Fluorescence Spectroscopy 9
      Selected Reading 10

Chapter 2
The Origin of Atomic Spectra 11

   1. The Nature of Electromagnetic Radiation 11
   2. Early Concepts 13
   3. The Balmer Equation 14
   4. From Balmer to Bohr 16
      4.1. Spectral Line Series for Hydrogen 21
      4.2. Energy Level Diagrams 21
   5. Modifications of the Bohr Theory 21
      5.1. Selection Rules for n and $k(l)$ 24
      5.2. Atoms with Two Valence Electrons 25
      5.3. Selection Rules and the Schrödinger Equation . . 26
   6. Alkali Metal Atom Spectra 28
      6.1. Doublet Structure of Alkali Metal Spectra . . . 28
      6.2. Electron Spin 29
   7. Alkaline Earth Atomic Spectra 30

8.	Spectral Series and Spectroscopic Term Symbols	31
9.	Zeeman and Stark Effects	31
10.	Spectral Line Intensities	33
	10.1. Statistical Weight	33
	10.2. The Boltzmann Distribution Factor	34
11.	Transition Probabilities—Oscillator Strengths	35
12.	Spectral Linewidths	37
13.	Atomic Fluorescence	38
14.	Metastable States—Laser Action	39
	14.1. Laser Action	39
15.	Molecular Spectra (Band Spectra)	41
	Selected Reading	45

Chapter 3
Filters, Prisms, Gratings, and Lenses ... 47

1.	Filters	47
	1.1. Absorption Filters	47
	1.2. Interference Filters	49
	1.3. Circular Variable Filters	51
2.	Prisms	52
	2.1. Dispersion of a Prism	53
	2.2. Resolving Power of a Prism	55
	2.3. Prism Materials	57
	2.4. Types of Prisms	58
3.	Interferometers	59
4.	Diffraction Gratings	60
	4.1. Dispersion of a Grating	63
	4.2. Resolving Power of a Grating	63
	4.3. Production and Characteristics of Gratings	65
	4.4. Grating Replicas	66
	4.5. Concave Gratings	66
	4.6. Holographic Gratings	67
	4.7. Echelle Gratings	67
5.	Lenses	70
	5.1. Uses of Lenses	70
	5.2. Lens Defects	72
	Selected Reading	74

Chapter 4
Spectrometers ... 75

1.	Prism Spectrometers	75
	1.1. The Cornu Prism Spectrometer	76

CONTENTS   xi

       1.2. The Littrow Spectrometer 76
  2. Plane Grating Spectrometers 78
       2.1. The Ebert Spectrometer 78
       2.2. The Czerny–Turner Spectrometer 81
       2.3. The Two-Mirror, Crossed-Beam, Plane Grating Spectrometer 82
       2.4. The Double-Grating Spectrometer 83
  3. Concave Grating Spectrometers 83
       3.1. The Rowland Spectrometer 84
       3.2. The Paschen–Runge Spectrometer 85
       3.3. The Eagle Spectrometer 86
       3.4. The Wadsworth Spectrometer 88
       3.5. The Grazing Incidence Spectrometer 89
       3.6. The Seya–Namioka Spectrometer 90
       3.7. Vacuum Spectrometers 90
  4. Direct Reading Spectrometers 92
  5. Selection of a Spectrometer 93
  6. Adjustment and Care of Spectrometers 94
       6.1. Vertical Adjustment of the Entrance Slit . . . 94
       6.2. Focusing the Entrance Slit 95
           6.2.1. Prism Instruments 95
           6.2.2. Grating Instruments 95
       6.3. Final Adjustments 95
       6.4. General Care of Spectrometers 96
     Selected Reading 97

Chapter 5
Accessory Equipment for Arc and Spark Spectrochemical Analysis 99

  1. The Spectrometer Slit 99
  2. The Hartmann Diaphragm 101
  3. The Step Filter 101
  4. Rotating Sectors 103
  5. Excitation Sources 103
       5.1. The Direct Current Arc 104
       5.2. The Alternating Current Arc 105
       5.3. The Electric Spark 106
       5.4. The Plasma Arc 108
       5.5. The Laser Source 112
       5.6. Multiple Source Units 113
  6. Arc and Spark Stands 114

6.1.	Special Assemblies for the Arc–Spark Stand	116
	6.1.1. The Stallwood Jet	116
	6.1.2. The Petry Stand	117
	6.1.3. Rotating Disk Electrode Device	117
7.	Order Sorters	118
8.	Densitometers and Comparators	118
9.	Miscellaneous Accessory Equipment	121
	9.1. Electrodes	122
	Selected Reading	123

Chapter 6
Recording and Reading Spectra 125

1.	The Photographic Process	125
	1.1. Characteristics and Properties of the Photographic Emulsion	126
	1.2. The Characteristic Curve	126
	1.3. The Reciprocity Law	129
	1.4. The Intermittency Effect	130
	1.5. The Eberhard Effect	130
	1.6. Graininess and Granularity	131
	1.7. Resolving Power	132
	1.8. Spectral Sensitivity	134
2.	Processing of Spectroscopic Films and Plates	135
	2.1. The Developing Process	135
3.	Hadamard Transform and Fourier Transform Spectroscopy	139
4.	Light-Sensitive Phototubes	140
	4.1. Spectral Response Designation	140
	4.2. General Characteristics of Multiplier Phototubes	141
	4.3. Solar Blind Phototubes	143
5.	Resonance Detectors	144
6.	Vidicon Detectors	145
	Selected Reading	146

Chapter 7
Qualitative and Semiquantitative Arc–Spark Emission Spectrochemical Analysis 147

1.	Sample Excitation	148
2.	Wavelength Measurements	149
	2.1. Line Identification by Wavelength Measurement	149
3.	Comparison Spectra	151

CONTENTS

   4. Spectral Charts 155
   5. Wavelength Tables 156
   6. Some Special Problems and Techniques of Spectrochemical Qualitative Analysis 157
      6.1. Spectral Line Interferences 157
      6.2. Spectral Band Interferences 157
      6.3. Arc Continuum Interference 158
   7. Increasing Spectral Line Intensities 158
   8. Semiquantitative Spectrochemical Analysis 159
      8.1. Determination of a Concentration Level 160
      8.2. The Harvey Method of Semiquantitative Spectrochemical Analysis 160
      8.3. Matrix Effects 164
      8.4. The Wang Method of Semiquantitative Spectrochemical Analysis 165
   9. Some Special Spectrochemical Problems 166
      9.1. Microsamples 166
      9.2. Microarea Sampling 166
   Selected Reading 167

Chapter 8
Quantitative Spectrochemical Analysis 169

   1. Some General Considerations 169
   2. The Internal Standard 170
   3. Spectroscopic Buffers 172
   4. Excitation of the Sample 175
   5. Selection of Spectral Lines 177
   6. Comparison Standards 177
   7. Sample Preparation 179
   8. Emulsion Calibration and Analytical Working Curves 180
      8.1. Emulsion Calibration 180
      8.2. The Emulsion Calibration Curve 185
   9. The Working Curve 188
      9.1. Construction of a Typical Analytical Working Curve 188
  10. The Calculating Board 192
  11. Background Correction 193
  12. Multielement Analysis with Direct Read-Out . . . 194
  13. Types of Samples 195
      13.1. Liquid Samples 195
      13.2. Metallic Samples 196
      13.3. Powder Samples 197

		13.4.	Organic Samples	197
		13.5.	Special Samples	198
	14.	Some Special Techniques		198
		14.1.	Fractional Distillation	198
		14.2.	Carrier Distillation	199
		14.3.	Transfer Methods	200
		14.4	Laser Methods	200
		14.5.	Controlled Atmospheres	200
		14.6.	Cathode Layer Excitation	201
		14.7.	Gases	202
		14.8.	Radioactive Samples	202
	15.	Time-Resolved Spectroscopy		203
		15.1.	Time-Resolving Components	203
		15.2.	Some Characteristics of Time-Resolved Spectra	204
		15.3.	Analytical Applications	205
	16.	Chemical Preparation of Samples		205
	17.	Applications of Spectrochemical Analysis		206
		17.1.	Metals and Alloys	206
		17.2.	Geology	207
		17.3.	Oils and Water	207
		17.4.	Plants and Soils	208
		17.5.	Men and Animals	209
		17.6.	Environmental Studies	209
		17.7.	Some Other Applications	209
		Selected Reading		210

Chapter 9
Flame Emission Spectroscopy. 211

	1.	Flame Emission Instrumentation Requirements		212
	2.	The Analytical Flame		212
		2.1.	Burners and Aspirators	216
		2.2.	Fuel–Oxidant Control	218
	3.	The Excitation Process in the Flame		219
		3.1.	Flame Emission Spectra	220
	4.	Flame Emission Interferences		222
		4.1.	Spectral	222
		4.2.	Ionization	224
		4.3.	Cation–Anion Interferences	227
		4.4.	Cation–Cation Interferences	228
		4.5.	Oxide Formation	228
		4.6.	Chemiluminescence	228

4.7.	Physical Interferences	229
5.	Control of Interferences	231
5.1.	Spectral	231
5.2.	Ionization Interference Control.	233
5.3.	Cation–Anion Interference Control	234
5.4.	Control of Oxide Interference	234
5.5.	Control of Physical Interference	235
6.	Simultaneous Multielement Analysis	235
7.	Analytical Treatment of Data	237
7.1.	Establishment of a Working Curve	237
7.2.	Background Correction	239
7.3.	Sample Bracketing	240
7.4.	The Method of Standard Additions	241
	Selected Reading	242

Chapter 10
Analytical Atomic Absorption Spectroscopy   243

1.	The Atomic Absorption Process	243
2.	Instrumentation Requirements	247
3.	Radiation Sources	248
3.1.	Hollow Cathode Lamps	249
3.1.1.	High-Intensity Lamps	251
3.1.2.	Multiple-Element Lamps	251
3.1.3.	Demountable Lamps	253
3.2.	Gaseous Discharge Lamps	253
3.3.	Electrodeless Discharge Lamps	254
3.4.	Flame Emission Sources	258
3.5.	Continuous Sources	258
4.	Production of the Atomic Vapor	259
4.1.	Nebulization of the Sample	259
4.1.1.	Ultrasonic Nebulization	261
4.2.	Flame Systems	262
5.	Fuels and Oxidants	264
5.1.	Atomic Distribution in Flames	264
6.	Non-Flame Absorption Cells	268
6.1.	Hollow Cathodes	268
6.2.	L'vov Furnace	269
6.3.	Woodriff Furnace	269
6.4.	Delves Cup	271
6.5.	Carbon Rod Analyzers	272
6.6.	Tantalum Boat Analyzer	273

	6.7.	Other Non-Flame Cells	275
	6.8.	Special Systems	278
7.	Monochromators	280
8.	Detectors	281
	8.1.	Resonance Detection	282
9.	Amplifiers	283
10.	Read-Out Devices	284
11.	Interferences in Atomic Absorption	285
	11.1.	Spectral Interferences	285
	11.2.	Ionization Interferences	286
	11.3.	Chemical Interferences	287
	11.4.	Interferences with Flameless Sampling	288
12.	Control of Interferences	289
	12.1.	Spectral Interference Control	289
	12.2.	Ionization Interference Control	290
	12.3.	Chemical Interference Control	290
		12.3.1. Flame Temperature	290
		12.3.2. Fuel-to-Oxidant Ratio	291
		12.3.3. Flame Region	291
		12.3.4. Releasing and Chelating Agents	291
		12.3.5. Chemical Separations	292
		12.3.6. Background Correction	292
13.	Analytical Treatment of Data	294
	13.1.	The Working Curve	295
	13.2.	Analytical Procedures	297
14.	Simultaneous Multielement Analysis	297
	Selected Reading	298

Chapter 11
Atomic Fluorescence Spectroscopy 299

1. Theoretical Basis of Analytical Atomic Fluorescence Spectroscopy 299
2. Advantages and Limitations of Atomic Fluorescence . . 302
3. Instrumentation 303
   3.1. Excitation Sources 304
      3.1.1 Hollow Cathode Lamps 304
      3.1.2. Metal Vapor Lamps 304
      3.1.3. Electrodeless Discharge Lamps . . . 305
      3.1.4. Continuous Sources 305
      3.1.5. Laser Sources 306
4. The Sample Cell 307

CONTENTS

    4.1. Total-Consumption Aspirator Burners 307
    4.2. Laminar Flow Burners 307
    4.3. Non-Flame Sample Cells 308
5. Monochromators 308
6. Interferences in Atomic Fluorescence 309
    6.1. Spectral Interferences 309
    6.2. Chemical Interferences 310
    6.3. Physical Interferences 310
7. Analytical Procedures 312
    7.1. The Analytical Working Curve 313
    7.2. Organic Solvents 314
    7.3. Detection Limits 315
    7.4. Sample Preparation 315
8. Applications and Future Developments 317
   Selected Reading 318

Appendix   I. Some Basic Definitions, Physical Constants, Units, and Conversion Factors 319
Appendix   II. Spectral Lines, Arranged by Wavelength, with gf and Intensity Values 321
Appendix III. Spectral Lines, Arranged by Elements, with gf and Intensity Values 337
Appendix IV. Spectral Charts 347
Appendix   V. Absorbance Values Calculated from Percentage Transmittances 353
Appendix VI. Numerical Values of the Seidel Function . . . 355
Appendix VII. Four-Place Logarithm Table 357
Appendix VIII. Detection Limits by Flame Emission and Atomic Absorption 361
Appendix IX. Periodic Table of the Elements 365
Appendix   X. Relative Atomic Weights 367

Author Index 369

Subject Index 371

Chapter 1
Historical Introduction

The use of atomic spectroscopy in chemical analysis requires, in addition to knowledge concerning the origin and production of spectra, a knowledge of the intrumentation required to produce, record, and measure the wavelengths of spectral lines and their intensities. These requirements therefore make necessary a knowledge of (1) the basic laws of optics, including characteristics of lenses and mirrors, (2) spectral dispersion elements, such as prisms and gratings, (3) methods of recording spectral wavelengths and intensities, such as photographic plates, photomultipliers, and strip chart recorders, and (4) methods of producing, under controlled conditions, spectra of the desired substances, as may be accomplished with arc, spark, or flame excitation techniques. Thus, a short history of atomic spectroscopy should include the major historical developments in each of these areas.

1. EARLY DEVELOPMENTS

Probably the first optical device of present-day spectroscopy to be discovered and used was the lens. There is archeological evidence of the existence of convex quartz lenses many years B.C. Though no early writings describing such objects have been discovered, it seems most likely that the properties of such lenses to magnify an object and/or to gather light rays from some object such as the sun were known to ancient peoples. In fact the possibility of using a lens to concentrate rays from the sun was mentioned by Aristophanes about 423 B.C.

Euclid, about 300 B.C., treated mathematically the size relations of the object and the image for plane and spherical mirrors. He located the focal point of concave mirrors and laid a foundation for the study of convergence and divergence of light beams reflected from mirrors. About 100 B.C., Hero presumably laid the basis of the law concerning the equality of the angles of incidence and reflection from a plane mirror when he stated that light takes the shortest path possible between two points.

The earliest recorded rainbow spectrum using glass as the dispersing medium is that of Seneca, about 40 A.D., who observed that the colors produced by passing sunlight through a three-cornered piece of glass produced colors similar to those of the rainbow. About 100 A.D., Ptolemy summarized knowledge of optics to that time and measured angles of incidence and refraction between air/glass, glass/water, and water/air.

In about 1038 A.D., Alhazen treated, in more detail, reflection of light from curved mirrors and the refraction of light through two different media. He also was aware of the atmospheric refraction of light but did not detail the laws of light refraction.

Roger Bacon, about 1250 A.D., refined the circular concave mirrors into parabolic surfaces and determined their precise focal points. He also studied lenses and observed the magnification possibilities of convex lenses.

About 1600 A.D. combinations of lenses were used and the telescope was invented about 1609 in Holland. News of this development reached Galileo in Italy and he improved on the telescope and demonstrated his own in 1610. This is also about the time that Snell (1591–1626) made his observation of the sine relationship between the angles of incidence and reflection as light passed between two different media.

2. THE NEWTONIAN ERA

Sir Isaac Newton (1642–1727) performed many experiments on the nature of light, the best known being those using a glass prism to study color. He noted that blue light was refracted a greater amount than was red. Thus, when sunlight was allowed to pass through a small hole, through the prism, and fall on a screen, he observed the resulting series of colored images of the hole. This Newton called a *spectrum*. A second prism did not further disperse the light but merely caused additional refraction. Newton had proved that the light from the sun was a heterogeneous mixture of rays that could be separated by a dispersion device.

Newton also felt that the separation of the light rays from the sun could be improved by the use of a lens. By using a lens to focus the sun's rays from the entrance hole onto the screen, he was able to produce a solar spectrum about 10 in. long. Newton realized that the ability of the prism to sort out the colors in sunlight depended on the fact that different colors of light possessed different indices of refraction and he measured indices of refraction of various substances. He was aware of Snell's law of the index of refraction; thus this work was an extension of Snell's observations.

HISTORICAL INTRODUCTION 3

Newton's study of combinations of lenses in telescopes led him to the conclusion that he could compensate for spherical aberration but that chromatic aberration could not be corrected. As in many cases, Newton's prestige as a scientist thus probably delayed further developments in optics for a number of years.

3. THE EARLY 1800'S (TO KIRCHHOFF AND BUNSEN)

The early 1800's produced a number of important developments related to spectroscopy. In 1800 Herschel discovered the infrared spectrum by using thermometers to measure energy beyond the visible red of the spectrum. In 1802 Wollaston observed dark lines in the sun's spectrum but failed to explain their presence. Also in 1802 Thomas Young made the first wavelength measurements of light using the theory of interference of wave motion as a basis. His measurements showed the visible spectrum to extend from 675×10^{-6} to 424×10^{-6} mm, remarkably close to presently accepted limits.

Fraunhofer, about 1814–15, observed the dark lines in the sun's spectrum that had been observed earlier by Wollaston (1802). He observed the fixed positions of the dark lines and improved the optics used for his observations. His use of a narrow slit and a telescope mounted behind the prism allowed him to carefully observe and locate the lines. Fraunhofer catalogued over 700 dark lines in the sun's spectrum. His study, using very narrow slits, led to his observation of diffraction patterns and in 1821 Fraunhofer invented the diffraction grating. Subsequently, he used a grating to determine the wavelengths of the dark lines in the sun's spectrum; eventually, the lines were to be named after him. For example, Fraunhofer measured the wavelengths of the D lines of sodium as 588.6×10^{-6} and 589.0×10^{-6} mm. These lines are the doublet and are now assigned wavelengths of 589.0×10^{-6} and 589.6×10^{-6} mm, respectively. These measurements made possible, for the first time, results in terms of wavelength rather than angles of refraction, a property dependent on the nature of the dispersing medium, rather than a fundamental property of light.

In 1823, J. F. Herschel published the first pictures of emission spectra and in 1831 suggested for the first time that the colors of an alcohol flame, as solutions were placed on the wick, produced "a ready and neat way of detecting extremely small amounts of them." He described how an alcohol flame could be made very bright when certain materials were placed in the flame.

Talbot, in 1825, also observed an orange color in a flame and ascribed its appearance as due to strontium added to the flame. He pointed out that this was the same color observed by Herschel when Herschel added a different salt of strontium to the flame. Talbot suggested this technique could be used to detect the presence of certain substances that would "otherwise require laborious chemical analysis to detect."

Wheatstone observed spark-excited spectra as early as 1835 and eventually published a tabulation of his observations. He reported that metals could be distinguished from one another on basis of their spectral lines. Foucault, in 1848, observed that a sodium flame emitted the Fraunhofer D lines and would also absorb the same rays from an electric arc placed behind it. He thus made the first observations of atomic absorption spectra other than those produced in the spectrum of the sun.

4. THE LATER 1800'S

Numerous and important developments in spectroscopy occurred during the last half of the nineteenth century. Kirchhoff[1] announced his law stating that, "The relation between the powers of emission and the powers of absorption for rays of the same wavelength is constant for all bodies at the same temperature." Kirchhoff showed this effect utilizing the Fraunhofer D lines. He showed that those lines produced with sodium were identical to those from the sun. He then explained the Fraunhofer lines as being caused by the absorption of the elements in the cooler regions of the sun's atmosphere from the continuous spectrum produced by the hot interior of the sun. Kirchhoff produced a large map of the solar spectrum and the path was opened to chemical analysis of the sun's atmosphere, a problem almost immediately studied by Kirchhoff and Bunsen. It was only a few years later that Kirchhoff and Bunsen discovered the elements cesium and rubidium, which they soon isolated chemically. This work of Kirchhoff and Bunsen[2] is considered by many scientists as the founding of modern analytical spectroscopy.

Anders Ångström, in 1868, published a table of 1200 lines, by wavelength, of the solar spectrum. Of these, 800 could be identified as lines of elements known on earth. By use of a diffraction grating, he was able to give the wavelengths to six significant figures and expressed in terms of 10^{-8} cm, a unit now referred to as the angstrom unit (Å) in his honor. Ångström's work was widely used as a standard of comparison and it is a tribute to his careful measurements that agreement with present-day standards is within approximately 13 parts in 100,000, an error caused by an error in line spacing of his grating.

Extension of spectral studies into the ultraviolet region occurred in 1862 when Stokes discovered that quartz was transparent in that region. The

[1] G. KIRCHHOFF, *Monatsber. Ber. Akad. Wiss.*, **1859**, 783 (1859).
[2] G. KIRCHHOFF, and R. BUNSEN, *Ann. Chim. Phys.*, **62**, 452 (1861).

first wavelength measurements in the ultraviolet were made by Mascart,[3] who photographed the UV region of the sun's spectrum to about 2950 Å. Rowland extended the measurements to 2150 Å and Schumann, by using a vacuum system and specially prepared photographic plates, extended the UV region to approximately 1200 Å.

In the period 1860–1900 a number of practical advances in analytical emission spectroscopy occurred. In 1874, Lockyer stated that the length, brightness, thickness, and number of spectral lines were related to the quantity of the element present in the sample. Hartley[4] studied the spectra of metals at varying concentrations and proposed a method of analysis based on the "last line" principle. The most persistent lines were those visible at the lowest concentrations of the element and thus served to measure the lowest concentration of the element that produced spectral lines under controlled excitation conditions.

Developments also occurred during this time with respect to improvements in the determination of the wavelengths of spectral lines. Rowland's tables[5] provided new wavelength measurements from the spectrum of the sun. He based his measurements on a weighted average for the sodium D_1 line, the longer wavelength line of the sodium doublet.

Michelson,[6] using interferometric techniques, measured the wavelengths of three cadmium lines to eight significant figures. He found differences between his results and those of Rowland. After correction of the wavelengths of the D_1 line, Rowland's data were in much better agreement with those of Michelson. In 1907 a value of 6438.4696 Å for the cadmium red line was adopted as a primary standard. Michelson's value for the same line, corrected to 15°C and 760 mm pressure in dry air, was 6438.4695 Å. The present wavelength standard is an emission line of krypton 86 which has a wavelength of 6057.8021 Å in vacuum and was established in 1960 by international agreement.

Theoretical considerations of emission spectra were slow to develop, although they started in the later 1800's and extended into the twentieth century. Balmer's[7] equation for the Balmer series of lines of hydrogen started the search for an explanation for the origin of atomic spectra. Later Ritz (1908) noted that lines of hydrogen observed in the ultraviolet by Lyman (1904) fit the Balmer equation if the constant was changed. This work was extended by Rydberg, Kayser, Runge, and Paschen. It was the work of Bohr,[8] with his concept of the "astronomical" atom and certain postulates

[3] E. Mascart, *Compt. Rend.*, **58**, 1111 (1864).
[4] W. N. Hartley, *J. Chem. Soc.*, **43**, 390 (1883).
[5] H. A. Rowland, *Am. J. Sci.*, **33**, 182 (1887).
[6] A. A. Michelson, *Phil. Mag.*, **34**, 280 (1892).
[7] J. J. Balmer, *Ann. Physik*, **25**, 80 (1885).
[8] N. Bohr, *Phil. Mag.*, **26**, 476 (1913).

concerning it, that placed the origin of spectra on a firm theoretical foundation. Following Bohr, Sommerfeld[9] proposed elliptical orbits, and later the Schrödinger wave equation permitted calculations of the quantum states necessary to describe the energy positions of electrons in atoms.

5. ARC AND SPARK EXCITATION

Many analytical applications of atomic spectroscopy produce their spectra by arc or spark excitation techniques and these methods form the basis for much of the present practice in the field. The historical development in this area is most difficult to document since almost from the start, after observations of the spectrum from the sun, the attempt was to utilize high-energy sources. This led immediately to arc and spark methods. The present-day applications of the arc or spark are improvements of the early work with attempts to better stabilize and control excitation conditions within the arc or spark in an effort to improve analytical data derived from the spectra. These techniques will be discussed in Chapter 5, which deals with accessory equipment for arc and spark spectrochemical analysis.

6. FLAME EMISSION SPECTROSCOPY

The early use of a flame as an excitation source for analytical emission spectroscopy dates back to Herschel[10] and Talbot,[11] who identified alkali metals by flame excitation. The work of Kirchhoff and Bunsen also was basic to the establishment of this technique of atomic excitation. One of the earliest uses of flame excitation was for the determination of sodium in plant ash (1873) by Champion, Pellet, and Grenier.[12] Thus use of the flame paralleled that of arc and spark excitation in the 1800's.

The work that gave principal impetus to flame excitation in terms of modern usage was that of Lundegårdh. In his method the sample solution was sprayed from an atomizer, under controlled conditions, into a condensing chamber and then into an air–acetylene flame. Lundegårdh[13] used photographic recording of spectra and measured the optical densities of selected analytical lines. His careful work led to relative accuracies of about 5–7% and his extensive report of this work laid the base for rapid, accurate analyses

[9]A. SOMMERFELD, *Ann. Physik.*, **51**, 1 (1916).
[10]J. F. W. HERSCHEL, *Trans. R. Soc. Edin.*, **9**, 445 (1823).
[11]W. H. F. TALBOT, *Brewster's J. Sci.*, **5**, 77 (1826).
[12]P. CHAMPION, H. PELLET, and M. GRENIER, *Compt. Rend.*, **76**, 707 (1873).
[13]H. LUNDEGÅRDH, *Lantbruks–Hogskol. Ann.*, **3**, 49 (1936).

of trace element constituents in soil extracts, plant ash, and other biological samples.

Since spectra produced by flames are much simpler (fewer lines) than those produced by arc and spark emission, simple devices for spectral isolation could be used. Developments in Europe in the 1930's led to simpler burners and made use of colored glass filters for spectral isolation. Read-out systems composed only of a photocell connected directly to a galvanometer were used. Such instruments, many still in use, were adequate for simple liquid samples for the determination of the alkali metals.

In 1939 Griggs[14] and Ells and Marshall[15] introduced Lundegårdh's method into the United States. In 1945 Barnes et al.[16] reported on a simple filter photometer for alkali metal determinations. The Perkin-Elmer Corporation apparently was the first American corporation to market such an instrument, which was similar to that developed by Barnes. In 1948 Beckman Instruments, Inc. made a flame excitation attachment available for their spectrophotometer.

Since the early developments in instrumentation in this country and the success of flame methods for the alkali metals there have been improvements in instrumentation that permit flame excitation to be used for other elements. These improvements have dealt primarily with the use of more efficient aspirators, hotter flames, monochromators (gratings or prisms) for spectral isolation, photomultipliers for increased sensitivity, and recorders for reading spectral line intensities. Today some 45–50 elements can be determined by use of flame excitation procedures.

7. ATOMIC ABSORPTION SPECTROSCOPY

The scientific basis for atomic absorption spectroscopy was laid when the origin of the Fraunhofer lines in the sun's spectrum was explained. The lines, first observed in 1802 by Wollaston, and studied more thoroughly by Fraunhofer (1814), were ultimately explained by Kirchhoff and Bunsen in 1859. They proved that the D lines in the Fraunhofer spectrum of the sun corresponded exactly to lines in the spectrum of sodium and thus that the absorption lines in the sun's spectrum were due to sodium atoms in the cooler, outer regions of the sun. The relation between emission and absorption spectra thus was formulated by Kirchhoff and summarized in the law bearing his name, which states that, "Matter absorbs light at the same wavelengths at which it emits light."

---

[14] M. A. GRIGGS, Science, **89**, 134 (1939).
[15] V. R. ELLS, and C. E. MARSHALL, Soil Sci. Soc. Am. Proc., **4**, 131 (1939).
[16] R. B. BARNES, D. RICHARDSON, J. W. BERRY, and R. L. HOOD, Ind. Eng. Chem., Anal. Ed., **17**, 605 (1945).

One of the first uses of atomic absorption for analytical purposes was by Woodson,[17] when he reported on a method of detecting mercury vapor in air. However, the techniques he described did not appear promising as a general analytical method for a variety of elements. In 1955 papers by Walsh[18] and by Alkemade and Milatz[19] both indicated the great potential of atomic absorption spectral techniques to the field of analytical chemistry. The work of Walsh appeared to be more detailed and thorough in its review of the potential in the field, so that he is now considered to be the originator of modern analytical atomic absorption spectroscopy. Walsh recognized the importance of a high-intensity, narrow linewidth source for high sensitivity and suggested the use of hollow cathode discharge lamps as meeting these requirements. Walsh made use of a flame to serve as a device to produce neutral atoms through which the energy emitted by the hollow cathode could be passed to produce the absorption signal. Both the above-mentioned items are still commonly used in modern atomic absorption instrumentation.

Since the foundations of atomic absorption spectroscopy were laid by Walsh a number of improvements in instrumentation and techniques have been made. Russell, Shelton, and Walsh[20] modulated the hollow cathode signal and used an amplifier tuned to the modulating frequency so measurements could be made without interference from flame emission. Sullivan and Walsh[21] developed very high-intensity hollow cathode lamps that led to lower detection limits. Willis[22] proposed the use of nitrous oxide–acetylene flame as a means of overcoming certain interferences and produce a higher population of free atoms in the flame.

Multiple-pass devices have been constructed so the effective optical path length through the flame could be increased, thus increasing the sensitivity of the method. Various attempts have been made to use a sample cell other than a flame to produce the atomic vapor in the optical path of the spectrophotometer. The graphite furnace devised by L'vov[23] in Europe and by Woodriff[24] in the United States are examples. More recently carbon rod analyzers and tantalum boat devices have been used. These devices depend on high temperatures to vaporize and dissociate the sample into the optical path to produce the absorption signal. Some absolute detection limits in the 10^{-12}–10^{-14} g range are reported using these techniques.

---

[17] T. T. WOODSON, *Rev. Sci. Instrum.*, **10**, 308 (1939).
[18] A. WALSH, *Spectrochim. Acta*, **7**, 109 (1955).
[19] C. T. J. ALKEMADE, and J. M. W. MILATZ, *Appl. Sci. Res., Sec. B*, **4B**, 289 (1955).
[20] B. J. RUSSELL, J. P. SHELTON, and A. WALSH, *Spectrochim. Acta*, **8**, 317 (1957).
[21] J. V. SULLIVAN, and A. WALSH, *Spectrochim. Acta*, **21**, 721 (1965).
[22] J. B. WILLIS, *Nature*, **207**, 715 (1965).
[23] B. V. L'VOV, *Spectrochim. Acta*, **17**, 761 (1961).
[24] R. WOODRIFF, R. W. STONE, and A. M. HELD, *Appl. Spectrosc.*, **22**, 409 (1968).

HISTORICAL INTRODUCTION

High-intensity sources other than hollow cathode tubes also have been investigated. Most promising are the microwave-excited electrodeless discharge lamps. These lamps produce very high-intensity, sharp line spectra but suffer stability problems and generally have short lifetimes compared to hollow cathode tubes.

8. ATOMIC FLUORESCENCE SPECTROSCOPY

Atomic fluorescence was studied as early as 1902 by Wood,[25] and Nichols and Howes[26] looked at fluorescence in flames, but neither of these reports dealt with analytical applications of atomic fluorescence. Alkemade[27] discussed resonance fluorescence in flames, considering modes of excitation. Winefordner and Vickers[28] then investigated the possibilities of using atomic fluorescence as a practical analytical technique. They used metal vapor discharge tubes as sources and were able to obtain sensitivities of better than 1 $\mu g/ml$ for mercury, zinc, cadmium, and thallium in an acetylene–oxygen flame. Later, Veillon et al.[29] were able to demonstrate the usefulness of a continuous source and added another dozen elements to the list of possibilities.

In England, West et al. (1966–1967) studied the same technique. They made use of microwave-excited electrodeless discharge sources, preparing tubes of over 30 elements. They considered a propane flame superior to either the acetylene–oxygen or hydrogen–oxygen flames. The method has been reported to be relatively free of interference effects.

Atomic fluorescence is the most recent development in analytical atomic spectroscopy; thus it has not had time to be evaluated as well as other techniques. Further developments in this field with respect to optimizing sources and sample cells, together with improvements in instrumental parameters and development of readily available commercial instrumentation, should lead to this technique serving in the area of analytical spectral methods to supplement the already well-established arc and spark emission, flame emission, and atomic absorption spectroscopy.

---

[25] R. W. Wood, Phil. Mag., 3, 128 (1902).
[26] E. L. Nichols, and H. L. Howes, Phys. Rev., 23, 472 (1924).
[27] C. T. J. Alkemade, in Proc. Xth Colloq. Spectros. Int., Spartan Books, Washington, D.C. (1963).
[28] J. D. Winefordner, and T. J. Vickers, Anal. Chem., 36, 161 (1964).
[29] C. Veillon, J. M. Mansfield, M. L. Parsons, and J. D. Winefordner, Anal. Chem., 38, 204 (1966).

SELECTED READING

Baly, E. C. C., *Spectroscopy*, Vol. I, Longmans, Green and Co. (1924).
Clark, George L., Editor, *The Encyclopedia of Spectroscopy*, Reinhold Publishing Co. (1960).
Hermann, R., and Alkemade, C. Th. J., *Flammenphotometrie*, Springer-Verlag (1960).
Kayser, H., *Handbuch der Spectroscopie*, Vol. I, S. Hirzel, Leipzig (1900).
Sarton, G., *A History of Science*, Vol. I, Harvard University Press (1959).
Slavin, W., *Atomic Absorption Spectroscopy*, Interscience Publishers (1968).

Chapter 2

The Origin of Atomic Spectra

Although detailed knowledge of atomic spectra is not required to use spectra for analytical purposes, it is most helpful for the analyst to have a basic understanding of the origin and the properties of spectra. The development of analytical methods using spectra as a basis for acquiring analytical data requires some understanding of such things as spectral line intensities, linewidths, temperature and pressure effects on spectral intensities, and characteristics of spectral lines suitable for spectroscopic analysis. Factors such as spectral line reversal, spectral line overlap, absorption of spectral energy, excited-state lifetimes, metastable states, degenerate states, and resonance line characteristics all are important to the analytical spectroscopist.

This chapter presents an elementary treatment of the origin of spectra suitable for those interested in applying atomic spectroscopy to the development of analytical procedures and in understanding the basis of spectroscopic methods. The development is historical and does not include the rigorous mathematics required for a detailed treatment of atomic spectra. For such a thorough discussion of the origin of atomic spectra the reader is referred to one of the many excellent books on the subject in the field of physics. Several are given in the list of selected reading at the end of this chapter.

1. THE NATURE OF ELECTROMAGNETIC RADIATION

Light is electromagnetic radiation to which the retina of the eye is sensitive, but electromagnetic radiation also includes regions identified as infrared, ultraviolet, x-ray, gamma-ray, and radiofrequency radiation. Atomic spectroscopy utilizes primarily the ultraviolet and visible regions of the electromagnetic spectrum. It is important therefore to understand some of the basic characteristics and properties of such radiation.

Electromagnetic radiation travels in straight lines in a uniform medium, has a velocity of 299,792,500 m/sec in a vacuum, and possesses properties of a wave motion and also of a particle (photon). The wave theory of light can satisfactorily account for properties such as *wavelength, interference, refraction, polarization, reflection,* and *diffraction*.

Wavelength can be visualized by reference to Figure 2-1. In the figure the distance *AB* is one wavelength. Associated with wavelength is frequency. The *frequency* is the number of waves passing a fixed point, such as *c*, in a unit length of time. Wavelength and frequency are related by the relation

$$\lambda v = c$$

where λ is wavelength, v is the frequency, and c is the velocity of light.

Two light waves can combine to *interfere* with one another, either constructively or destructively. Consider the two waves shown in Figure 2-1. The two light waves have the same wavelength, frequency, and amplitude but are *out of phase by 180°*. When combined they destructively interfere to cancel exactly one another. If they were *in phase*, they would constructively interfere to produce a wave having twice the amplitude of each single component.

Since light travels at a different speed in a transparent medium than it does in a vacuum, light can be *refracted*. The index of refraction of a medium is the ratio of the speed of light in a vacuum to the speed of light in the medium. The index of refraction is also a function of wavelength; thus longer wavelengths (red light) have a smaller index of refraction in a transparent

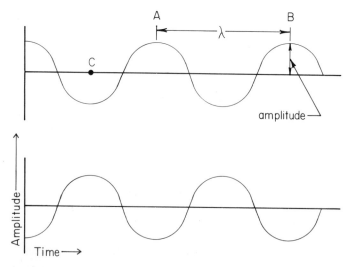

FIGURE 2-1. Some characteristics of electromagnetic radiation.

medium than do shorter wavelengths (violet). When light of a particular wavelength enters an optically dense transparent medium the speed of light decreases but its frequency remains constant; therefore it follows that the wavelength in the medium is less than in a vacuum.

A *polarized* beam of light is one in which the changes in amplitude that occur with time are all in the same plane. Polarization can be produced by selective absorption of light as it passes through certain substances. Polaroid films can cause this phenomenon to occur. If two Polaroid films are properly oriented, polarized light will pass through both films. If they are adjusted so there is a 90° difference in their orientations, light will not pass through.

When light strikes some surfaces it is *reflected*. The angle of reflection from a plane surface equals the angle of incidence. Reflection of light by plane and concave mirrors frequently is used in atomic spectroscopy to direct a light beam in a desired direction.

If the path of a light beam from its source is partially obstructed by an opaque object, light penetrates into a region beyond the object that cannot be reached by following a straight line from the source. This phenomenon is called *diffraction* and is important in understanding the principle of the diffraction grating; it is used in Chapter 3 to develop the principles of the diffraction grating.

When light is emitted or absorbed it is necessary to consider that light possesses particle properties. The quantum concept was initially introduced by Max Planck and was later used by Niels Bohr (1913) to account for atomic spectra. This concept of light is therefore important in our understanding of the origin of atomic emission and absorption spectra since it postulates that emission and absorption of light occurs discontinuously or in isolated "quanta." The quantum is a definite unit of radiant energy given by the expression

$$E = h\nu$$

where E is the energy of the quantum, h is Planck's constant, and ν is the frequency of the radiation.

2. EARLY CONCEPTS

The practice of analytical spectroscopy preceded the development of the theories concerning the origin of spectra by a number of years. Bunsen and Kirchhoff studied spectra produced by salts and salt solutions heated in flames and noted the characteristic spectral line emissions of a number of elements. They also observed that substances absorb energy most strongly at the same wavelengths at which emission occurs. Their results led Kirchhoff to state that the power of emission is equal to the power of absorption for all bodies which are in temperature equilibrium with their radiations. This statement, known as Kirchhoff's law, explained the Fraunhofer lines in the sun's spectrum that had been observed earlier. The basis was thus laid for the use of spectral lines, both emission and absorption, for analytical purposes.

During the late 1800's, after the work of Bunsen and Kirchhoff, many line spectra of the elements were identified and recorded. Spectra were classified by element, wavelength, line intensities, and by series, depending on line characteristics. Four series were described for alkali metals and called sharp, principal, diffuse, and fundamental. The influence of this terminology is still retained in modern spectroscopic notation as S, P, D, and F to identify different atomic energy levels. Early attempts to formulate some mathematical relationship between lines of a single element included harmonic or overtone relations similar to sound waves. Stoney, in 1871, suggested a reciprocal wavelength relation for certain hydrogen lines but further efforts were unable to attach any real significance to his results.

3. THE BALMER EQUATION

However, Balmer, in 1885, reported on an empirical relation that very precisely fit the wavelengths of a series of lines of hydrogen. The wavelengths of the lines converged toward a limit at approximately 3646 Å, and were described by the equation

$$\lambda = \lambda_0 \frac{n^2}{n^2 - 4} \qquad (2\text{-}1)$$

where n is an integer greater than two. λ_0 is a constant, which Balmer measured as 3645.6 Å. The Balmer equation gave wavelengths for the first four lines of the hydrogen spectrum that agreed to within 0.10 Å of Rowland's experimental wavelengths.

The Balmer equation, written in terms of wavenumbers, becomes

$$\bar{v} = R_H \left(\frac{1}{4} + \frac{1}{n^2} \right) \qquad (2\text{-}2)$$

where the constant R_H has a value of 109,678 cm^{-1}. Wavenumber is defined as

$$\bar{v} = 1/\lambda \qquad (2\text{-}3)$$

where λ is the wavelength expressed in cm. The wavelength λ and frequency v of electromagnetic radiation are related as follows:

$$v = c/\lambda \qquad (2\text{-}4)$$

THE ORIGIN OF ATOMIC SPECTRA

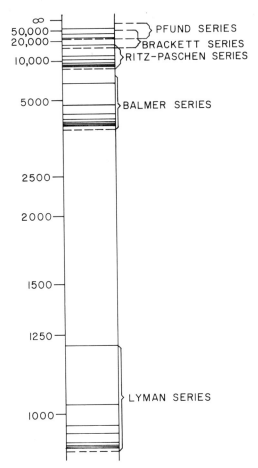

FIGURE 2-2. Schematic representation of the spectrum of the hydrogen atom.

where c is the velocity of light. The wavenumber, therefore, also is given by

$$\bar{v} = v/c \qquad (2\text{-}5)$$

from which equation (2-2) is derived.

Soon after the development of the Balmer equation another series of spectral lines of hydrogen in the ultraviolet region were reported by Lyman. These lines were found to fit the Balmer equation (2-2) if the first term in the parentheses is 1/1 rather than 1/4. It appeared therefore that a more general form of the Balmer equation would be

$$\bar{v} = R_H \left(\frac{1}{n_1^2} - \frac{1}{n_2^2} \right) \qquad (2\text{-}6)$$

with the Lyman series having a value of $n_1 = 1$.

Analysis of equation (2-2) for the Balmer series of hydrogen lines indicates that the spectral emission lines are given by the difference between $R_H/4$ and R_H/n^2. The ratio $R_H/4$ is called a "fixed" term, while R_H/n^2 is called a "running" term. Similar treatment of the Lyman series produces similar results if the fixed term is $R_H/1$, a fact suggested by Ritz. Thus, the wavenumbers of lines of any series are the results of differences between two terms, one of them being of "fixed" value.

Equation (2-6) led to the identification of other series of the lines for hydrogen, including the Paschen series ($n_1 = 3$), the Brackett series ($n_1 = 4$), and the Pfund series ($n_1 = 5$). The Balmer series is in the visible region of the spectrum, the Lyman series is in the ultraviolet, and the Paschen, Brackett, and Pfund series appear in the infrared. Their distribution is shown in Figure 2-2. Equation (2-6), which accounts for all presently known lines of hydrogen, led Ritz (1908) to propose his combination principle, that the wavenumbers of all lines in a series are the result of the difference in energy between a "fixed" and a "running" term.

4. FROM BALMER TO BOHR

The wave theory of electromagnetic radiation can explain a number of observed phenomena associated with light, such as diffraction, refraction, and interference, but fails to explain other properties. These include such things as the photoelectric effect and the emission and absorption of radiation by bodies. Instead, those phenomena involving interaction of light with matter are explained by utilizing the "corpuscular" character of electromagnetic radiation.

Planck (1901) was the first to discard classical mechanics to explain "blackbody" radiation when he proposed that an oscillator could only acquire and lose energy in discrete units, called quanta. The magnitude of the quantum of energy was not fixed, but depended on the change in the oscillator energy E_n according to the equation

$$E_n = nh\nu \qquad (2\text{-}7)$$

where E_n is the quantized energy to which the oscillator is restricted, ν is the quantum frequency, n is an integer, and h is Planck's constant (6.62 × 10^{-27} erg-sec). Einstein (1905) provided a strong confirmation for Planck's concept of quantized energy when he used the idea to explain the photoelectric effect.

Attempts to explain the origin of sharp line spectra of the elements were not successful on the basis of the Rutherford concept of the atom. In this model negative electrons were proposed to revolve around a positive nucleus and were held in position by a balance between Coulombic attraction and centripetal force. Since the electron is centripetally accelerated, classical electromagnetic radiation theory would require the accelerating electrical charge to continuously emit radiation. The result would be that electrons would emit or lose energy and follow a spiral course into the nucleus. The atom, according to the Rutherford concept, would be unstable and continuously radiate energy. However, neither of these conditions exist.

It remained for Bohr (1913) to break with classical mechanics and apply the quantum concept of Planck to atoms and atomic spectra. The model proposed by Bohr and the equation he developed for the spectrum of atomic hydrogen were based on a set of new postulates, including the following:

(1) An atom can only absorb or emit discrete units of energy called quanta (or photons), characteristic of each atom. No energy is absorbed or emitted when the electrons remain in discrete energy states.
(2) Absorption or emission of energy by an atom occurs when the electron changes from one stationary energy state to another. One photon of energy is absorbed or emitted as a result.
(3) The energies of the allowed states can be calculated by classical mechanics.
(4) The allowed energies are such that the angular momentum of the system is an integral multiple of $h/2\pi$.

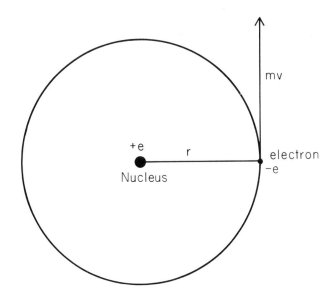

FIGURE 2-3. Bohr model of the hydrogen atom.

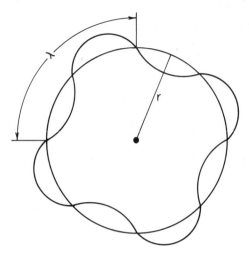

FIGURE 2-4. Wave motion restriction of an electron in orbit.

The first and second postulates are consistent with the line spectra of hydrogen and the Ritz combination principle. The third and fourth postulates provide a method to solve for the energy levels of an atom.

The Bohr development of these postulates follows below and gives rise to the Bohr equation for calculation of the wavelengths of the spectral lines of hydrogen. Consider hydrogenlike ions, with one electron revolving about a nucleus of $+Ze$ charge; the electron carries a charge of $-e$. The mass of the nucleus is large compared with the mass of the electron, permitting the approximation that the electron revolves around the nucleus rather than about the center of mass of the system. This system is shown in Figure 2-3.

The total energy E of the atom is the sum of its electrostatic energy V and its kinetic energy T, which are defined by

$$V = -Ze^2/r \quad \text{and} \quad T = \tfrac{1}{2}mv^2$$

where m is the mass of the electron, v is its velocity, Z is the atomic number, and r is the radius of its orbit. For a stable orbit the centripetal force must equal the electrostatic force; thus

$$\frac{mv^2}{r} = \frac{Ze^2}{r^2} \quad \text{or} \quad mv^2 = \frac{Ze^2}{r} \tag{2-8}$$

so

$$T = \frac{Ze^2}{2r} \tag{2-9}$$

THE ORIGIN OF ATOMIC SPECTRA

The total energy thus is

$$E = T + V = -\frac{Ze^2}{2r} \tag{2-10}$$

The negative sign is due to the convention used, that the potential energy of the system is decreased from zero as the electron moves from infinity to distance r from the nucleus.

The fourth Bohr postulate requires that the angular momentum of the electron in allowed orbits be restricted to integral multiples of $h/2\pi$. De Broglie, in 1925, pointed out that the fourth Bohr postulate requires that the wavelength properties of the electron be similar to that shown in Figure 2-4, where r is the radius of the electron orbit and λ is an associated wavelength.

For circular orbits, as proposed by Bohr, the angular momentum is mvr, so

$$mvr = nh/2\pi \tag{2-11}$$

By combining equations (2-8) and (2-11) to eliminate v and solving for r, we obtain the result

$$r = \frac{n^2 h^2}{4\pi m Z e^2} \tag{2-12}$$

Equation (2-12) permits calculation of the radius of the hydrogenlike ions for various values of n since all other quantities in the equation are known. For $n = 1$ and $Z = 1$, the ground state for hydrogen, $r = 0.529$ Å, a value in close agreement with those given by other methods for the determination of the radius of the hydrogen atom. The radii of electron orbits in a hydrogenlike atom increase as the square of n and decrease with increasing nuclear charge.

The energy associated with each allowable orbit (value of n) may be calculated by substitution of the value of r [from equation (2-12)] into the expression for the energy [equation (2-10)], to obtain

$$E = -\frac{2\pi^2 m e^4 Z^2}{n^2 h^2} \tag{2-13}$$

The most stable (lowest) state for hydrogen ($n = 1$) is called the *ground* state of the atom. Other states are less stable and therefore are referred to as *excited* states. When energy is supplied to the atom the energy may be absorbed during a *transition* (for example, from $n = 1$ to $n = 2$), going from a lower state to a higher state. Conversely, energy will be emitted in a transition from a higher state to a lower state. The frequency of the emitted energy (photon) is given by the Planck equation, $E = h\nu$.

The energy of an electronic transition must be the energy difference between state 1 and state 2; thus

$$E_{\text{transition}} = E_2 - E_1$$

or

$$E_{\text{transition}} = \left(-\frac{2\pi^2 me^4 Z^2}{n_2^2 h^2}\right) - \left(-\frac{2\pi^2 me^4 Z^2}{n_1^2 h^2}\right)$$

$$= \frac{2\pi^2 me^4 Z^2}{h^2}\left(\frac{1}{n_1^2} - \frac{1}{n_2^2}\right) \tag{2-14}$$

When converted to frequency, this becomes

$$\nu_{\text{transition}} = \frac{2\pi^2 me^4 Z^2}{h^3}\left(\frac{1}{n_1^2} - \frac{1}{n_2^2}\right) \tag{2-15}$$

The quantity $2\pi^2 me^4 Z^2/h^3$ corresponds to R, the empirical Rydberg constant, and can be evaluated on the basis of the numerical values of the constants involved. For the Rydberg constant with $Z = 1$

$$R = \frac{2\pi^2 me^4 Z^2}{h^3}$$

$$= \frac{2(3.1416)^2(9.1085 \times 10^{-28})(4.8029 \times 10^{-10})^4}{(6.6252 \times 10^{-27})^3}$$

$$= 3.2898 \times 10^{15} \text{ cycles/sec} \tag{2-16}$$

This quantity, expressed in wavenumbers $\bar{\nu}$ [from equation (2-5)], gives

$$R = 109{,}737 \text{ cm}^{-1}$$

which may be compared with the experimentally determined value of 109,677.6 cm^{-1}, a remarkable agreement.

Equation (2-15) assumes an infinite mass for the nucleus. When two bodies, such as the electron and the proton of hydrogen, interact, the nucleus does not maintain a fixed position and both the nucleus and the electron revolve about their common center of mass. The result is that the mass of the electron should be replaced by the *reduced mass* μ, defined as

$$\mu = \frac{m}{1 + (m/M)} \tag{2-17}$$

where m is the mass of the electron and M is the mass of the nucleus. The Rydberg constant for hydrogen, in wavenumbers, thus becomes

$$R = \frac{2\pi^2 e^4}{h^3 c}\frac{m}{1 + (m/M)} = \frac{2\pi^2 \mu e^4}{h^3 c} \tag{2-18}$$

Since μ is slightly smaller than m, this substitution into the Rydberg constant produces even closer agreement between the calculated and experimental values of the constant.

4.1. Spectral Line Series for Hydrogen

Equation (2-15) permits the calculation of the frequency of the photon emitted for any electronic transition of hydrogenlike ions; consequently, the wavelength of the transition can be calculated from the expression $\lambda = c/v$. The calculated wavelengths and experimental wavelengths agree closely in all cases for all the series identified for hydrogen.

If the condition $n = 1$ (ground state for hydrogen) and $n_2 = \infty$ are used in equation (2-15), the ionization energy for hydrogen can be calculated and equals 13.595 eV (electron volts). At higher energies a continuum occurs since energy above the ionization energy is distributed as translational energy with levels so close together they cannot be resolved and appear as a continuum. Thus, careful measurement of the wavelength at which the continuum begins is a measure of the ionization energy of the atom if the electron originates from the lowest energy level (ground state of the atom). Properly locating the inception of the continuum is difficult since the series of spectral lines that converge to the limit of $n_2 = \infty$ cannot easily be resolved. In spite of such experimental difficulties, spectroscopic measurements are among the most accurate to determine ionization energies of atoms.

4.2. Energy Level Diagrams

Grotrian (1928) developed a graphical method for presenting atomic energy levels and electronic transitions that is almost universally used. The diagram permits the representation of spectral terms and transitions graphically, as shown in Figure 2-5 for hydrogen. The vertical axis is an energy axis and the energy levels (terms) are shown as horizontal lines. For hydrogen the terms are equal to R_H/n^2. The separations of the levels decrease as the value of n increases. The continuum (ionization) starts at $n = \infty$.

A spectral emission line results from a transition from a higher value of n to a lower value. The vertical distance representing a transition is a measure of the energy of the transition. The width of the vertical line representing a transition is an approximate measure of the line intensity.

5. MODIFICATIONS OF THE BOHR THEORY

The Bohr theory was useful in interpreting atomic spectra and laid the foundation for further development and refinement. The theory could only be applied to a one-electron system and a circular electron orbit. Limitations of the Bohr concept were discovered soon after the theory was developed. Sommerfeld (1916) proposed the concept of elliptical orbits to meet, in part,

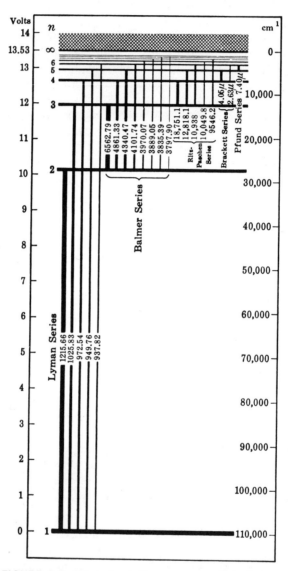

FIGURE 2-5. Energy level diagram for atomic hydrogen. [From G. Herzberg, *Atomic Spectra and Atomic Structure*, Dover Publications (1944). Used by permission of Dover Publications.]

THE ORIGIN OF ATOMIC SPECTRA

some of the observed discrepancies in atomic spectra of elements other than hydrogen. Such orbits were known to be dynamically stable if the nucleus was located at one focus. This extension of the Bohr theory provided an explanation of the series of lines observed with alkali metals. These series had been observed before the Bohr theory was developed and had been designated as sharp, principal, diffuse, and fundamental series. To explain these spectra Sommerfeld therefore proposed elliptical orbits but restricted them to certain ratios of minor to major axes. He introduced a second quantum number k, called the azimuthal quantum number, with n/k equal to the ratio of major axis to minor axis. Thus, k can take any value from 1 to n, and when $n = k$ a circular orbit results.

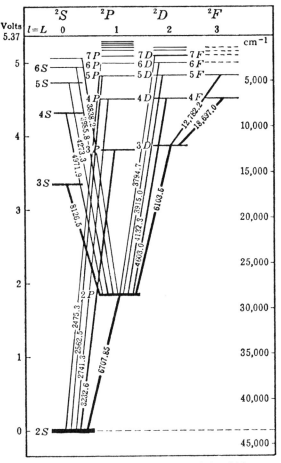

FIGURE 2-6. Energy level diagram of the lithium atom. [From G. Herzberg, *Atomic Spectra and Atomic Structure*, Dover Publications (1944). Used by permission of Dover Publications.]

In the case of hydrogen all elliptical orbits have *almost* equal energies with circular orbits. For other elements the outer electron (valence electron), which is primarily responsible for the observed spectrum of the element, is under the influence of inner (core) electrons. Elliptical orbits in this situation have different energies depending on the degree of ellipticity. For each value of n, therefore, there will be n different energy levels for k.

Letter designations for spectral series were used by spectroscopists before the development of the quantum theory and are still in use, that is, for $k = 1, 2, 3$, and 4 the letter designations are s, p, d, and f, respectively, and refer to the sharp, principal, diffuse, and fundamental series. A Grotrian diagram for lithium is shown in Figure 2-6 illustrating these spectral series. The spectral lines that originate from S and drop to a P level form the sharp series, those from P to S give the principal series, etc. The Grotrian diagram in this case must show the values of k as well as those for n. Values of n are shown vertically as in Figure 2-5. The values for k are then shown plotted horizontally for $k = 1, 2$, or 3, and the connecting lines, for example from $2P$ to $1S$, indicate electronic transitions. The energy released in an electronic transition from a higher to lower level is the *vertical displacement* of the energy states and not the length of the connecting line.

Also shown in the diagram, immediately below values of k, is a quantum number identified as L. The number L is a *resultant* angular momentum of all electrons; l is used for the angular momentum for *one* electron. Values of L are 0, 1, 2, and 3 with respective atomic states S, P, D, and F. For alkali metals, with only one valence electron, the resultant angular momentum of the inner electrons is zero since all closed shells have net zero angular momentum. For lithium, the angular momentum of the atom, therefore, is the same as for a single electron, that is, $L = l$. The quantum number L will be discussed further in a later section.

5.1 Selection Rules for *n* and *k(l)*

If one calculates the total number of spectral lines that should occur involving all values of n and k and compares this number with the number of spectral lines actually observed, it is obvious that certain transitions do not occur. The analysis of spectra led to the conclusion that some transitions were "permitted" and some were "forbidden." The following rules were formulated as a result.

(a) n can change by any integral number, including 0.
(b) $k(l)$ must change by ± 1.

5.2. Atoms with Two Valence Electrons

The Bohr–Sommerfeld theory provided a reasonably satisfactory explanation of the spectra of atoms having only one valence electron. With two or more electrons discrepancies occur and certain arbitrary selection rules for atomic transitions were required. In an attempt to solve this problem, several persons, including de Broglie, Schrödinger, Heisenberg, and Born, combined quantum mechanical and wave mechanical concepts. A detailed account of these efforts is beyond the scope of this book. However, some qualitative understanding of these concepts will be helpful and thus a brief account is included.

De Broglie suggested that an electron (or any other corpuscle) has associated with it a wave motion whose wavelength is given by the expression

$$\lambda = h/mv \tag{2-19}$$

where h is Planck's constant, m is the mass of the corpuscle, and v is its velocity. To understand an electron, the wave behavior associated with the electron must be investigated. De Broglie limited the wave behavior to standing waves having an integral number of waves in an orbit, applying the restriction that

$$n\lambda = 2\pi r \tag{2-20}$$

where $n = 1, 2, 3$, etc. This assumption leads to the Bohr equation for circular orbits and to discrete stationary states for each value of n (see Figure 2-4).

Schrödinger's wave equation can be used to calculate the stationary wave states and can be written as

$$\frac{\partial^2 \psi}{\partial x^2} + \frac{\partial^2 \psi}{\partial y^2} + \frac{\partial^2 \psi}{\partial z^2} + \frac{8\pi^2 m}{h^2}(E - V)\psi = 0 \tag{2-21}$$

where E is the total energy, V is the potential energy, m is the mass of the electron, h is Planck's constant, and ψ is the amplitude of the wave. When ψ is finite, single-valued, continuous, and zero at infinity, the equation can be solved for certain values of E. The calculated values of E are called eigenvalues and the corresponding wave functions are called eigenfunctions and represent the stationary states for the wave motion. The experimentally determined energies of an atom as measured from its spectrum appear as eigenvalues of the Schrödinger wave equation.

If the wave mechanical model (the Schrödinger equation) for the hydrogen atom is solved, the possible energy values are those obtained from the Bohr theory. The Schrödinger equation also yields, for each stationary energy state, a degenerate set of eigenfunctions differing in angular momentum and spatial orientation. These are designated as l and m, which are always

TABLE 2-1
Possible Quantum Number Values for n, l, and m

n	1	2		3		
l	0	0	1	0	1	2
m	0	0	−1, 0, +1	0	−1, 0, +1	−2, −1, 0, +1, +2

integers. The quantum number l corresponds to the Bohr–Sommerfeld quantum number k, the azimuthal quantum number. For any value of n, the quantum number l takes on values of $0, 1, 2, \ldots, n-1$. The relation between k and l then is $l = k - 1$. This relation also is shown at the top of Figure 2-6. Present practice is to use l as the azimuthal quantum number rather than k.

The quantum number m is called the magnetic quantum number and is related to the spatial quantization of the angular momentum of the electron in its orbit. The spatial orientation of the angular momentum is restricted by the condition that the component of angular momentum of an atom in a magnetic field is an integral multiple of $h/2\pi$. Allowed values of m are $-l$, $-l+1$, $-l+2, \ldots, +l$. In the absence of a magnetic field energy levels differing only in the value of m are of equal energy and are called *degenerate* states. In the presence of a magnetic field a "splitting" of the degenerate states occurs with each different value of m producing a single nondegenerate state. This behavior is known as the Zeeman effect when using an external magnetic field and the Stark effect when using an external electrical field.

The possible combinations of n, l, and m are shown in Table 2-1. What were assumptions of the Bohr–Sommerfeld theory for n, k, and m become, through the solution of the Schrödinger wave equation, n, l, and m. Thus all three quantum numbers arise from the solution of the Schrödinger equation.

5.3 Selection Rules and the Schrödinger Equation

The selection rules given earlier for n and k were experimentally determined. The Bohr–Sommerfeld model for the atom did not account for them. Calculations based on the wave mechanical model indicate that n can change by any integer amount, as $\Delta n = 0, 1, 2, 3$, etc., and that l must change by ± 1. This corresponds to the change in k that was determined experimentally. Since the wave mechanical model of the atom gives solutions in terms of statistical probability, a more accurate statement of the selection rule for l would be that the *probability* of a transition other than $\Delta l = \pm 1$ is very small.

THE ORIGIN OF ATOMIC SPECTRA

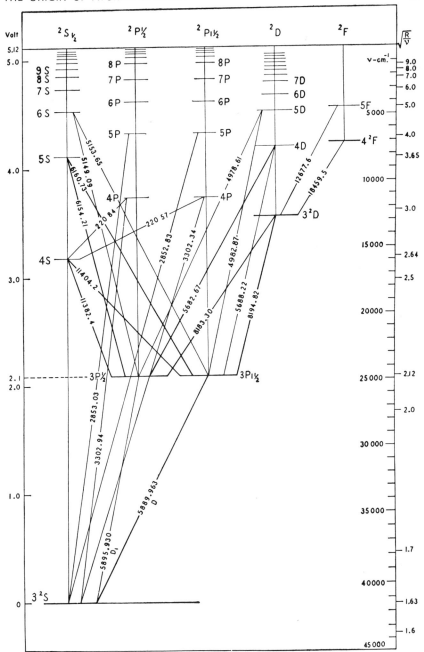

FIGURE 2-7. Energy level diagram of the sodium atom. [From W. Brode, *Chemical Spectroscopy*, 2nd ed., Wiley, New York (1943). Used by permission of Wiley, New York.]

6. ALKALI METAL ATOM SPECTRA

An energy level diagram for sodium is shown in Figure 2-7. To understand the atomic spectra of the alkali metals, it is necessary to consider the four sets of energy levels designated by the S, P, D, and F notations given earlier. These levels exist in hydrogen but can only be distinguished by an extremely high-resolution spectrograph. The ground state of any alkali metal is an S state, and in the case of sodium is designated as $3S$ since the lowest level is $n = 3$. The sharp series (S) starts with $L = 0$, the P with $L = 1$, etc. In Figure 2-7 those transitions with an upper level in the S state dropping to a lower P level form the sharp series. When the upper state is a P state and the lower an S state a principal series is formed. Note in Figure 2-7 that L changes by ± 1. The wavelengths of each transition also are shown.

The energies associated with the $4S$, $4P$, $4D$, and $4F$ terms are not equal. This behavior is due to the change of eccentricity of the elliptical orbits and is contained within a "Rydberg correction." The Rydberg correction is the deviation from the value of R for hydrogen and is necessary if the electron penetrates the atom core, since the "effective" nuclear charge thus is changed. The Rydberg correction is greatest for S terms and is progressively smaller for P, D, and F terms. The spectrum of sodium is very similar to that of hydrogen if due consideration is given the Rydberg correction. This structure similarity therefore results in the conclusion that the alkali metals have an external structure similar to hydrogen, that is, one outer electron.

6.1. Doublet Structure of Alkali Metal Spectra

Inspection of the $3P$ energy state in Figure 2-7 (and all other P states) reveals that it consists of two states of almost equal energy. Examination of the spectra of other alkali metals reveals similar behavior. Each of the spectral lines involving the S series is a doublet arising from a single S level dropping to a "split" P level. The line splitting increases as the atomic weight of the alkali metal increases. One of the most familiar cases of doublet formation (line splitting) is that of the sodium doublet at 5895.93 and 5889.86 Å, known as the D lines of sodium, and is shown in Figure 2-7 as arising from electrons dropping from the "split" $3P$ levels to the ground state, or $3S$ level. Line splitting must therefore arise from term splitting, in this case, a splitting of the $3P$ state into two states of nearly equal energy. The S terms are single and the P terms are split into two levels. The splitting of the D terms is so small (see Figure 2-7) that experimentally they are treated as if they were single. The splitting of the D terms becomes more noticeable for the alkali elements of higher atomic number than potassium.

THE ORIGIN OF ATOMIC SPECTRA

TABLE 2-2
J Values for Various Doublet Terms

Term	L	J
S	0	1/2
P	1	1/2, 3/2
D	2	3/2, 5/2
F	3	5/2, 7/2

The existence of the doublet systems in the alkali metal elements requires another quantum number. This number was called the *inner quantum number* by Sommerfeld and is designated by the letter J and is used to distinguish the components of doublets. The S terms all have a J index of $\frac{1}{2}$; the P terms of $\frac{1}{2}$ and $\frac{3}{2}$; and the D terms of $\frac{3}{2}$ and $\frac{5}{2}$. Table 2-2 lists the J values for doublet terms. Analysis of the spectra of the alkali metals leads to the selection rule that $\Delta J = 0$ or ± 1, but also that the transition from $J = 0$ to $J = 0$ is excluded.

6.2. Electron Spin

To account for the doublet character of the spectral lines of the alkali metals, it is necessary to introduce another degree of internal freedom to the atom. Goudschmidt and Uhlenbeck (1925) suggested this was electron spin, designated as the quantum number s (not to be confused with the term s electrons). The spin angular momentum is limited to the two values of $\pm \frac{1}{2} h/2\pi$, or $s = \pm \frac{1}{2}$, corresponding to opposite spins for the electron.

The origin of the quantum number J can now be explained. The total angular momentum is equal to $J \times h/2\pi$ and results from the combination of electron orbital angular momentum and electron spin angular momentum vectorially added. For alkali metals, with one valence electron, $S = +\frac{1}{2}$ or $-\frac{1}{2}$, so J can have two values ($J = +\frac{1}{2}$ and $J = -\frac{1}{2}$). The vector addition of L and S must produce integral differences. If $L = 1$ and $S = \frac{1}{2}$, J can be $\frac{3}{2}$ and $\frac{1}{2}$. The number of possible J values is $J = 2S + 1$. For a two-electron system, such as an alkaline earth element, $S = 1$ or 0, and for a three-electron system $S = \frac{3}{2}$ and $\frac{1}{2}$ and these values may combine with possible L values.

The *multiplicity* of a term is given by $M = 2S + 1$. If $S = \frac{1}{2}$, as for an alkali metal, $M = 2$ and a doublet occurs. For alkaline earth elements (two-electron systems) $S = 0$ or 1. This results in two sets of terms, a singlet and a triplet system.

7. ALKALINE EARTH ATOMIC SPECTRA

A Grotrian diagram of the energy levels of calcium is shown in Figure 2-8. The spectrum of calcium consists of a singlet and a triplet system, as predicted, since the values of S are 0 and 1. Terms within each system combine according to the selection rules already given. Transitions between levels of the two systems to produce intercombination lines do not occur except in a few cases and those that occur are very weak. This behavior gives rise to the spin selection rule $\Delta S = 0$. The existence of the two systems in calcium (the

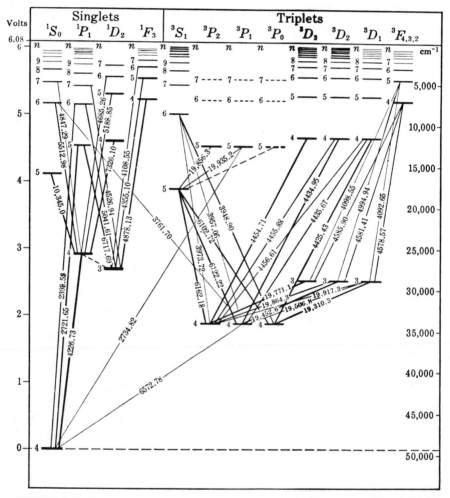

FIGURE 2-8. Energy level diagram of calcium. [From G. Herzberg, *Atomic Spectra and Atomic Structure*, Dover Publications (1944). Used by permission of Dover Publications.]

singlet and triplet systems) occurs with all alkaline earth elements as well as helium and all two-electron configurations. This behavior constitutes strong evidence for the existence of two electrons in the valence shell of each of these elements.

8. SPECTRAL SERIES AND SPECTROSCOPIC TERM SYMBOLS

As shown with hydrogen, the emission spectrum of an element consists of several series of lines arising from transitions from some higher energy state to a common lower state. In hydrogen these series are quite well separated, with only minimal overlap. Reference to Figure 2-2 will make this clear.

With the more complex spectra of heavier elements the different series have considerable overlap, as shown in figure 2-9. The upper spectrum is that of atomic sodium. Below it are shown the principal, sharp, diffuse, and fundamental series sorted out of the complete sodium spectrum. The doublet character of the separate series also is shown. The typical characteristic of each series to converge to a series limit also is clear.

Notation to describe a spectral energy level using the quantum numbers n, L, S, and J has been devised as follows. The terms are written as $n^M L_J$. Here n is the principal quantum number; M ($M = 2S + 1$) is the multiplicity; L is written as a letter, that is, for $L = 0$ the letter used is S, for $L = 1$ it is P, for $L = 2$ it is D, and for $L = 3$ it is F. Both J and M are written as numbers. For example, the upper levels of the two D lines of sodium arise from a one-electron system where $n = 3$, $L = 1$, $S = \pm\frac{1}{2}$, and $J = \frac{1}{2}$ or $\frac{3}{2}$. The term symbols for these two levels of sodium are $3^2 P_{1/2}$ and $3^2 P_{3/2}$, respectively.

The common lower level for these two sodium lines arises from one electron where $n = 3$, $l = 0$ and $s = \frac{1}{2}$, so $L = 0$, $S = \frac{1}{2}$, $M = 2$, and $J = \frac{1}{2}$, so the term symbol is $3^2 S_{1/2}$. Therefore these two transitions are written as

$$3^2 S_{1/2} - 3^2 P_{1/2} \quad \text{and} \quad 3^2 S_{1/2} - 3^2 P_{3/2}$$

The lower level of a transition is written first. The lines are part of the principal series of sodium and their positions in the series can be seen in Figure 2-9, where the series notations also are shown along the vertical axis.

9. ZEEMAN AND STARK EFFECTS

The splitting of a degenerate energy level into states of slightly different energies in the presence of an external magnetic field is known as the Zeeman effect. The total angular momentum J is quantized along the magnetic axis

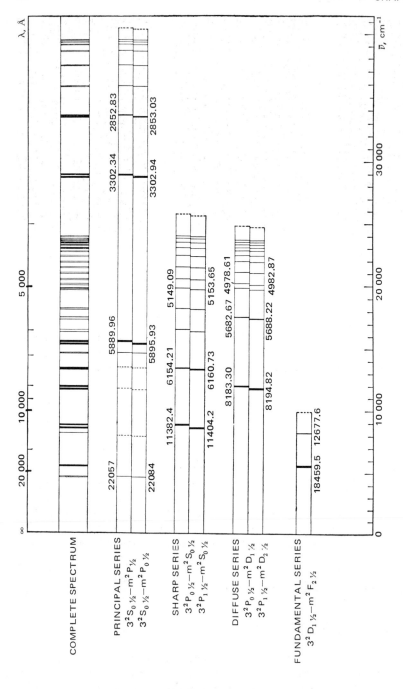

FIGURE 2-9. Spectral series of sodium. [From W. Brode, *Chemical Spectroscopy*, 2nd ed., Wiley, New York (1943). Used by permission of Wiley, New York.]

THE ORIGIN OF ATOMIC SPECTRA

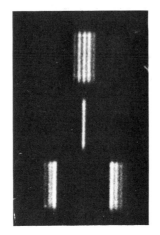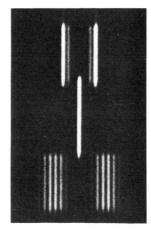

FIGURE 2-10. Zeeman spectra of spectral lines of scandium. [From Ph.D. dissertation by Lorin Neufeld, Kansas State University.]

and has values $M_J = J, J-1, \ldots, -J$ as possible projections of J along the axis. Since J is determined by L and S, the amount of Zeeman splitting is affected by their values. Zeeman splitting derives from the interaction of the external magnetic field with J. The magnitude of the splitting is proportional to the strength of the magnetic field.

Since the level is split only in the presence of a magnetic field, the components of the line are degenerate in the absence of the field. Since only discrete orientations are permitted, the angular momentum vector J must be space-quantized in a magnetic field. The number of components of M that are possible is $2J + 1$: thus if $J = 2$, then $M = 5$, and if $J = \frac{5}{2}$, $M = 6$. Examples of the Zeeman effect are shown in Figure 2-10 for several spectral lines of scandium.

If splitting can occur in a magnetic field, an analogous effect should be observable in an electric field. Experimental evidence verifies that a similar splitting can be produced electrically; this is known as the Stark effect.

10. SPECTRAL LINE INTENSITIES

10.1 Statistical Weight

The intensity of a spectral line is related to the total number of states in the level from which the transition originates as given by the expression $M = 2J + 1$. If a population is assigned to each state of a level (a condition verified by experiment), the level of higher M will produce a more intense line in the absence of a magnetic field. If a simple level (with $J = 0$) is assigned a statistical weight of 1, then a level with angular momentum J has a statistical weight of $2J + 1$.

For two levels of different J values the probability of finding an atom in one or the other of the two states is

$$\frac{2J_1 + 1}{2J_2 + 1} = \frac{g_1}{g_2}$$

providing both states are of approximately equal energy.

Consider the example of the relative intensities of the sodium D_1 and D_2 lines. Both lines have a common lower level of $3^2S_{1/2}$. The two upper levels, of almost equal energies, are $3^2P_{1/2}$ and $3^2P_{3/2}$. The numbers of degenerate states are, respectively, $2(1/2) + 1 = 2$ and $2(3/2) + 1 = 4$. This calculated relative intensity ratio of $1:2$ is in accord with experimental spectral line intensity observations.

10.2 The Boltzmann Distribution Factor

If thermal equilibrium exists among the atoms being considered, and if E_n is the energy of state n, the number of atoms in state n is proportional to $e^{-E_n/kT}$ if n is a nondegenerate state. This exponential term is known as the Boltzmann distribution term, where k is the Boltzmann constant (1.381×10^{-16} erg/°C) and T is the absolute temperature.

If state n is degenerate, the probability of finding an electron in state n equals the Boltzmann term multiplied by the statistical weight g_n, or is proportional to $g_n e^{-E_n/kT}$. The ratio of atoms found in two excited states is then given by

$$\frac{N_n}{N_m} = \frac{g_n e^{-E_n/kT}}{g_m e^{-E_m/kT}} \tag{2-22}$$

If state m is the ground state $E_m = 0$, equation (2-22) reduces to

$$N_n = N_m \frac{g_n}{g_m} e^{-E_n/kT} \tag{2-23}$$

and the intensity of the electronic transitions from state n to the ground state will be proportional to N_n. Equation (2-23) indicates that the population of the excited state is exponentially related to the absolute temperature and that states of higher energy have smaller populations; thus, allowed low-energy transitions are more intense than allowed high-energy transitions.

THE ORIGIN OF ATOMIC SPECTRA

11. TRANSITION PROBABILITIES— OSCILLATOR STRENGTHS

Spectral emission lines originate from electronic transitions between energy levels of atoms or ions. The frequency of the photons emitted is related, by Planck's constant, to the energy difference between the upper and lower energy states, or

$$hv = E_2 + E_1 \tag{2-24}$$

where v is the frequency, h is Planck's constant, and E_2 and E_1 are the energies of states 2 and 1, respectively.

Einstein proposed that the number of transitions occurring between states 2 and 1 ($dN_{2\to1}$) per unit time is governed by a law similar to radioactive decay, or

$$-dN_{2\to1}/dt = A_{2\to1} N_2 \tag{2-25}$$

where N_2 is the number of atoms in state 2 and $A_{2\to1}$ is a transition probability coefficient usually referred to as the Einstein coefficient or the Einstein transition probability of spontaneous emission.

The emission power P of the spectral line is the energy emitted per unit time and is the product of the number of photons $dN_{2\to1}$ and the energy of the photons, or

$$P_{2\to1} = dN_{2\to1} hv = hv A_{2\to1} N_2 \tag{2-26}$$

Classical radiation theory originally proposed a quantity called *oscillator strength*, which was related to the intensity of emission of a given radiation. It was originally defined as the number of electrons per atom responsible for the radiation. This explanation no longer is valid but the term oscillator strength f is still used and is defined as follows:

$$A_{2\to1} = \frac{8\pi^2 e^2}{\lambda^2 mc} f_2 \tag{2-27}$$

where e is the charge on the electron, m is the mass of the electron, c is the velocity of light, and λ is the wavelength of the transition from state 2 to state 1.

Combining equations (2-26) and (2-27) gives

$$P_{2\to1} = \frac{8\pi^2 e^2 h}{\lambda^3 m^2} N_2 f_2 \tag{2-28}$$

From equation (2-28) it is evident that the emitted radiant power of a particular transition is proportional to the oscillator strength if the population of atoms in upper excited state is constant. Further, if N_2 is known, the oscillator strength can be determined from the spectral line intensity. Corliss and Bozman[1] have determined the oscillator strengths of 25,000 atomic lines of 70 elements. They used a copper arc at 5100 ± 110°K as an excitation source, and obtained relative intensity values which then were converted to absolute values by calibration against known absolute gf values as normalized from a number of sources. The statistical weight g refers to the upper level of the two states involved in the transition.

Corliss and Bozman list, for each line, the values of gA, gf, and $\log gf$. These data can be of value to analytical chemists in their choice of spectral lines for analytical purposes and also for qualitative comparisons of line intensities of elements. Appendixes II and III of this book present data on the more intense spectral lines of the elements abstracted from the Bureau of Standards Tables.

The oscillator strength of the emission transition from state 2 to state 1 is not equal to the oscillator strength of the same transition when energy is absorbed by the atom in moving from state 1 to state 2, or $f_{2 \to 1} \neq f_{1 \to 2}$. The quantitative relation between absorption of energy and emission for the same transition is given by

$$g_2 f_{2 \to 1} = g_1 f_{1 \to 2} \tag{2-29}$$

since the statistical weight must be taken into account. This relation is useful when atomic absorption processes are being considered.

Equation (2-28) frequently is modified to relate the spectral line intensity to the population of atoms in the lower energy state (N_1) of the transition being considered. The expression relating the populations of the two states is

$$N_2 = N_1 \frac{g_2}{g_1} e^{-E_{2 \to 1}/kT} \tag{2-30}$$

where N_1 is the population of the lower state, N_2 is the population of the upper state, $E_{2 \to 1}$ is the energy of the transition, k is the Boltzmann constant, T is the absolute temperature, and g_1 and g_2 are the statistical weights of the two states.

Combining equations (2-28) and (2-30) gives

$$I_{2 \to 1} = \frac{8\pi^2 e^2 h}{\lambda^3 m} N_1 \frac{g_2}{g_1} f_{2 \to 1} e^{-E_{2 \to 1}/kT} \tag{2-31}$$

From equation (2-31) it is evident that the intensity of a spectral line for a particular transition is proportional to its oscillator strength, and varies exponentially with the absolute temperature and inversely exponentially with the energy of the transition.

---

[1] C. H. CORLISS, and W. R. BOZMAN, *Experimental Transition Probabilities for Spectral Lines of Seventy Elements*, National Bureau of Standards Monograph 53 (1962).

12. SPECTRAL LINEWIDTHS

The width of a spectral line is determined by several factors, including the following:

1. The *natural* width of a spectral line is determined by the Heisenberg uncertainty principle and the lifetime of the excited state. The Heisenberg principle states that the position and velocity of an electron cannot be specified with complete accuracy. Since the lifetime of an excited electronic state is of the order of 10^{-8}–10^{-10} sec, there must therefore be associated with the electron sufficient energy uncertainty to provide for slight broadening of the spectral line. The natural linewidth is very small and of the order of 10^{-4} Å.

2. *Doppler broadening* is due to the random kinetic motion of the atom toward and away from the point of observation during the emission process. Doppler broadening, therefore, is temperature dependent; the higher the temperature, the greater the line broadening. Doppler broadening in an electric arc source can be from 0.01 to 0.05 Å. Doppler line shift has been used to determine the velocity of stars in relation to the earth.

3. *Pressure broadening* is caused by collisions with foreign atoms in the plasma, and is sometimes called Lorentz broadening. If an atom is in collision at the time it is emitting, the frequency of the emitted radiation will be changed slightly. If two atoms with electric dipole moments approach one another, the combination has a potential energy curve characteristic of a molecule. If one of the atoms is excited, the emitted radiation is slightly different from what would have occurred if the atom was free from the presence of other atoms or ions. The frequency shift that occurs is usually to the red. Since this type of line broadening is proportional to the particle concentration, it is called pressure broadening. Pressure broadening can be of the order of 0.05 Å.

4. *Resonance broadening* results from collisions with similar atoms. Since the concentration of emitting atoms is usually very small compared with the concentration of nonemitting species, resonance broadening normally is very small.

5. *Stark broadening* occurs if there are a considerable number of ions and electrons in the excitation source. Excited atoms therefore are subject to strong electrical fields from nearby ions and electrons and may collide with them. Spectral radiation emitted under these conditions may be sufficiently affected to cause the line to broaden. Since the local electrical fields are continually changing and are not of uniform intensity, a broad, diffuse line results. This effect was observed experimentally by early spectroscopists and led to their identification of certain spectral line series as being "diffuse" and "fundamental."

Spectral linewidths can be calculated if temperature and pressure are known. The natural linewidth and usually the resonance broadening can be neglected. In low-temperature sources, such as a flame, Stark broadening also can be neglected. Doppler broadening and pressure broadening therefore are the most important factors to consider at low temperatures. Spectral linewidths under these conditions range from about 0.02 to 0.04 Å. In higher temperature sources linewidths may increase to 0.1–0.2 Å.

13. ATOMIC FLUORESCENCE

Atomic fluorescence occurs when an atom reaches an excited state by the absorption of electromagnetic radiation and then returns to a lower state, or a ground state, by the emission of energy. There are four distinct types of atomic fluorescence processes and these are illustrated in Figure 2-11. Figure 2-11a is an example of resonance fluorescence. In this case the emitted energy is the same as the absorbed energy.

Stepwise line fluorescence is shown in Figure 2-11b. In this case the atom is excited to a state *above* the first excited state. This is followed by a radiationless deactivation to the first excited state followed by emission of energy $h\nu_2$ to return to the ground state.

Fluorescence of the type shown in Figure 2-11c is called direct line fluorescence. Excitation raises the electron to an excited state above the

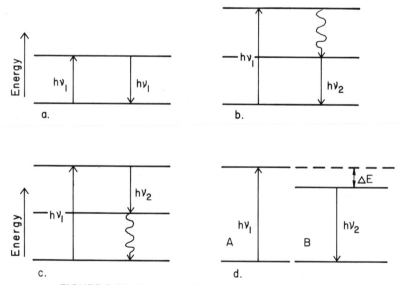

FIGURE 2-11. Four types of atomic fluorescence processes.

lowest available state just as in the case of stepwise line fluorescence. The excited electron returns to the ground state by first emitting radiation corresponding to hv_2 to return to an intermediate energy state, followed by a radiationless transition to the ground state.

Sensitized fluorescence is shown in Figure 2-11d and involves a transfer of excitation energy from one atom (A) to another (B). A mixture of atoms A and B is excited with radiation of energy equal to hv_1. If the population of B is such that collisions between A and B will occur within the lifetime of the excited state of A, some energy will be transferred to B followed by emission of energy (hv_2) by B. The energy difference between the excited state of A and the excited state of B, indicated as ΔE in the figure, will be absorbed as kinetic energy by atoms of both A and B.

Analytical uses of atomic fluorescence have been developed in recent years. Most of the methods utilize resonance fluorescence, but other types of fluorescence also are useful. For example, spectral emission lines of mercury have been used to produce fluorescence of elements such as iron, thallium, chromium, and magnesium. The instrumentation and techniques for analytical applications of atomic fluorescence are described in Chapter 11.

14. METASTABLE STATES—LASER ACTION

A metastable state is defined as an excited state of an atom which does not (according to the selection rules) combine with the ground state of the atom. For example, the 2^3S_1 (the lowest lying level of the helium triplet system) will not combine with the 1^1S_0 (ground) state of helium. Thus, the 2^3S_1 state is said to be metastable. The 2^3S_1 state of helium lies about 19.7 eV above the ground state and frequently is used as a source of collisional excitation energy to study excited atomic and molecular states as well as chemical reactions. The 2^1S_0 state of helium also is metastable since the selection rule that $\Delta L = \pm 1$ does not allow for the transition to the ground state. Figure 2-8, the energy level diagram for calcium, shows a similar situation; the 5^1S_0 cannot return to the 4^1S_0 state, and the 4^3P_2 state cannot drop to the 4^1S_0 ground state. The excited states for which the probability of returning to the ground state by emission of energy is very low also cannot be reached from the lower state by the direct absorption of energy. These states can be populated by an indirect process, such as energy transfer by collisions with high-energy particles.

14.1. Laser Action

The helium–cadmium ion gas laser furnishes an excellent example of one mechanism of laser excitation and emission. Figure 2-12 presents the

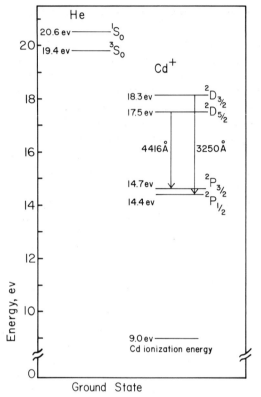

FIGURE 2-12. Energy level diagram of a helium–cadmium ion laser system.

basic energy level diagram of this system. The helium metastable states ($^1S_0$ and $^3S_1$) are populated by utilizing radiofrequency energy for ionization and excitation. Some electrons are sufficiently energetic to produce excited helium atoms by collisional transfer of energy. The metastable helium atoms lose their energy by collision with ground state cadmium atoms. The collisions result in the formation of excited cadmium ions by a process that can be described by the reaction

$$\text{He}^* + \text{Cd}^0 \rightarrow \text{He}^0 - \text{Cd}^{+*} + e^- + \Delta E$$

where ΔE is the energy difference between that possessed by the excited helium atoms and the excited cadmium ions. The $^2D_{3/2}$ and $^2D_{5/2}$ excited states of Cd$^+$ can be stimulated to emit energy by laser action as they drop to the $^2P_{1/2}$ and $2P_{3/2}$ states, resulting in emission at 3250 and 4416 Å, respectively. The approximately 20 eV of energy possessed by the excited states of helium is more than sufficient to ionize and excite the ground state cadmium atoms.

The stimulated emission of a laser is in contrast to the spontaneous emission that produces fluorescence, and results in emission with unique properties. The light that is produced (1) is of very high power density, (2) is monochromatic, (3) travels in a narrow, parallel beam, and (4) is coherent, that is, all the emitted radiation is in phase.

Solid substances also are used to produce laser action. The most common of these is the ruby laser, Al_2O_3 containing about 0.05% Cr_2O_3. The electronic levels of the Cr^{3+} ion are used to produce laser action in a manner similar to that described for the helium–cadmium ion system. In the ruby laser, energy is produced at a wavelength of 6943 Å.

Tunable lasers have been produced using an organic dye in the laser. A dye has a broad, continuous fluorescence spectrum rather than a single-line fluorescence as occurs with fixed-frequency lasers. A continuous emission spectrum (rather than single-line emission) is produced by the laser-active material. Because most dye molecules are large, they possess many bonds and internal degrees of freedom. The opportunity therefore exists for many vibrational and rotational levels, all very close together. The energy thus produced is a virtually continuous energy level band.

The dye laser can be "tuned" by changing the concentration of the dye. However, if the laser cavity is arranged to include a diffraction grating or prism, the bandwidth can be dramatically reduced with little loss of power. Rotation of the grating or prism changes the frequency of the wave that is parallel to the laser cavity. Bandwidths as narrow as 1–2 Å can be achieved, and with a Fabry–Perot interferometer bandwidths of 0.01 Å can be obtained.

Because of the unique characteristics of their emitted energies, lasers have been used for sample vaporization and excitation sources in atomic emission spectroscopy. They also have been used as sources for atomic absorption and atomic fluorescence analysis. Their application in these areas will no doubt increase as lasers become cheaper and more readily available.

15. MOLECULAR SPECTRA (BAND SPECTRA)

Although not of primary concern to atomic spectroscopists, a limited background concerning molecular spectra is important. Band spectra produced by molecular emission are sometimes used analytically, while some other band spectra appear at wavelengths that interfere with important atomic emission lines.

Molecular spectra are more complex than atomic spectra and thus are more difficult to analyze. Molecules, like atoms, can emit or absorb energy

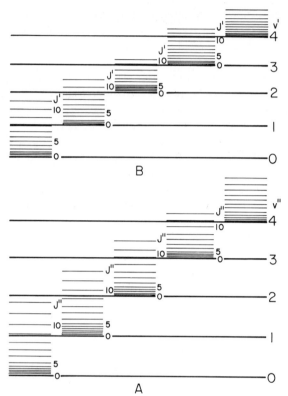

FIGURE 2-13. Typical energy level diagram for a molecule.

only in discrete amounts corresponding to certain energy states of the molecule. This energy is related to the Planck equation

$$\Delta E = h\nu \tag{2-32}$$

where ΔE is the difference between two energy states and h and ν have their usual meanings. Energy is absorbed in a transition from a lower to a higher energy state and emitted when the transition occurs from a higher to a lower state.

Molecules may absorb or emit energy in three different ways. These include electronic transitions, vibrational transitions, and rotational transitions. Molecules also possess translational energy but it is not observed with present instrumentation. The total quantized energy of a molecule is therefore given by the expression

$$E_{\text{total}} = E_e + E_v + E_r \tag{2-33}$$

THE ORIGIN OF ATOMIC SPECTRA

Transitions between molecular electronic levels occur in the visible and ultraviolet regions of the spectrum. Since both vibrational and rotational transitions are of much lower energy, they occur in the infrared, with the purely rotational levels being in the far-infrared and microwave regions ($> 20 \mu$m).

When electronic transitions occur in molecules they are accompanied by vibrational and rotational transitions which are superimposed on the electronic transitions. The result is a complex spectrum which includes the combinations of energy levels shown in Figure 2-13. A typical band system (that for CN) is shown in Figure 2-14 which was obtained using a dc arc, graphite electrodes, and photographic recording. Figure 2-15, obtained using a plasma arc and photographic recording followed by microphotometer tracing of the spectrum, shows a series of bands in the useful atomic emission spectral region.

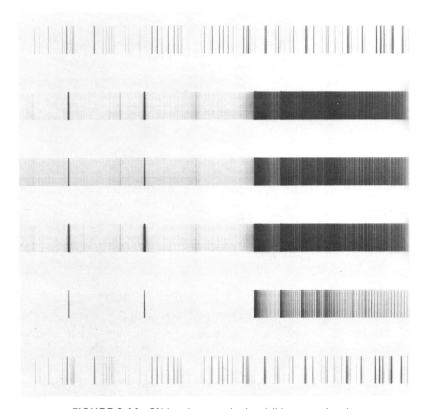

FIGURE 2-14. CN band spectra in the visible spectral region.

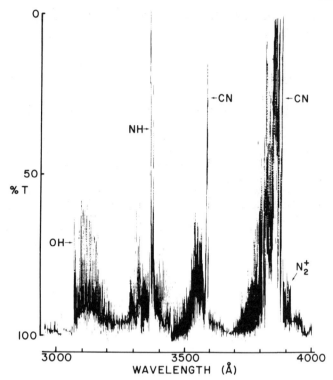

FIGURE 2-15. Microphotometer tracings of some band spectra between 3000 and 4000 Å. [From Ph.D. dissertation by S. E. Valente, Kansas State University.]

These spectra appear as bands or very closely spaced lines because the energy differences between vibrational and rotational states are small. In addition, the translational (kinetic) energies of the molecules also broaden the spectral lines. This is a case of Doppler broadening applied to molecules. Collisional broadening also has some effect on linewidth. The line structure of electronic spectra of molecules can only be studied by use of an extremely high-resolution spectrometer.

Some molecular spectra have been used by atomic spectroscopists for analytical purposes. These include the electronic band spectra of CaO, MgO, S_2, and C_2. More often the band spectra encountered in analytical atomic spectroscopy adversely affect atomic spectroscopy since they tend to mask or obscure useful atomic spectral lines. Bands such as those produced by CN, N_2, NH, and OH are particularly troublesome in the region from 3000 to 4000 Å, a region containing many useful atomic lines. For example, the most sensitive lines of copper fall in the OH band region of the spectrum.

THE ORIGIN OF ATOMIC SPECTRA

Some of the methods used to minimize effects of band spectra on atomic spectra will be dealt with in later chapters.

Selected Reading

Condon, E. U., and Shortley, G. H., *The Theory of Atomic Spectra*, Cambridge University Press, New York (1957).

Herzberg, G., *Atomic Spectra and Atomic Structure*, Dover Publications, New York (1944).

Kuhn, H. G., *Atomic Spectra*, 2nd ed., Longmans, Green and Co., London (1969).

Mitchell, A. C. G., and Zemansky, M. W., *Resonance Radiation and Excited Atoms*, Cambridge University Press (1934; reprinted 1961).

Pauling, L., and Goudsmit, S., *The Structure of Line Spectra*, McGraw-Hill, New York (1930).

Schwalow, A. L., *Lasers and Light*, W. H. Freeman, New York (1969).

Webb, J. Pierce, Tunable Organic Dye Lasers, *Anal. Chem.*, **44**, 30A (1972).

White, H. E., *Introduction to Atomic Spectra*, Cambridge University Press and McGraw-Hill, New York (1929).

Chapter 3
Filters, Prisms, Gratings, and Lenses

The optical system of any instrumentation utilizing radiant energy for analytical purposes includes some device to isolate the spectral region desired together with other optical devices to direct and focus the radiant energy. Spectral isolation is achieved by using a filter, a prism, or a grating, depending on the degree of wavelength resolution required. The collection and focusing of the light from the source are usually accomplished by using lenses and/or mirrors. Since careful control of the light path is needed to achieve maximum efficiency, an optical bench frequently is used to mount the parts of the optical system.

1. FILTERS

1.1 Absorption Filters

The simplest device used to isolate a spectral region is a color filter. The three main requirements of a filter are (1) high transmittance in the desired wavelength region, (2) low transmittance at all other wavelengths, and (3) a narrow wavelength band in the high-transmittance region. A number of other interdependent factors are involved in a useful color filter, including the spectral response curve of the detector, the intensities and bandwidths of the emissions to be measured, and the possibility of interferences in the spectral region transmitted by the filter from emission of elements other than the desired one.

Commonly available color filters have relatively wide spectral bandwidths. Widths of 350–400 Å are common, measured at a transmittance of one-half the maximum. The most suitable filter is that which has its maximum transmission at the desired wavelength and a small spectral bandwidth at

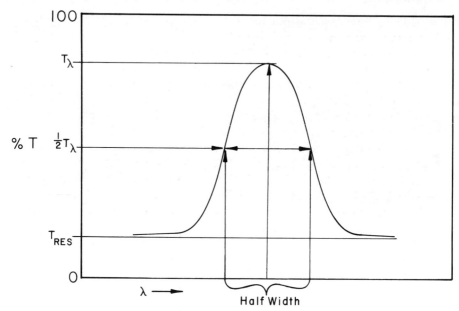

FIGURE 3-1. Characteristics of a light filter.

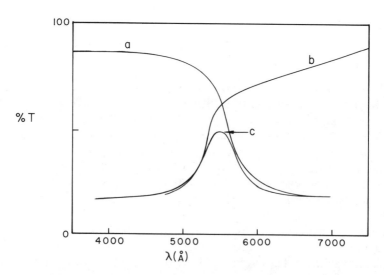

FIGURE 3-2. Combination of two light filters to isolate a spectral region.

one-half the maximum transmission as well as at the low transmission points. These terms are illustrated in Figure 3-1.

Commonly used colored filters consist of colored glass, colored films, or solutions of colored substances. Since most colored substances have broad transmission curves, it is necessary to combine colored substances to provide a suitable filter. Figure 3-2 illustrates, in an idealized way, how this can be accomplished. Substance a transmits below about 6000 Å and absorbs above this region. Substance b transmits above about 5800 Å and absorbs below. The combination of these two substances into a single filter produces a light filter with maximum transmission at about 5900 Å with a relatively sharp cutoff both above and below this wavelength. A filter with characteristics such as those in Figure 3-2 would be well adapted to isolate the sodium D lines at 5890 and 5896 Å from spectral emission above 6000 Å and below 5500 Å. Among most colored glasses or gelatins, sharp cutoffs on the long-wavelength side are less common. Stability of the colored compounds used to make filters also is a problem. The substances used should be stable and long-lived to produce a satisfactory filter.

1.2. Interference Filters

The interference filter can provide better spectral isolation than a color filter. The bandwidth at one-half transmittance may be as low as 100 Å. The principle of the interference filter is shown in Figure 3-3. The filter itself is composed of two parallel, half-silvered (semitransparent) surfaces I and J, carefully spaced by a transparent material. The half-silvered surfaces are separated by one-half of the desired wavelength.

As light of wavefront AB enters the filter, a fraction is reflected from surface J to surface I. If the spacing is such that distance CD equals one-half the

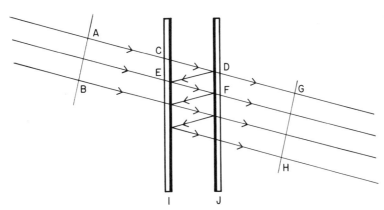

FIGURE 3-3. Light path through an interference filter.

wavelength $\lambda/2$ and DE also equals one-half the wavelength $\lambda/2$, the reflected ray $CD + DE$ will be in phase with the ray entering directly at E at wavelength λ. Therefore, light of wavelength λ will be reinforced and emerge from the filter. If the entering beam contains components slightly shorter or slightly longer than λ, there will be some constructive interference but of lower intensity than at λ. A spectral band therefore is passed by the filter.

In Figure 3-3 the entering beam is drawn at an oblique angle to illustrate the principle of the filter. The filter is used with the incident light beam normal to the filter surface. If radiant energy enters the filter at some angle other than normal, the spectral band passed by the filter shifts toward shorter wavelengths.

Frequencies harmonically related to the fundamental frequency of the interference filter also will be reinforced. A filter with maximum transmittance at 6000 Å also will transmit at 3000 and 2000 Å, corresponding to a path length for CD plus DE of 2λ and 3λ, respectively. If these harmonically related frequencies are a source of difficulty in the use of the interference filter, they can be removed by combining the interference filter with a color filter with peak transmittance at the same wavelength as the fundamental of the interference filter. Such a combination will provide the better spectral isolation of the interference filter with removal of harmonically related transmission peaks.

The interference wedge is a variation of the interference filter and is illustrated in Figure 3-4. In the interference wedge the transparent filler between the half-silvered surfaces is wedge-shaped. Thus, referring to Figure 3-4, light beam 1 passes through a narrow portion of the wedge to reinforce short-wavelength light, and light beam 2, at the thicker end of the wedge, reinforces at a longer wavelength. The interference wedge therefore is a continuously variable light filter. By use of an entrance slit, any particular wavelength maximum can be selected by moving the proper portion of the wedge into the optical path. The interference wedge will respond to harmonically related wavelengths just as the interference filter does.

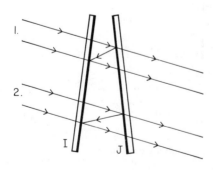

FIGURE 3-4. Light paths through an interference wedge.

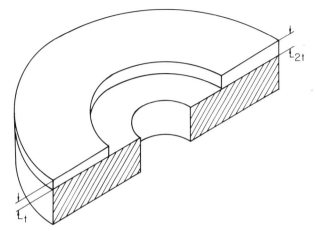

FIGURE 3-5. Schematic diagram of a circular variable filter. [From V. L. Yen, Circular variable filters, *Optical Spectra* **3**, No. 3, 78 (1969). Used by permission of *Optical Spectra*.]

Interference filters provide the best wavelength selectivity of any filters available. It is not possible to provide the necessary resolving power required for more complex spectral isolation. The filters therefore are primarily useful for simple systems where passage of a spectral band will meet spectral isolation requirements.

1.3. Circular Variable Filters

A circular variable filter is an interference filter constructed so rotation of the filter changes the spectral bandpass of the filter. The filter can be constructed so the wavelength change is a linear function of the rotation angle.

Figure 3-5 represents a typical circular variable filter design. As the filter is rotated, the wavelength of the bandpass of the filter changes from that produced by thickness t to $2t$ (at 180°) and then back to t as rotation continues. Wavelength coverage can be varied. Filters have been made to cover 2–4 μm and also 4–8 μm over 180° rotation. The minimum wavelength produced has been 0.4 μm and the maximum 25 μm, so the filters are of most use in the infrared spectral region. The spectral bandpass of the filters is about 1% of the center frequency at the half-intensity level.

Circular variable filters can be used in scanning monochromators; they are easily calibrated, have constant resolution, are rugged, and require little maintenance. The filters also have application in radiometers, where they have replaced sets of filters of fixed wavelengths. Another application is as order sorter for spectrophotometers.

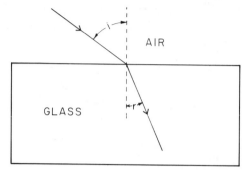

FIGURE 3-6. Refraction of light entering a denser medium.

2. PRISMS

Refraction of light as it passes from a medium of one density to a medium of a different density was known for centuries, but it remained for Snell (1620) to formulate the law that quantitatively describes the phenomenon. Snell observed that the ratio of the sine of the angle of incidence to the sine of the angle of refraction is constant. Referring to Figure 3-6, with the angle of incidence in air given as i and the angle of refraction as r, Snell's law is expressed as

$$\frac{\sin i}{\sin r} = \text{const} \tag{3-1}$$

However, the index of refraction also is a property of the substances involved. Thus in determining the index of refraction of a substance, the relation that exists between light traveling through a vacuum and through the denser medium is determined. In addition, the index of refraction is a function of wavelength, so the wavelength also must be specified. Snell's law then becomes

$$\frac{\sin i}{\sin r} = n_\lambda \tag{3-2}$$

where i is measured in a vacuum and r in the denser medium. Then n_λ is the index of refraction of the denser medium at wavelength λ. Another definition of index of refraction is that it is the ratio between the velocity of light through a vacuum and that through the denser medium.

The action of a prism to form a spectrum depends on the fact that the index of refraction is a function of wavlength, with the index of refraction increasing as the wavelength decreases. Cauchy first suggested the empirical formula

$$n = A + \frac{B}{\lambda^2} \tag{3-3}$$

FILTERS, PRISMS, GRATINGS, AND LENSES

to express the relation between wavelength and index of refraction. The constants A and B vary for different media. Knowing A and B permits calculation of n at any given wavelength.

Another useful empirical formula, if applied to a limited wavelength region, was given by Hartmann as

$$n = n_0 + \frac{c}{\lambda - \lambda_0} \tag{3-4}$$

Both formulas are useful in determining the index of refraction of a transparent substance at any given wavelength.

2.1. Dispersion of a Prism

Figure 3-7 shows a trace of the path of a monochromatic beam of light through a prism at the angle of minimum deviation. At minimum deviation the path through the prism is parallel to the base of the prism and angles $i_1 = i_2$ and $r_1 = r_2$.

Snell's law gives

$$n = \frac{\sin i_1}{\sin r_1} \quad \text{and} \quad n = \frac{\sin i_2}{\sin r_2} \tag{3-5}$$

The angle θ by which the beam is deviated by the prism is $i_1 - r_1 + i_2 - r_2$. Since $A = r_1 + r_2$, then $\theta = i_1 + i_2 - A$. At minimum deviation

$$\theta = 2i - A \quad \text{and} \quad i = (\theta + A)/2 \tag{3-6}$$

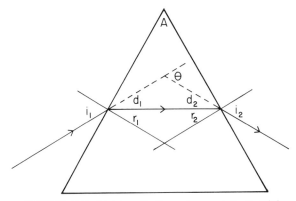

FIGURE 3-7. Light path through a prism at minimum deviation.

Substitution of equations (3-5) and (3-6) into Snell's law gives

$$n = \frac{\sin i}{\sin r} = \frac{\sin[(\theta + A)/2]}{\sin(A/2)} \qquad (3\text{-}7)$$

This relation holds for the condition of minimum deviation.

Since the dispersion of a prism is $d\theta/dn$, it is necessary to differentiate equation (3-7) and obtain its reciprocal,

$$\frac{d\theta}{dn} = \frac{2\sin(A/2)}{\cos[(\theta + A)/2]} \qquad (3\text{-}8)$$

Since

$$\sin^2 \frac{\theta + A}{2} + \cos^2 \frac{\theta + A}{2} = 1 \quad \text{or} \quad \cos \frac{\theta + A}{2} = \left(1 + \sin^2 \frac{\theta + A}{2}\right)^{1/2}$$

and

$$\sin \frac{\theta + A}{2} = n \sin \frac{A}{2}$$

we have

$$\frac{d\theta}{dn} = \frac{2\sin(A/2)}{[1 - n^2 \sin^2(A/2)]^{1/2}} \qquad (3\text{-}9)$$

is obtained as an expression which relates the change in angular dispersion as the refractive index changes.

The change in angular dispersion with wavelength $d\theta/d\lambda$ may be obtained as follows:

$$\frac{d\theta}{d\lambda} = \frac{d\theta}{dn} \frac{dn}{d\lambda}$$

$d\theta/dn$ has already been evaluated [(3-9)]. The quantity $dn/d\lambda$ can be evaluated by differentiating the Hartmann equation (3-4), which gives

$$\frac{dn}{d\lambda} = \frac{c}{(\lambda - \lambda_0)^2} \qquad (3\text{-}10)$$

The expression for the angular dispersion of a prism as the wavelength changes therefore is

$$\frac{d\theta}{d\lambda} = \frac{-2c\sin(A/2)}{(\lambda - \lambda_0)^2 [1 - n^2 \sin^2(A/2)]^{1/2}} \qquad (3\text{-}11)$$

The equation shows that as the wavelength increases, the angular dispersion decreases. Thus the angular dispersion of red light (longer wave-

FILTERS, PRISMS, GRATINGS, AND LENSES

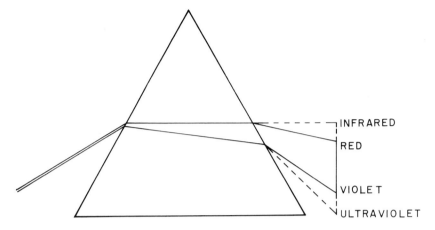

FIGURE 3-8. Path of a heterogeneous (white) beam of light through a prism.

length) is less than that for violet light (shorter wavelength). Figure 3-8 illustrates this relation for white light as it is dispersed as it passes through a prism.

2.2. Resolving Power of a Prism

The theoretical resolving power of a prism (or grating) is given by

$$R = \lambda/\Delta\lambda \qquad (3\text{-}12)$$

where $\Delta\lambda$ is the wavelength difference of two lines that are just resolved. The

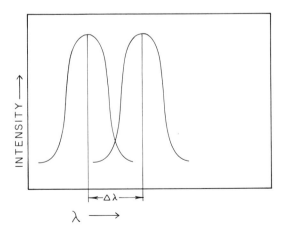

FIGURE 3-9. Rayleigh criterion for resolution of spectral lines.

criterion used by Lord Rayleigh is most commonly used for this purpose. Rayleigh stated that two lines could be resolved if the maximum intensity of one line is just "over" the minimum of the line near it. Figure 3-9 shows this separation, as used by Rayleigh.

Figure 3-10 shows how light from an infinitely narrow slit is focused to cover the face of the prism at the angle of minimum deviation. Two beams emerge of wavelengths λ and $\lambda + \Delta\lambda$, which just meet the Rayleigh criterion for resolution. The beams are separated by the angle $\Delta\theta$. Thus, $\Delta\theta = \lambda/a$, where a is the beam width. Consequently

$$R = \frac{\lambda}{\Delta\theta}\frac{d\theta}{d\lambda} = a\frac{d\theta}{d\lambda} \tag{3-13}$$

Since

$$\frac{d\theta}{d\lambda} = \frac{d\theta}{dn}\frac{dn}{d\lambda}$$

we have

$$R = a\frac{d\theta}{dn}\frac{dn}{d\lambda} \tag{3-14}$$

Substituting the expression for $d\theta/dn$ [(3-9)] into (3-14) at minimum deviation gives

$$R = a\frac{2\sin(A/2)}{[1 - n^2\sin^2(A/2)]^{1/2}}\frac{dn}{d\lambda} \tag{3-15}$$

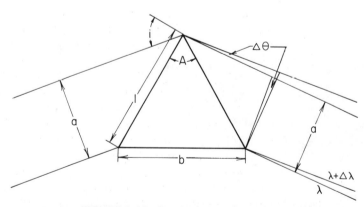

FIGURE 3-10. Resolving power of a prism.

FILTERS, PRISMS, GRATINGS, AND LENSES

Since

$$\left(1 - n^2 \sin^2 \frac{A}{2}\right)^{1/2} = \cos i$$

we have

$$R = a \frac{2\sin(A/2)}{\cos i} \frac{dn}{d\lambda} \qquad (3\text{-}16)$$

From Figure 3-10, $a = l \cos i$, and $b = 2l \sin(A/2)$, so

$$R = b \, dn/d\lambda \qquad (3\text{-}17)$$

Equation (3-17) for the theoretical resolving power of a prism uses the Rayleigh criterion for resolution, and is based on the use of an infinitely narrow entrance slit and on an optical system that makes use of the entire entrance face of the prism. If the entire face of the prism is not used, b, the length of the base of the prism, must be replaced by some smaller quantity which would be the "effective" base for the ray of light as it passes through the prism. Since the resolving power depends on the total effective thickness of the prism, two prisms, used in sequence, could produce greater resolution. Such arrangements are not common, however, since they are optically difficult to adjust.

The change of index of refraction with change in wavelength $dn/d\lambda$ is a property of the prism material and thus cannot be changed except by a change in the prism itself as used in the spectrometer.

2.3. Prism Materials

Prisms are made of a variety of transparent materials. Most common are glass, for the visible region, and quartz, for the ultraviolet and visible regions. Ideally a prism material should be transparent over a wide wavelength range, have a large change in index of refraction with wavelength $dn/d\lambda$, and have a low temperature coefficient for its index of refraction dn/dT.

A wide variety of glasses possessing different indices of refraction is available for use in the visible region. For example, a light crown glass has an index of refraction of 1.5170 at 5893 Å and a dense flint glass has an index of refraction of 1.6499 at the same wavelength. Salt (NaCl) crystals are used frequently since they are transparent in the infrared, but they are subject to severe deterioration unless carefully protected from moisture.

Liquid prisms have been used for special purposes, the liquid being enclosed in a hollow prism. Most liquids, however, possess high temperature coefficients for their indices of refraction and temperature gradients within the prism can be troublesome.

In the ultraviolet region, lithium fluoride, fused silica, and crystalline quartz all have been used. Crystalline quartz has an index of refraction of 1.54426 at 5893 Å, somewhat lower than glass, but is transparent to much lower wavelengths than glass. Quartz has the disadvantage of being doubly refracting; thus the entering light beam is split into two beams polarized perpendicular to each other. If the prism is cut so the beam travels along the optical axis of the quartz crystal, the effect of double refraction disappears.

Quartz also will rotate the plane of polarization of a beam of plane polarized light. This effect occurs even along the optical axis of the quartz, but can be compensated in the construction of the prism.

2.4. Types of Prisms

The two most common types of quartz prisms, as used in emission spectrographs, are the Cornu and Littrow types. The Cornu prism is made of two pieces of quartz as shown in Figure 3-11. Two pieces of quartz, one left-handed (A) and the other right-handed (B), are joined as shown in the figure. The optical axis is along C, which is parallel to the base. The prism is a 60° prism, all angles being the same. The two prisms that combine to make the Cornu prism have their joining surfaces in close contact to prevent light loss due to reflection. If the surfaces are made optically flat, no cement is needed to hold the two pieces together. The Cornu prism thus provides transparency into the ultraviolet region and compensates for the rotation of polarized light in the prism.

The Littrow-mounted quartz prism is a 30°, 60°, 90° prism, being essentially one-half of the Cornu prism. The back surface of the prism has a reflecting surface, either silver or aluminum, such that the light is reflected through the prism as shown in Figure 3-12. The effect thus is the same as

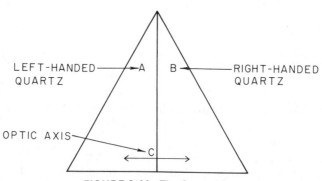

FIGURE 3-11. The Cornu prism.

FILTERS, PRISMS, GRATINGS, AND LENSES

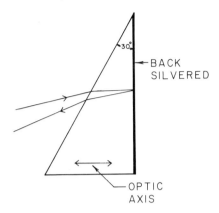

FIGURE 3-12. The Littrow prism.

light passing through a 60° prism and the optical rotation as the light enters the prism is compensated for as the ray returns through the prism. The optical axis is parallel to the base.

Glass prisms do not require the special attention given quartz since optical rotation and birefringence do not occur in glasses. However, glass prisms for spectrometers commonly are made similar to those of quartz. The 60° prism is common, as is a Littrow (half-prism) type. Prisms usually are used near the position of minimum deviation. For most uses the 60° prism appears to be close to the optimum design.

3. INTERFEROMETERS

The interferometer was developed by Michelson in 1887 and used by him to measure the wavelength of light. Figure 3-13 shows the essential features of an interferometer. It consists of two plane mirrors and a beam splitter. The beam splitter transmits approximately half of all incident radiation to a movable mirror and reflects half to a stationary mirror. Each component reflected by the two mirrors returns to the beam splitter, where the waves are combined.

If the two mirrors are equidistant from the beam splitter, the amplitudes of the two beams combine constructively. If one mirror is moved a distance of $\lambda/4$, the emerging beam will be the result of two beams 180° out of phase and they will combine destructively. If the entering beam from the source is monochromatic, the detector will observe a cosine signal whose amplitude is a function of the mirror position. If the entering beam is polychromatic, the signal observed at the detector is a summation of all interferences as each

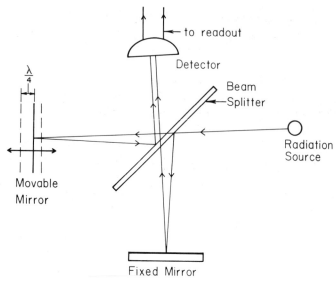

FIGURE 3-13. The principle of an interferometer.

wavelength component interacts with each other component. The resulting signal is called an interferogram and is a complex pattern of amplitude as a function of the distance the mirror is moved. Proper mathematical analysis of the interferogram permits recovery of the spectrum of the radiation source. Therefore an interferometer, in addition to its use to measure the wavelength of electromagnetic radiation, also can be used to obtain a spectrum of the source radiation.

4. DIFFRACTION GRATINGS

Fraunhofer usually is credited with the discovery of the diffraction grating. His grating was made of parallel wires wound back and forth between the threads of two long screws. Rowland and Michelson improved on the grating and found it to be an excellent device to use as a replacement for the prism to produce spectra. It was found especially useful in the ultraviolet and infrared regions of the spectrum, where prisms encounter transparency problems.

In developing the theory of grating action as a spectral dispersion medium, it is useful to observe the behavior of a monochromatic beam of light passing through a single narrow slit.

FILTERS, PRISMS, GRATINGS, AND LENSES

When a beam of monochromatic light with a parallel wavefront passes through a slit ab (see Figure 3-14), most of the light continues along the same path to point A on the screen. Light from edge a of the slit travels the same distance as from edge b and therefore arrives at A "in phase" to produce a highly intense image of the slit. Some light energy, however, is diffracted at the slit edges a and b and strikes the screen at other points. Light striking at point B follows two paths, aB and bB. If path bB is $\lambda/2$ longer than path aB, destructive interference occurs at B and a region of minimum intensity occurs. Similarly, at point C, where bC is one λ longer than aC, reinforcement occurs and a second maximum is produced. Thus a "diffraction pattern" is produced on the screen with an intensity distribution as shown.

A diffraction grating is made up of a series of narrow slits that are parallel and equally spaced. Some gratings are "transmission" gratings, through which light passes. Other gratings are called "reflection" gratings, where the light is reflected from closely spaced parallel grooves back to the side of the grating from which the entering light beam originated. In either case the same principle of grating action applies. To simplify the diagram of grating action (Figure 3-15), a transmission grating will be used.

When light passes through many parallel, equally spaced slits the diffraction pattern becomes very sharp. These maxima are the spectrum produced by a diffraction grating. In Figure 3-15, A–E represents a plane grating with equally spaced, parallel openings for transmission of light. The entering wavefront AF is parallel, and strikes the grating at an angle θ_1 from the normal. After passing through the grating the diffracted light is scattered as through a series of single slits.

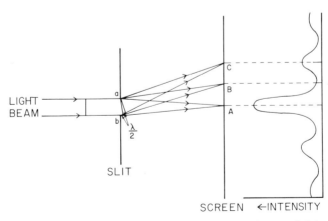

FIGURE 3-14. Diffraction of a monochromatic beam of light through a single slit.

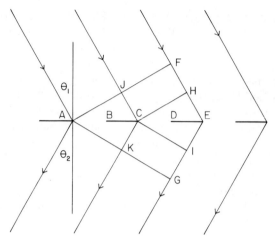

FIGURE 3-15. Diffraction of a plane transmission grating.

Consider a beam of light AF entering the grating and emerging with wavefront AG at angle θ_2. The part of the beam passing point C has a longer path length than that passing point A by the distance JCK, and that passing point E has a longer path by the distance FEG. If the distance JCK is one wavelength, then distance FEG equals 2λ of the light along the wavefront AKG. The three waves will be in phase for wavelength λ and reinforcement will occur.

The distance AC is a measure of the grating spacing, usually designated by the letter d. The condition for reinforcement of light passing through the grating to produce a bright image of the source is $JC + CK = \lambda$. Since $JC + CK = d(\sin\theta_1 + \sin\theta_2)$, then

$$\lambda = d(\sin\theta_1 + \sin\theta_2) \tag{3-18}$$

It is obvious that reinforcement of the light also can occur for multiples of the wavelength, so the general equation for reinforcement of light through a grating is

$$n\lambda = d(\sin\theta_1 + \sin\theta_2) \tag{3-19}$$

The equation holds for all values of θ_1 and θ_2 and is valid for reflection gratings as well as transmission gratings.

If $n = 1$, the spectrum is said to be a first-order spectrum, if $n = 2$, it is second order, etc. This relationship indicates that a first-order 6000 Å spectral line, a second-order 3000 Å line, and a third-order 2000 Å line would be superimposed on one another. Another deduction from equation

FILTERS, PRISMS, GRATINGS, AND LENSES

(3-19) concerns the condition that exists when the incident light is normal to the grating surface ($\theta_1 = 0$). In this case equation (3-19) reduces to $n\lambda = a \sin \theta_2$. Reference to this relation as well as Figure 3-15 indicates that the short wavelengths are dispersed least and long wavelengths most from the grating normal. This is in contrast to the spectrum produced by a prism, in which the longer wavelengths are deviated least.

4.1. Dispersion of a Grating

The angular dispersion of a grating is defined as $d\theta/d\lambda$ ($d\theta_2$ of Figure 3-15) and is the angular separation between lines of unit wavelength difference. An expression for the angular dispersion of a grating can be obtained by differentiation of equation (3-19), holding θ_1 constant, to give $n \, d\lambda = d \cos \theta_2 \, d\theta_2$, or

$$\frac{d\theta_2}{d\lambda} = \frac{n}{d \cos \theta_2} \tag{3-20}$$

Common usage of a grating calls for observing the spectrum on a normal to the grating ($\theta_2 = 0$). For the "normal" spectrum

$$\frac{d\theta_2}{d\lambda} = \frac{n}{d} \tag{3-21}$$

For a "normal" spectrum the angular deviation of spectral lines from the normal is very small and $d\theta_2/d\lambda = \text{const}$; thus there is a linear relation between angular dispersion and wavelength for the "normal" spectrum. It also should be noted that the dispersion increases as the "order", that is, a second-order spectrum has twice the dispersion of one of first order, etc.

As commonly used, dispersion is the wavelength separation of two spectral lines which are a unit distance apart at the focal plane of the spectrometer. This is actually *reciprocal linear dispersion* or $d\lambda/f \, d\theta_2$, where f is the distance from the grating to the focal plane of the spectrometer. Reciprocal linear dispersion has the dimensions of angstrom units per millimeter.

4.2. Resolving Power of a Grating

The resolving power of a grating has the same dimensions as for a prism and use is again made of the Rayleigh criterion for resolution. The numerical value of the resolving power is given by $\lambda/\Delta\lambda$, where $\Delta\lambda$ meets the Rayleigh definition.

Line AD of Figure 3-16 represents a grating from which wavelength λ emerges at angle θ. The just-resolved beam, of wavelength $\lambda + \Delta\lambda$, emerges

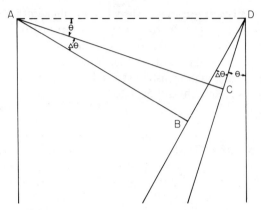

FIGURE 3-16. Resolving power of a grating.

at angle $\theta + \Delta\theta$, thus differing in emerging angle by $\Delta\theta$, with $\Delta\theta = \lambda/AB$, where AB is the beam width. Now

$$\frac{\lambda}{\Delta\lambda} = \frac{\lambda}{\Delta\theta} \frac{d\theta}{d\lambda}$$

and $AB = AD \cos\theta = Nd \cos\theta$, where N is the number of lines on the grating surface between A and D, and d is the line spacing. From the dispersion formula (3-20), we have

$$\frac{d\theta}{d\lambda} = \frac{n}{d \cos\theta} \qquad (3\text{-}22)$$

Thus, by substitution, we obtain

$$\frac{\lambda}{\Delta\lambda} = (Nd \cos\theta)\frac{n}{d \cos\theta} = Nn \qquad (3\text{-}23)$$

The theoretical resolving power of a grating thus depends on the product of the spectral order and the number of lines ruled on the grating. It *does not* depend on wavelength or grating spacing. If the entire face of the grating is not covered by the entering beam of light, resolution is decreased accordingly. This situation is similar to that described earlier for a prism, where maximum resolution occurs when the entire face of the prism is utilized by the entering light beam.

The above formula also assumes use of an infinitely narrow entrance slit, as does the formula for theoretical resolving power of a prism. The relation between theoretical and attainable resolution can be illustrated from Michelson's measurement of the wavelength of the green mercury line at 5641 Å. Operating in the sixth order on a 10-in. ruled surface Michelson attained a resolution of 600,000, compared with a theoretical resolution of 660,000.

4.3. Production and Characteristics of Gratings

Almost all gratings presently in use are reflection gratings and are produced by ruling grooves on the grating surface. To produce a high-quality grating requires special ruling machines to produce uniform, parallel grooves, 15,000–30,000/in., over a 6–12 in. surface. The grooves must be of the same depth and shape. Because of the extremely close tolerance required in producing a grating, it is understandable why gratings produced some years ago were not considered to be as desirable as prisms.

If errors arise in groove spacing, errors will occur in the spectra produced. The two most frequently occurring errors are periodic, depending on (1) each turn of the screw of the ruling engine and (2) regular intervals between two repetitive cycles of the ruling engine. The first condition produces "Rowland ghosts" and the second "Lyman ghosts".

Rowland ghosts can be easily identified since the "parent" line will be produced along with pairs of symmetrically spaced ghost lines on opposite sides of the parent. An intensely exposed spectrum of an element can be examined by observation of a strong line of the element and the region to either side of the line. In severe cases several ghost pairs will occur with the pair closest to the parent line being most intense. The ghost intensity increases with the order of the spectrum. Manufacturers of gratings commonly hold Rowland ghost intensities to less than 1/1000 in the first order for a grating of 15,000 lines/in. and will furnish guarantees on their gratings.

Periodic errors in the ruling engine produce Lyman ghosts, which are far removed from the parent line. Their wavelengths of appearance are normally a simple fraction of the wavelength of the parent line, such as 2/5, 3/5, 7/5, etc. Lyman ghosts are not troublesome in analytical applications using modern gratings since their intensities are usually held to less than 1/10,000 the intensity of the parent line.

The "blaze" of a grating is an important characteristic. By proper shaping of the groove during the ruling of a grating, it is possible to increase the maximum intensity of the spectrum at a predetermined angle and hence a predetermined wavelength region. Thus one grating may be blazed for maximum intensity at 4000 Å and another for 7000 Å.

Rowland is usually credited with developing the first successful mechanical engine for ruling gratings and his design is still the basic design in use today. The grating blank is mounted on a table moved by a precision screw mechanism. The ruling device is a diamond stylus mounted so it can be accurately moved in a direction at right angles to the movable table. As the grating is moved forward the diamond stylus is raised from the grating blank surface, then lowered into place for ruling the next groove.

Harrison has devised an interferometric control of groove spacing in a grating. An interferometer is used to control spacing by using interferometer fringes as the controlling device. Harrison has achieved a Rowland ghost intensity of about 1/20,000 in the first order of a 15,000 lines/in. grating and has produced some large gratings with theoretical resolving powers of about one million.

4.4. Grating Replicas

Replicas of gratings can be made successfully and can provide excellent resolution and high reflection efficiency and definition. They usually have somewhat higher scattered light properties than do originals. Most producers of replica gratings do not make the details of their processes available. One method, however, is to prepare a collodion solution, flow it over a grating, allow it to set, strip it off, and mount it on a permanent base.

Newer techniques and use of other polymerizable substances have led to production of replica gratings of high quality. The manufacturer can select a very high quality original as a master grating from which superior replica gratings can be produced. Improvements in techniques and the availability to the manufacturer of high-quality originals has made the use of replica gratings in most modern spectrometers very common.

4.5. Concave Gratings

Concave gratings have the advantage of focusing the light beam as well as producing a spectrum. Rowland was responsible for their development and produced concave gratings with a high degree of accuracy. Concave gratings are particularly useful in the vacuum ultraviolet region, where the usual optics employed with spectrometers would absorb a large portion of the useful electromagnetic radiation.

In recent years, however, plane gratings have become almost universal in spectrometers, for the following reasons:

(1) Simple adjustment of wavelength range is possible.
(2) Higher orders can easily be obtained, with associated higher resolution and dispersion.
(3) Most plane gratings are highly stigmatic.
(4) Instrument design can be simpler.
(5) Mechanical adjustment of the spectrometer is simpler.
(6) The gratings can be made and tested more easily.

4.6. Holographic Gratings

The first holographic grating was developed in the 1940's utilizing light fringes produced by two interfering beams from a high intensity mercury lamp. More recently holographic gratings have been produced using laser beams to produce much sharper interference beams than had been produced previously. In 1965 Shankoff[1] obtained a patent on the production of holographic gratings that exhibited blaze properties.

Holographic gratings can be produced by focusing interference fringes from a light beam on a photosensitive emulsion. The developed photographic image will be a transmission grating if the emulsion graininess is less than the distance between the interference fringes. Reflection gratings can be produced by use of a photosensitive resist, exposure and development, and evaporation of a metal on the image.

The techniques involved in producing a high quality holographic grating are considerable. Exposure time is long if an image of low graininess is required. Changes in temperature or air pressure and the presence of air currents during exposure can decrease the uniformity of the interference fringes. The optics of the focusing system must be of exceedingly high quality. The physical size of the grating is determined by the optical system.

Holographic gratings have been prepared with 6000 grooves/mm that have excellent resolution and appear to be free of ghosts and spurious lines. Holographic gratings, at present, however, are relatively inefficient and exhibit polarization effects. With unpolarized light the efficiency varies from 10 to 40%, increasing at longer wavelengths. With polarized light, peaks and valleys appear in the efficiency curves as plotted against wavelength. Thus spectral intensity measurements are difficult to make using holographic gratings.

Holographic gratings are still expensive to prepare but if the polarization effects can be remedied and if a practical method to blaze them can be developed they may become very useful, especially if their size can be increased.

4.7. Echelle Gratings

In 1949 Harrison[2] described the production of a grating intermediate between the Michelson echelon and the Wood echelette, which he called an echelle. The echelle grating possesses properties different from those of other gratings, particularly as to the method by which high resolution and linear dispersion are achieved.

---

[1] A. SHANKOFF, U.S. Patent 3,567,444 (1965).
[2] G. R. HARRISON, *J. Opt. Soc. Am.*, **39**, 522 (1949).

The general equation for the resolving power of a grating is given by

$$\frac{\lambda}{\Delta\lambda} = \frac{2Nd \sin \theta}{\lambda} \tag{3-24}$$

where N is the number of grooves illuminated with radiation, d is the groove spacing, and θ is the angle between the normal to the grating and incident radiation. From equation (3-24) it is evident that resolving power is increased at high angles using a large (wide) grating.

The linear dispersion of a grating is given by

$$\frac{dl}{\Delta\lambda} = \frac{2f \tan \theta}{\lambda} \tag{3-25}$$

where f is the focal length. Dispersion therefore may be increased by increasing the focal length or increasing the grating angle. The echelle grating uses a coarsely ruled grating (8–80 grooves/mm) at a large angle (60°–70°) to obtain high resolution and dispersion. Operation at high orders (40–120) for the visible–ultraviolet range also is required.

The echelle has broad, flat grooves ruled with high precision. A typical echelle is shown in Figure 3-17. The grooves are ruled with the width several times greater than the height. Since each step of the echelle is many times greater than the wavelength of the incident light, many different wavelengths will coincide; thus, the use of high orders of the echelle is necessary.

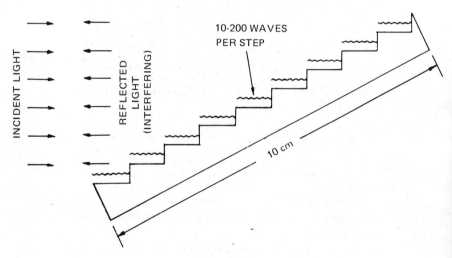

FIGURE 3-17. Reflection echelle grating. [From G. J. Matz, Recent Advances in Echelle Spectrometer Analysis, *American Laboratory*, **4** (3), 75, (1973). Used by permission of International Scientific Communications, Inc.]

FILTERS, PRISMS, GRATINGS, AND LENSES

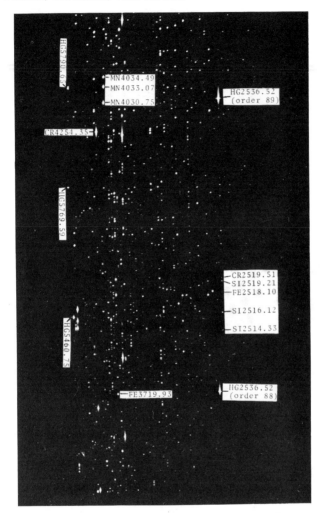

FIGURE 3-18. Echelle–prism spectrum of stainless steel with mercury lamp precision wavelength reference. [From G. J. Matz, Recent Advances in Echelle Spectrometer Analysis, *American Laboratory*, **4** (3), 75. Used by permission of International Scientific Communications, Inc.] Entrance slit, 25 μm wide by 200 μm high. Polaroid type 55 fine-grain film was used.

Since spectra of different orders may coincide or overlap, some method of separating orders is required. Separation of orders is achieved by use of a second dispersing element, either a prism or conventional grating, mounted so its dispersion is in a direction perpendicular to that of the echelle. This combination of dispersing elements produces a two-dimensional spectral pattern. An example of such a pattern is shown in Figure 3-18, which was obtained using the Spectra Metrics, Inc., echelle spectrometer, having a 0.75-m focal length, to produce a reciprocal linear dispersion of 0.65 Å/mm and a resolution of 0.006 Å at 2000 Å. Photographic recording of spectra can be used for qualitative or quantitative data, or photoelectric detection is possible with a limited number of elements.

5. LENSES

Efficient use of a spectrometer requires that the light energy entering the entrance slit be uniform in intensity and selected from the proper portion of the excitation source. To take full advantage of the spectrometer, the illumination of the slit should be of maximum intensity. To achieve these results requires the use, in most cases, of lenses and/or mirrors, in various combinations.

Optimum resolution of the spectrometer is obtained when the dispersing element is filled with light. Maximum intensity of the spectrum occurs when the entrance slit is uniformly illuminated for the full height of the dispersing element. Often it is not possible to place the source at the optimum position in front of the entrance slit. In this case a lens or combination of lenses can be used to accomplish the same effect.

5.1. Uses of Lenses

Lenses are usually used with spectrometers to serve four functions: (1) as a condensing system, (2) to collimate (make parallel) a light beam, (3) to focus the beam at some predetermined point, and (4) to enlarge or decrease the image size of the source. A lens may serve two or more of these functions at the same time.

The relation of the focal length of a lens and the distances of the object and the image is given by

$$\frac{1}{d_o} + \frac{1}{d_i} = \frac{1}{f} \qquad (3\text{-}26)$$

The linear ratio of object size to image size is the ratio of d_o/d_i. If d_o is twice d_i the object will have twice the linear dimensions of the image. Equation (3-26) shows also that if $d_o = \infty$ (light rays parallel), the focal length is equal to the image distance.

A condensing lens frequently is used in spectroscopy to increase the intensity of illumination of the slit. Figure 3-19 illustrates this effect. In Figure 3-19a, only a small amount of the total energy emitted from the source falls on the slit. If the source is a point, the solid angle intercepted by the slit

FILTERS, PRISMS, GRATINGS, AND LENSES

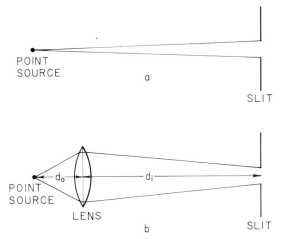

FIGURE 3-19. A condensing lens used to focus radiant energy on a spectrometer entrance slit.

is very small. If a condensing lens is used, as shown in Figure 3-19b, a much larger fraction of the total energy from the source can be directed to the entrance slit of the spectrometer. The condensing lens also is acting as a focusing device in this case. It is useful to recall the lens equation [equation (3-26)] in a case such as shown in Figure 3-19, as well as the size ratio of image to object. For example, if the image size is equal to the object size, $d_o = d_i$. Other ratios of d_o to d_i can be used to form images either larger or smaller than the object as required.

In some cases it is useful to collimate the light beam and use two lenses in an optical system. This arrangement is particularly useful if the source is at some distance from the entrance slit of the spectrometer and is illustrated in Figure 3-20. The collimating lens produces a parallel light beam from the

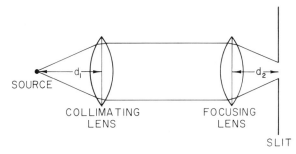

FIGURE 3-20. Use of a collimating lens and a focusing lens in a spectrometer optical system.

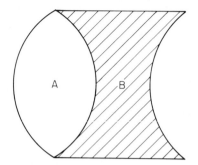

FIGURE 3-21. Compound lens for correction of chromatic aberration.

source, which is focused on the slit by the focusing lens. Distance d_2 equals the focal length of the focusing lens. If the focal lengths of the two lenses are equal, the image and object will be the same size. If, for example, the focal length of the focusing lens is twice that of the collimating lens, the image will be a $2\times$ linear magnification of the object.

5.2. Lens Defects

There are three defects in lenses that are important in their use in spectroscopy. These are (1) chromatic aberration, (2) spherical aberration, and (3) coma.

Chromatic aberration occurs because the index of refraction of glass is a function of wavelength. As a result, all colors do not focus at the same focal point. Chromatic aberration can be minimized by use of compound lenses, made of two or more different glasses, with different indices of refraction, cemented together. In Figure 3-21 the positive lens A is made of a less dispersive material and the negative lens B of a more dispersive substance. Such a compound lens can be designed to correct chromatic aberration at two wavelengths and is called an achromatic lens. It also is possible to use three elements in a lens and correct at three different wavelengths.

Spherical aberration is failure of a lens to bring to a focal point the rays passing through various parts of the lens. For a spherical, double convex lens, as shown in Figure 3-22, the rays farthest from the center of the lens focus nearer the lens than those rays passing through near the center of the lens. This effect can be serious and for use in spectrometers it frequently requires correction. Two approaches to the problem are possible. First, use of only the central part of the lens is possible, but this method also reduces the light-gathering power of the lens. Second, the lens shape may be changed. Such lenses are much more difficult to produce than spherical lenses, and, consequently, are much more expensive.

FILTERS, PRISMS, GRATINGS, AND LENSES

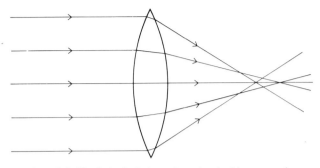

FIGURE 3-22. Spherical aberration of a double convex lens.

Coma is the spherical aberration of light rays passing through the lens at an oblique angle, as is illustrated in Figure 3-23. The coma type of defect produces a spot that has a "tail" on it. It will have an appearance similar to a comet with a tail. The effect may be most noticeable in a collimating lens. Coma can be reduced by using a smaller central part of the lens, resulting in reduced light intensity. Lenses should be mounted so light enters the lens along the optical axis of the lens if at all possible.

Lenses are commonly constructed of glasses. A variety of glasses are available with different indices of refraction. Of even more importance in spectroscopy is the transparency of the lens material at different wavelengths. Any lens causes some loss of light by reflection and by absorption. Coated lenses can reduce loss of light by reflection but have no effect on absorption. In the visible region glass lenses are quite suitable. Glass is not transparent

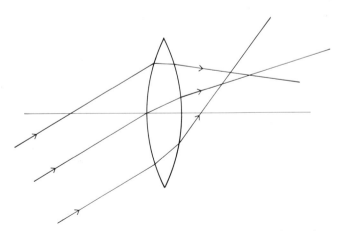

FIGURE 3-23. The "coma" effect of light rays entering a lens obliquely.

in the ultraviolet region. Fused quartz lenses extend the region further, but to go to 2000 Å requires lenses of crystalline quartz and such lenses are used in spectrometers designed to operate to this wavelength. Below 2000 Å the problem becomes increasingly difficult and may require vacuum systems and use of front-silvered mirrors to produce usable spectra.

SELECTED READING

Born, M. and Wolf, E., *Principles of Optics,* 3rd ed., Pergamon Press, New York (1965).
Brown, E. B., *Modern Optics,* Reinhold, New York (1965).
Clark, G. L., Editor, *The Encyclopedia of Spectroscopy,* Reinhold, New York (1960).
Grove, E. L., Editor, *Analytical Emission Spectroscopy,* Vol. 1, Part 1, Marcel Dekker, New York (1971).
Harrison, G. R., Lord, R. C., and Loofbourow, J. R., *Practical Spectroscopy,* Prentice-Hall, New York (1948).
Robertson, J. K., *Introduction to Physical Optics,* 3rd ed. D. Van Nostrand, New York (1941).
Sawyer, R. A., *Experimental Spectroscopy,* Prentice-Hall, New York (1944).
Slavin, M., *Emission Spectrochemical Analysis,* Wiley–Interscience, New York (1971).

Chapter 4

Spectrometers

To use atomic spectra for analytical purposes, regardless of the application, certain basic instrumentation is required. Included are a spectral isolation device—filter, prism, or grating; a slit to permit the radiation to strike the monochromator as a narrow beam; a device to allow observation of the spectrum; and the necessary optics, lenses, mirrors, etc. to collect and focus the incident light beam. These devices, assembled into one instrument, constitute the monochromator. Some lenses and mirrors, an optical bench, and the instrumentation for observing the spectrum are or may be external to the basic monochromator.

For some purposes, the necessary spectral isolation may require only the isolation of a band of frequencies. For such cases a filter may be used. Most situations, however, require a spectral resolution much better than can be achieved with a filter. A prism or a grating is necessary for such applications. There are numerous methods and optical arrangements used in monochromators. They each have advantages and disadvantages and many are designed for a particular use. The following discussion describes the more common monochromators, their characteristics, and some of their applications.

1. PRISM SPECTROMETERS

Two basic designs of prism spectrometer are commonly used for spectrochemical analytical purposes, the Cornu and the Littrow types. Many of these instruments are presently in use, although grating spectrometers are gradually replacing prism instruments.

1.1. The Cornu Prism Spectrometer

This type of spectrometer is shown in Figure 4-1. The Cornu prism is made of two 30°–60°–90° prisms, one of "left-handed" quartz, the other of "right-handed" quartz, cemented together to form a 60°–60°–60° prism. A quartz collimator lens is used to render the beam from the entrance slit parallel and to completely fill the face of the prism. Another quartz lens is mounted near the exit side of the prism to bring the dispersed light from the prism to a focus at a suitable viewing position. Frequently a photographic plate or film is mounted along the focal position to record the spectra. Usually this instrument uses collimator and lenses of about 0.5 m focal length at 2100 Å. A spectral range of 2100–7000 Å can be recorded on a 10-in. spectrum plate. The instrument normally has only the one fixed position for recording of spectra. A linear reciprocal dispersion of 32 Å/mm at 3000 Å is obtained with the spectrometer using 0.5-m focal length lenses.

1.2 The Littrow Spectrometer

A schematic diagram of a Littrow spectrometer is shown in Figure 4-2. The Littrow spectrometer utilizes a 30°–60°–90° quartz or glass prism with the back surface silvered to reflect the light beam back on a path almost parallel to the entering beam but vertically displaced from the entering beam. The lens is used as both a collimating and a focusing lens. The light rays are brought to a focus at the viewing position. A photographic plate usually is used to record the spectrum. The photographic plate must be mounted along the focal plane of the spectrometer and must be slightly curved to be in focus along the length of the plate.

The large Littrow-type instrument has a focal length of 1.8 m and can be used over a spectral range of 2000–8000 Å with a quartz prism. This spectral range requires the use of 30 in. of photographic plates. The spectral range can be increased to 10,000 Å by use of a glass prism.

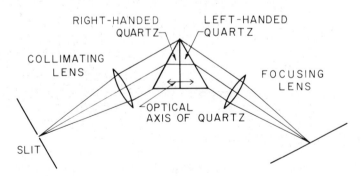

FIGURE 4-1. Schematic diagram of a Cornu prism spectrometer.

SPECTROMETERS

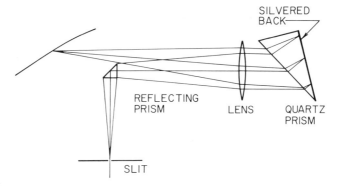

FIGURE 4-2. Schematic diagram of a Littrow prism spectrometer.

To permit the photographing of the entire spectral range, the large Littrow spectrometer has adjustable prism and photographic plate positions. The prism can be rotated to select the proper wavelength range and it also can be positioned laterally, along with the lens to permit proper focusing as the wavelength changes. The photographic plate holder also must be tilted along the optical axis to permit focusing at each prism setting. The prism and lens are mounted on a removable table so rapid exchange of prisms is possible.

The two most-used large Littrow spectrometers are those produced by Bausch and Lomb and by Adam Hilger, Ltd. The instruments have similar specifications; however, the Hilger unit combines all three wavelength

FIGURE 4-3. Photograph of a large Littrow spectrograph (no longer being manufactured). [Courtesy Bausch and Lomb Optical Company.]

adjustments into one operation. They are adjusted independently in the Bausch and Lomb spectrometer. The linear reciprocal dispersion is 13.6 Å/mm at 5500 Å for the glass prism. Many of these instruments are still in use and, when properly adjusted, provide excellent spectra. Figure 4-3 is a photograph of the Bausch and Lomb large Littrow spectrograph with its optical bench.

2. PLANE GRATING SPECTROMETERS

Improvements in the ruling of gratings have led to their increasing use in present-day monochromators. Plane grating spectrometers (1) have simple wavelength range adjustments, (2) have higher orders, permitting higher resolutions and dispersion, and (3) are highly stigmatic. The earliest plane grating spectrometers used a mounting similar to the Littrow prism spectrometer. The spectrometers were difficult to focus and adjust over a suitable wavelength range. Therefore, they were abandoned for other mounting arrangements.

2.1. The Ebert Spectrometer

The mounting for a plane grating and associated optics as designed by Ebert is shown in Figure 4-4. It consists of an entrance slit S, a concave

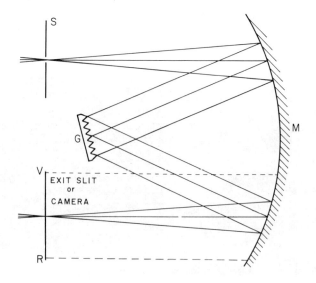

FIGURE 4-4. Schematic diagram of an Ebert plane grating spectrometer.

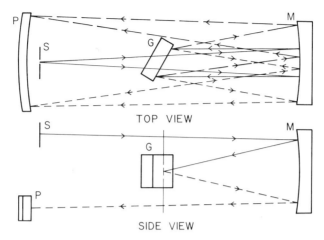

FIGURE 4-5. Schematic diagram of an Ebert–Fastie plane grating spectrometer.

mirror M, a grating G, and an exit slit or a position for recording the spectrum. The grating can be rotated to observe different portions of the spectrum. The focal length of the concave mirror determines the length of the optical path. The mounting is achromatic, as will be any plane grating spectrometer utilizing mirrors for focusing elements. The entrance rays lie to one side of the grating and the exit rays on the other, a "side-by-side" arrangement. A large mirror is required to cover an extended wavelength range since rays for different wavelengths are divergent.

Fastie suggested a modification of the Ebert spectrometer as shown in Figure 4-5. The grating is moved forward toward the mirror and an "over-and-under" optical path is used. The entering rays from the entrance slit pass below the grating to the concave mirror and above the grating to the exit slit or photographic plate holder. A smaller mirror can be used than with the original Ebert system and can include a wide wavelength range.

The Ebert–Fastie design has been used in several commercial instruments with various focal length mirrors. One popular commercially available unit is the 3.4-m Ebert–Fastie spectrometer produced by the Jarrell-Ash Company. This instrument utilizes a gear and worm drive to rotate the grating with a wavelength scale attached. The grating may be rotated by motor drive to provide spectral scanning. Provision also is made for photographic recording of spectra. The plate holder is 20 in. long, to hold two 10-in. plates or 20 in. of photographic film. The 20-in. camera permits photographing a 2550-Å range of the spectrum operating in the first order of a grating of 15,000 grooves/in.

FIGURE 4-6. Photograph of a 3.4-m Ebert–Fastie spectrograph. [Courtesy Jarrell–Ash Division, Fisher Scientific Co.]

The Jarrell-Ash instrument also can be used with a photomultiplier tube mounted at an exit slit for scanning and recording spectra on a strip-chart type of recorder. It also is possible to provide for direct reading of several elements by locating exit slits and photomultiplier tubes mounted at proper wavelength positions. Gratings are easily replaceable and a series of gratings of different blazes and rulings is available. The linear reciprocal dispersion of the 3.4-m instrument with a 15,000 grooves/in. grating in the first order is 5.1 Å/mm. A photograph of this spectrometer is shown in Figure 4-6 along with its support base, multisource power supply, optical bench, and arc–spark stand.

The Ebert optical system also is popular for smaller spectrometers, those of about 1 m to $\frac{1}{4}$ m dimensions. The Ebert mount, with its smaller dimensions, places much lower requirements on the concave collimating and focusing mirror. The design of most of these smaller spectrometers includes provision for scanning a spectral region, using a mechanical drive mechanism to rotate the grating and an exit slit, and a photomultiplier tube. Some also have photographic capability. Many of these units find application in flame emission, atomic absorption, and atomic fluorescence spectroscopy as well as other applications that do not require the higher resolving power of the larger instruments.

2.2 The Czerny–Turner Spectrometer

The Czerny–Turner spectrometer optics is shown in Figure 4-7. This design uses two concave mirrors, one for collimating the light beam and the other for focusing the beam at the exit slit or photographic plate. The second mirror can be made sufficiently large to cover the required field. The result is a "side-by-side" optical path with the added advantage of a large focusing concave mirror.

Several commercial instruments using the Czerny–Turner optical system are available. Jarrell-Ash employs this arrangement for scanning spectrometers in 2-m, 1-m, and $\frac{3}{4}$-m versions. Provision for photographing spectra is also available, as is a variety of gratings. Applied Research Laboratories has a 1.5-m unit with provisions for photographic recording, scanning, or direct reading of spectral line intensities.

The McPherson Instrument Corp. use the Czerny–Turner design in a spectrometer with vacuum capability. The instrument case can be evacuated to extend the lower spectral range to about 1050 Å. It can be used as a scanning spectrometer, with photographic recording, or with up to six side-window photomultiplier tubes mounted to view selected wavelengths. Direct read out of wavelength is provided with a counter, and gratings may be interchanged as needed. A linear reciprocal dispersion of 16.6 Å/mm is produced with a grating of 1200 grooves/mm. Figure 4-8 is a photograph of this unit.

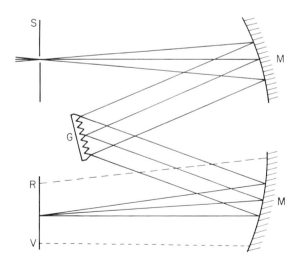

FIGURE 4-7. Schematic diagram of a Czerny–Turner plane grating spectrometer. [Courtesy McPherson Instrument Corp.]

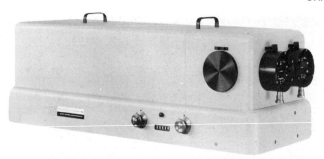

FIGURE 4-8. Photograph of a Czerny–Turner plane grating spectrograph. [Courtesy McPherson Instrument Corp.]

2.3. The Two-Mirror, Crossed-Beam, Plane Grating Spectrometer

The McPherson Instrument Corp. has a compact plane grating monochromator using a patented optical system as shown in Figure 4-9. This arrangement, used in a 0.3-m monochromator, provides a compact unit with minimum off-axis aberration. With a grating of 1200 grooves/mm, a linear reciprocal dispersion of 26.5 Å/mm is obtained. The spectrometer can be evacuated for use to about 1050 Å and spectral scanning is available both manually and mechanically. Gratings can easily be replaced as needed.

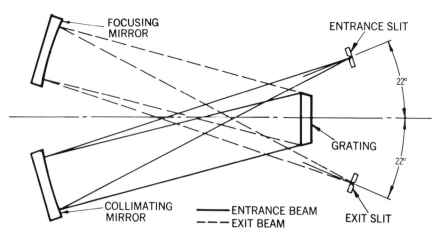

FIGURE 4-9. Schematic diagram of a crossed-beam, plane grating spectrometer. [Courtesy McPherson Instrument Corp. U.S. Patent No. 3,409,374.]

SPECTROMETERS

FIGURE 4-10. A double-grating spectrometer. [Courtesy Baird-Atomic, Inc.]

2.4. The Double-Grating Spectrometer

The use of two gratings and two optical paths in series provides a spectrometer with exceptionally low scattered light and high dispersion. Such properties permit detection of weak spectral lines that may lie close to a strong line. The Spex Industries double monochromator uses two complete Czerny–Turner mounting units. The gratings are mechanically connected and turn simultaneously. Provision is made to permit use of only one grating if desired. Gratings may be interchanged and scanning is either electrical or manual.

The Baird-Atomic, Inc., dual-grating instrument, shown in Figure 4-10, uses an Ebert mounting. The two gratings are independently adjustable. A focal length of 2 m provides a reciprocal linear dispersion of 4 Å/mm with 1200 grooves/mm gratings.

3. CONCAVE GRATING SPECTROMETERS

Concave gratings have the advantage of providing their own focusing. No auxiliary lenses or mirrors are required to bring the image of the entrance slit to a focus at the exit slit or camera position. They are therefore especially useful for working at short wavelengths. A variety of mountings have been used for concave gratings, all except the Wadsworth, by applying variations of principles first stated by Rowland. The source, the concave grating, and the spectrum all lie on a circle to which the grating is tangent, and which has a diameter equal to the radius of curvature of the concave grating.

3.1. The Rowland Spectrometer

Figure 4-11 is a diagram of the Rowland mounting for a spectrometer. The concave grating and the exit slit or plate holder are mounted at opposite ends of a rigid bar across the diameter of the Rowland circle. Heavy metal tracks extend from G to S and from S to P. The angle GSP is a right angle, so the entrance slit S also lies on the circumference of the Rowland circle. Small carriages are located under the grating and the plate holder and each is free to move along its own rail. Thus the grating can be made to move along GS and the plate holder along PS. The grating and the plate holder continue to be at a fixed position with respect to each other; the relative position of the slit changes, but the angle GSP remains a right angle. Figure 4-11 illustrates this effect by comparing the solid Rowland circle with the dashed position. The angle of the light ray incident upon the grating has been changed; thus a different portion of the spectrum will be observed at the position of the photographic plate. The spectral dispersion remains practically constant since spectra are always observed normal to the grating.

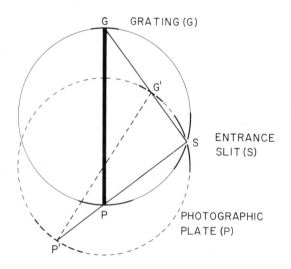

FIGURE 4-11. Schematic diagram of the Rowland mounting for a concave grating.

SPECTROMETERS

The Rowland mounting is basic to the use of the concave grating but is difficult to use and adjust. For the research that Rowland planned, i.e., measurements of spectral line wavelengths, it was ideal at that time. Other techniques are now available for wavelength measurements, so the purpose of the original Rowland mounting no longer is important and no instruments of this type have been constructed in recent years.

3.2. The Paschen–Runge Spectrometer

A schematic diagram of the Paschen–Runge spectrometer is shown in Figure 4-12. This has been a popular mounting, especially for large concave gratings, and also is used in some smaller units. The Rowland circle is the basis of the design, with the grating and entrance slit at fixed positions in the circle to give a fixed angle of incident radiation striking the grating. The light from the entrance slits focuses along the Rowland circle and photographic plates or phototubes are mounted along the circle to permit examination of a wider spectral range than would otherwise be possible. The Paschen–Runge mounting does not provide linear dispersion except near the grating normal. This disadvantage, however, must be weighed against the advantage of its wide spectral range.

The slit position and its relation to the angle of incidence should be chosen to take advantage of the highest reflectivity of the grating. The spectral order desired also must be considered. Some Paschen–Runge spectrometers have two fixed slits. By using angles of incidence of about 20° and 50°, both

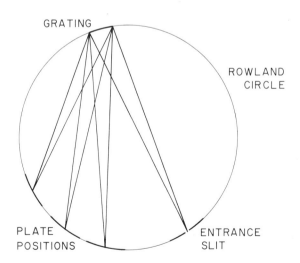

FIGURE 4-12. Schematic diagram of the Paschen–Runge mounting for a concave grating.

FIGURE 4-13. The "Quantometric Analyzer." [Courtesy Applied Research Laboratories.]

on the same side of the normal to the grating, a wide spectral range is possible utilizing the best angle of reflectivity of the grating.

Two commercial variations of the Paschen–Runge mounting are shown in Figures 4-13 and 4-14. Figure 4-13 shows the Applied Research Laboratories' Quantometric Analyzer. A fixed entrance slit and grating are used with exit slits at various wavelength positions on the Rowland circle. The spectral line intensities are obtained using photomultiplier tubes mounted at each desired wavelength.

A variation of the same optical system, as shown in Figure 4-14, is used in the Consolidated Electrodynamics direct reading spectrometers. The optical system is "folded," with the use of a "folding mirror," to make the instrument more compact. Using a 30,000 lines/in. grating provides a linear reciprocal diversion of about 2.78 Å/mm. Spectral intensities are determined by using photomultiplier tubes mounted at the exit slits.

3.3. The Eagle Spectrometer

The Eagle mounting for a concave grating is shown in schematic form in Figure 4-15. The mounting places the grating entrance slit and photographic plate near one another and the optical path is quite similar to that of the Littrow prism spectrometer. The incident beam and the reflected beam of light are displaced vertically. The grating, entrance slit, and photographic plate positions are on the Rowland circle, or the path length is equal to that of the Rowland circle. Some Eagle mountings utilize a reflecting prism to

SPECTROMETERS

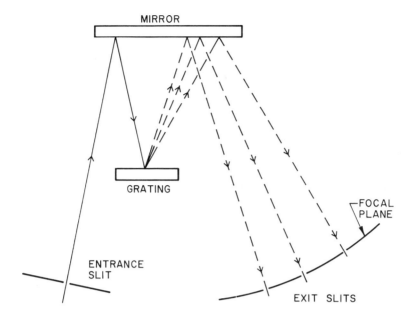

FIGURE 4-14. Optical diagram of a "folded" Paschen–Runge spectrometer.

bend the incident beam at right angles to the reflecting beam (see Figure 4-15), but the total path length must remain as dictated by the Rowland circle. In some Eagle mountings the incident beam is not bent but enters the spectrometer slit mounted immediately above or alongside the photographic plate.

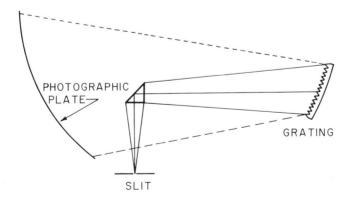

FIGURE 4-15. Schematic diagram of the Eagle mounting for a concave grating.

Changing from one wavelength region to another with the Eagle spectrometer requires three precise adjustments: (1) grating rotation, (2) motion of the grating on a line toward the photographic plate, and (3) photographic plate rotation. These adjustments allow the grating entrance slit and photographic plate to lie on the Rowland circle and still permit the angles of incidence and reflection to change. The Eagle spectrometer does not have linear dispersion since the photographic plate is not normal to the grating.

3.4. The Wadsworth Spectrometer

The Wadsworth spectrometer, shown schematically in Figure 4-16, does not make use of the Rowland circle; the focal distance is one-half the radius of curvature and another optical element, a concave collimating mirror, is introduced into the optical path. In this arrangement the photographic plate is mounted on a line normal to the concave grating, and thus stigmatic spectra are produced and the dispersion is linear.

The concave mirror is mounted at its focal distance from the entrance slit to provide a parallel light beam to the grating rather than the dispersed beam produced by the Rowland-circle-mounted spectrometers. The photographic plate and the grating are mounted on a bar. To change wavelengths, the bar is rotated away from the concave mirror on an axis under the grating and the photographic plate is moved along the axis to maintain focus.

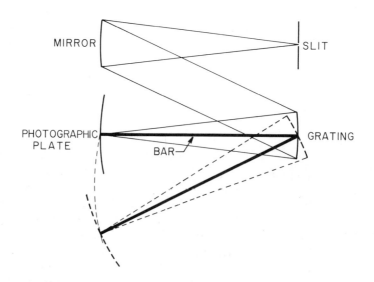

FIGURE 4-16. Schematic diagram of a Wadsworth mounting for a concave grating.

SPECTROMETERS

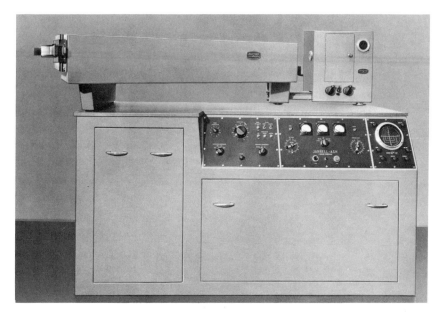

FIGURE 4-17. Photograph of a Wadsworth-mounted 1.5-m spectrometer. [Courtesy Jarrell–Ash Division, Fisher Scientific Co.]

The Wadsworth optical system is used in several commercially available spectrometers with a variety of focal lengths. A 1.5-m Wadsworth spectrometer produced by the Jarrell-Ash Division, Fisher Scientific Company is shown in Figure 4-17 along with an excitation source and arc–spark stand. Photographic film recording of spectra is used. With a grating of 590 lines/mm a linear reciprocal dispersion of 10.8 Å/mm is obtained in the first order.

3.5. The Grazing Incidence Spectrometer

Figure 4-18 is a schematic diagram of the optical system of a grazing incidence spectrometer. Under these conditions, when the incident light strikes the grating at an angle of 82°–88° from the normal, the wavelengths below about 1000 Å will be almost totally reflected and observations to wavelengths as low as 10 Å *are possible*. The astigmatism of such an optical system is high. The dispersion is nonlinear and varies with the grazing incidence angle and with the wavelength. Used with a vacuum system, however, the grazing incidence spectrometer gives a spectral region not available when using other more conventional metrometers.

The optical system shown in Figure 4-18 is used in the McPherson Instrument Corp.'s 2-m grazing incidence vacuum spectrometer. It has a

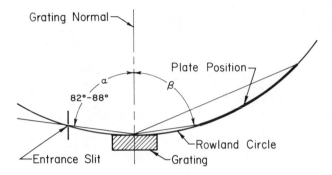

FIGURE 4-18 Schematic diagram of a grazing incidence, concave grating spectrometer. [Courtesy McPherson Instrument Corp.]

useful wavelength range of 10–2500 Å. Gratings can be interchanged and photographic plate or phototube read-out of spectra is available. With an angle of incidence of 82°, a grating of 300 lines/mm, and at a wavelength of 1000 Å, the reciprocal linear dispersion is 4.2 Å/mm.

3.6. The Seya–Namioka Spectrometer

Figure 4-19 is a schematic diagram of spectrometric optics as proposed by Seya and Namioka. They suggest that if the incident ray and emergent ray produce an angle of 70°15′, then rotating the grating produces only very slight defocusing. The system thus is excellent to use for scanning the spectrum by rotating the grating while the entrance and exit slits remain in a fixed position. The design is very useful in the vacuum ultraviolet. Several commercial variations of this instrument are available.

Figure 4-20 is a photograph of the McPherson Instrument Corp.'s version of the Seya–Namioka monochromator. This instrument is a $\frac{1}{2}$-m scanning monochromator with vacuum capability. With a grating of 1200 grooves/mm a reciprocal linear dispersion of 16.6 Å/mm is obtained with a resolution of 0.5 Å using 10-μm slits. The wavelength range is from 500 to 3000 Å with a 1200 grooves/mm grating. Twelve scanning speeds are available and a vacuum as low as 10^{-7} Torr can be maintained in the monochromator.

3.7. Vacuum Spectrometers

If it is necessary to observe wavelengths below approximately 1850–1900 Å, air must be removed from the spectrometer. Displacing air with nitrogen extends the low-wavelength region to about 1750 Å. However, if a vacuum system is used, the minimum wavelength is further extended.

SPECTROMETERS

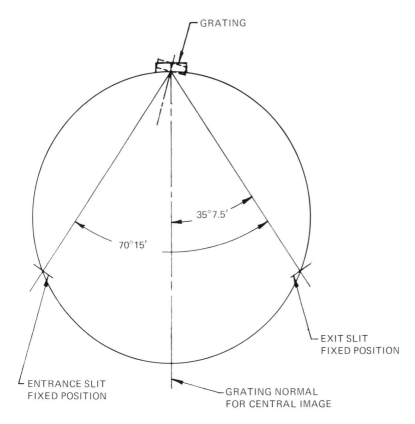

FIGURE 4-19. Schematic diagram of the Seya–Namioka concave grating spectrometer. [Courtesy McPherson Instrument Corp.]

Almost any spectrometer can be used in the vacuum ultraviolet (about 1100–2000 Å) by properly designing the enclosure and utilizing a vacuum pumping system. Vacuum pumping systems usually include a mechanical pump with an oil diffusion pump. A cold trap provides protection of the grating and the internal optics from long-term oil contamination. A vacuum as low as 10^{-7} Torr frequently is used.

Vacuum spectrometers require windows over the entrance and exit slits to maintain the vacuum. They are usually made of lithium fluoride or calcium fluoride. If photographic plates are used to record spectra, special arrangements are required to either place the plate inside the monochromator case or provide a system to remove the plate and replace it without breaking the vacuum. Both arrangements are used commercially.

FIGURE 4-20. Photograph of Seya–Namioka spectrometer with vacuum capability. [Courtesy McPherson Instrument Corp.]

The usual exit slit also may be dispensed with, if photomultiplier tubes are used to record spectral line intensities, by mounting the phototubes internally.

Vacuum spectrometers are available from several manufacturers and in a wide variety of focal lengths. The Czerny–Turner spectrometer is frequently used as a vacuum system. The Wadsworth mounting also is used since the instrument can be smaller than many others because the optical path is folded by the use of a collimating, concave mirror. Grazing incidence spectrometers require a vacuum system to be useful in the spectral range for which they are designed. The Seya–Namioka design also is very useful in the vacuum-ultraviolet range if a scanning-type spectrometer is suitable.

4. DIRECT READING SPECTROMETERS

If a spectrometer is to be used for quantitative spectrochemical analysis for a selected number of elements, it is possible to speed up the process of analysis by using a direct reading system. In fact, the time required for analysis once the sample is prepared may be only 1 or 2 min. For direct reading spectrometers, photomultiplier tubes are used to measure the intensity of radiation at selected wavelengths instead of a photographic film or plate. Exit slits are carefully positioned along the focal curve to allow light of specific wavelengths to strike the phototubes mounted behind each slit. One slit and one phototube are required for each element. The most common system is to collect the electrical output of the phototube during the exposure period as a charge on a capacitor. After the exposure is completed the capacitor is discharged and the collected electrical energy measured.

The results can be placed on any one of several read-out devices, a strip chart recorder, a digital voltmeter, or a printed report from a computer.

The internal standard principle can be applied to direct readers. If a phototube is positioned to receive energy from an internal standard line and another phototube is used to collect energy from an element of unknown concentration, a voltage ratio can be obtained. Then the voltage ratio is used to determine the unknown concentration. Direct reading spectrometers using photomultiplier tubes can be used to measure light intensities accurately over a range of four orders of magnitude; thus the use of photomultipliers greatly extends the usable concentration range over that obtained by photographic recording of spectra.

Almost all companies producing spectrometers offer direct readers. They vary from quite small instruments, capable of determining 6–12 elements simultaneously, to large dispersion instruments that can determine as many as 50 elements simultaneously. Many instruments also can interchange photographic recording and direct reading.

5. SELECTION OF A SPECTROMETER

The choice of a spectrometer depends on its intended use. Most present-day instruments use a grating as the spectral dispersing element. A wide variety of excellent gratings is currently available. They are supplied with certification as to their precision, blaze efficiency, and freedom from ghost lines, and are generally of high quality.

Some of the questions that should be answered before selecting a spectrometer for emission spectrochemical analysis are the following.

1. Is the spectrometer to be used entirely for repetitive, routine analysis, or is it also to serve a research purpose?
2. How quickly are analytical data required?
3. Is photographic recording of spectra satisfactory?
4. What type of samples are to be analyzed, i.e., solutions, solids, metals, soils, biological materials?
5. What resolving power will be required?
6. How much dispersion is needed?
7. How much space will the instrument require?
8. What are the power requirements of the excitation sources?
9. What kind of accessory equipment is needed?
10. Is the vacuum-ultraviolet spectral region needed?

Many spectrometers are available as packaged units. For example, small emission spectrographs are available as complete units including power source, arc–spark stand, etc. Those units used as direct readers include the phototubes and electronic components as part of the package. Vacuum spectrometers usually include the vacuum pumping system.

It is not possible to detail all the factors involved in the selection of a spectrometer, but the above short discussion should serve to focus attention on many of the more important questions.

6. ADJUSTMENT AND CARE OF SPECTROMETERS

There are many types of spectrometers, with a wide variety of mountings, focal lengths, resolutions, and light-gathering ability. Each of these requires its own set of particular adjustments for best performance. The small spectrometers, of focal lengths less than 1 m, are much less critical to adjust than the longer focal length instruments. Most smaller units are factory aligned and require only minimum adjustment on delivery. The larger units require much more careful alignment and most manufacturers send a company representative to the site to make initial adjustments of the monochromator, grating or prism, internal optics, and initial focusing. In addition, the manufacturers provide detailed instructions for alignment of their particular instruments. These instructions should be followed carefully if any realignment is attempted.

In general, adjustments of spectrometers require careful positioning of the monochromator element, and optical alignment of the entrance slit, optical elements within the monochromator, and photographic plate or exit slit. The image of the entrance slit must be adjusted so it is in sharp focus at the photoreceptor.

6.1. Vertical Adjustment of the Entrance Slit

It is important that the entrance slit edges be parallel to the apex of the prism or to the rulings on the grating. This adjustment is made by rotating the entrance slit after the prism or grating and internal optics are aligned. Preliminary adjustment of the slit may be made visually to determine if the spectral lines are perpendicular to the spectrum. A further refinement is to draw $X-Y$ coordinates on a translucent screen and view the spectrum at the focal position. The spectral lines should be parallel to the vertical coordinates when the horizontal coordinate is parallel to the spectrum. A final adjustment to properly align the slit is to take a series of spectra, adjusting the slit very slightly for each exposure. Examination of the spectra will permit selection of the best vertical position for the slit. Once adjusted, further vertical positioning of the slit should not be necessary except when the slit assembly is removed and replaced.

SPECTROMETERS

6.2 Focusing the Entrance Slit

6.2.1. Prism Instruments

Spectrometers using prisms as monochromator elements usually require two adjustments to focus the image of the entrance slit on the photographic plate: (1) the distance between the prism and the plate, and (2) the "tilt" of the plate. Preliminary focusing can be done visually, usually by the use of a ground glass screen at the position usually occupied by a photographic plate. A mercury or neon arc is a suitable source for visual observation. Once preliminary focus has been achieved, the final setting is determined by taking a series of spectra while varying the focus setting and then selecting the setting producing the sharpest focus.

If the "tilt" of the photographic plate also requires adjustment, the spectral lines at the center of the plate should first be placed in sharp focus. Then a series of spectra should be taken, varying plate tilt. The sharpest spectrum from one end of the photographic plate to the other then should be used. In the large prism spectrometers (such as the large Littrow) the photographic plate is bent to maintain focus over the entire spectral range covered by the plate. This adjustment is not variable and is built into the photographic plate holder.

6.2.2. Grating Instruments

Focusing of the entrance slit of a grating monochromator is similar to the procedure used for prism instruments. Visual adjustment should be followed by taking a series of test spectra. It also is important that the slit be previously adjusted to be parallel to the grating lines. The focusing of concave gratings usually is accomplished by movement of the photographic plate holder along the line normal to the grating. Since the photographic plate should lie along the Rowland circle, plate holders are made to hold the plate at the proper radius of curvature.

6.3. Final Adjustments

The final test of a spectrometer is to determine its effective resolving power. A good indication of resolution can be obtained by observation of the hyperfine structure of the green line of mercury at 5461 Å. Care should be exercised in making such measurements to use a narrow, accurately aligned entrance slit and minimum exposure to record the spectral lines. If spectral lines are too dense, the lines, as observed photographically, will be broadened and maximum available resolving power will not be observed.

6.4. General Care of Spectrometers

Spectrometers should have the care accorded any fine instrument. They should be protected from common chemical laboratory fumes. Air should be dust-free, and the room, especially for the larger spectrometers, should be air conditioned. The room should be equipped with adequate electric power and an exhaust system is necessary to carry any gases and any particulate matter from the excitation source. Sample preparation should be done in a separate room.

Lenses and mirrors should be protected from dust and dirt when not in use. They can be cleaned when necessary, but care must be exercised, especially with mirrors. Dust can be removed from lenses by wiping gently with lens paper or soft cloth. A solvent such as ethanol can be used to remove fingerprints. Front surface mirrors are easily scratched. Dust can be removed by brushing carefully with a soft brush. The surface can be washed with very dilute ammonium hydroxide followed by rinsing with distilled water. A 50–50 mixture of water and ethanol also may be used and should be followed by rinsing with distilled water.

It is best not to attempt to clean gratings, especially the larger, more expensive ones. Much more harm than good may occur if gratings are handled carelessly. If absolutely necessary, a grating can be washed by flowing ethanol or distilled water over the surface. If replaceable gratings are used in the monochromator, the unused gratings should always be protected from dust and laboratory fumes by storage in a tight container.

Variable-width spectrographic slits are constructed so they cannot be easily damaged, but slit width always should be adjusted carefully. If a slit jaw should jam, the micrometer threads could be damaged if attempts are made to force the micrometer mechanism. Dust particles on the edges of the slit interfere seriously with formation of the spectral image, producing narrow white lines through the spectrum. If the variable-width slit jaws are opened 1 mm or more, the jaw edges can be cleaned by running a match stick or small piece of wood along the jaw edges. Fixed slits are more difficult to clean, but ultrasonic cleaners seem to be effective. The fixed slit can be immersed in distilled water in an ultrasonic cleaner for a few minutes and then drained and dried carefully.

Most important to remember in taking care of spectrometers is to (1) follow manufacturers' instructions carefully and (2) proceed with caution, considering that the spectrometer is an expensive, precision instrument.

Selected Reading

Ahrens, L. H., and Taylor, S. R., *Spectrochemical Analysis*, 2nd ed., Addison-Wesley, Reading, Massachusetts (1961).
Brode, Wallace R., *Chemical Spectroscopy*, 2nd ed., Wiley, New York (1943).
Clark, George L., Editor, *The Encyclopedia of Spectroscopy*, Reinhold Publishing Co. (1960).
Grove, E. L., Editor, *Analytical Emission Spectroscopy*, Vol. 1, Part 1, Marcel Dekker, New York (1971).
Harrison, George R., Lord, Richard C., and Loofbourow, John R., *Practical Spectroscopy*, Prentice-Hall (1948).
Mavrodineanu, R., Editor, *Analytical Flame Spectroscopy*, Springer-Verlag, (1969).
Nachtrieb, Norman H., *Principles and Practice of Spectrochemical Analysis*, McGraw-Hill, New York (1950).
Sawyer, Ralph A., *Experimental Spectroscopy*, Prentice-Hall (1944).

Also see literature supplied by manufacturers of spectrometers and spectrometric equipment.

Chapter 5

Accessory Equipment for Arc and Spark Spectrochemical Analysis

To use a spectrometer to produce atomic spectra for spectrochemical analysis, a number of items are required that can be classed as accessory equipment. The spectrometer is used to produce the atomic spectra, but for analytical purposes the spectra must be obtained in a reproducible manner. This chapter describes the accessory equipment commonly used to produce emission spectra for analytical purposes.

1. THE SPECTROMETER SLIT

Although the entrance slit may be considered a basic part of the spectrometer, its importance in spectrochemical analysis requires that it be considered in this chapter. The entrance slit is a narrow, rectangular opening through which light enters the spectrometer. The spectral lines produced on the photographic plate are images of the entrance slit. It therefore is important that the slit have parallel edges, be free of imperfections, and be constructed so that light is not reflected from the slit edges. The slit should be narrow, so narrow spectral lines are produced. Since the theoretical resolving power of a spectrometer, as developed in Chapter 3, is based on infinitely narrow slit, an actual slit, with a finite width, cannot produce maximum theoretical resolution. There are several reasons for this, including (1) the finite width of the slit, (2) the response of the photographic plate or other photoreceptor, (3) the relative intensities of the diffraction maxima, (4) imperfections in the monochromator element, and (5) imperfections in other optical accessories.

Fixed slits or variable-width slits are available for spectrometers. The fixed slit widths are adjusted by the manufacturer. If fixed slits are used, the monochromator case has a mounting position in which interchangeable slits are mounted. Slits usually are constructed of stainless steel with the inner

FIGURE 5-1. Bilateral variable slit mechanism. [Courtesy Jarrell-Ash Division, Fisher Scientific Co.]

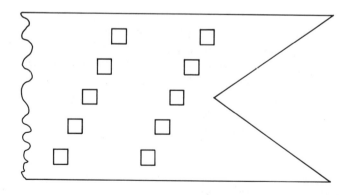

FIGURE 5-2. The Hartmann diaphragm.

edge beveled so that the outer edge, which determines the slit width, is very thin. Fixed slits provide constant slit width and are easily interchanged. They also provide a convenient means of ensuring that the slit is mounted parallel to the monochromator element.

Variable-width slits must be precision-constructed devices capable of resetting to a definite slit width with a high degree of accuracy. As the slit opens and closes, the two edges of the slit must remain parallel. Most variable-width slits have a bilateral motion, that is, both edges of the slit move during the adjustment of slit width. A micrometer screw type device is used to move the slit edges and a scale attached to the micrometer screw serves to measure slit width. A spring device is used to prevent damage to the micrometer screw if the slit is unintentionally closed too tightly. Figure 5-1 is a photograph of a bilateral slit showing the screw mechanism used to increase and decrease the slit width.

Slit mountings usually include arrangements to mount optical devices next to the slit. These devices may include neutral filters to limit exposure intensity, to limit the length of the spectral line, to view only a part of the total spectral line length, or to provide a gradation in line intensity over the length of the slit.

2. THE HARTMANN DIAPHRAGM

One of the most common devices used immediately adjacent to the entrance slit is the Hartmann diaphragm, shown in Figure 5-2. The diaphragm has a "V"-shaped wedge on the end and a series of rectangular openings spaced vertically and horizontally along the diaphragm. The diaphragm is constructed of metal and slides horizontally immediately in front of the slit. The "V" wedge can be used to limit the length of the slit and thus the length of the spectral line.

The rectangular openings in the Hartmann diaphragm can be used to record several spectrograms on one photographic plate without moving the plate holder. This prevents any possible lateral movement of the photographic plate between exposures or error in the slit adjustment.

3. THE STEP FILTER

Sometimes it is desirable to obtain spectral lines in which the light intensity has been varied in a regular manner. One way to accomplish this

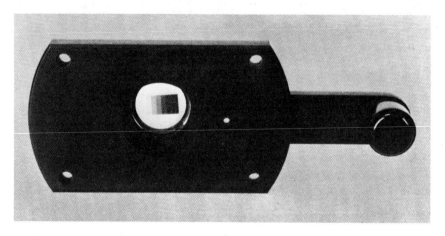

FIGURE 5-3. A seven-step neutral density filter. [Courtesy Jarrell-Ash Division, Fisher Scientific Co.]

is to mount a "neutral" step filter immediately in front of the entrance slit. A seven-step neutral filter is shown in Figure 5-3. A neutral filter attenuates light equally at all wavelengths. The filter shown in Figure 5-3 has seven steps with light transmissions of 100, 63.9, 39.8, 25.1, 15.9, 10.0, and 6.4% for each successive step. If it is mounted vertically immediately in front of the entrance slit, each spectral line will show the intensity of each step. A portion of a spectrum photographed in this manner is shown in Figure 5-4. Such spectra are very useful to obtain (1) a photographic emulsion calibration curve and (2) spectral line segments in a usable intensity range. The filter also can be mounted horizontally and any step positioned over the entrance slit, for uniform attenuation of the spectral lines is thus produced.

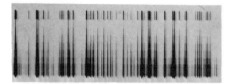

FIGURE 5-4. Photograph of a portion of an iron spectrum using a seven-step neutral filter.

4. ROTATING SECTORS

Rotating sector disks also are frequently used in emission spectroscopy. The three principal types of rotating sectors are shown in Figure 5-5. The sector disk, illustrated in Figure 5-5A, has a variable aperture and is used to reduce the intensity of exposure. The sector disk is mounted in the optical path in front of the entrance slit to the spectrometer and is rotated at a constant rate.

The sector disk in Figure 5-5B can be used for the same purposes as a step filter. The disk is cut to form a series of circular steps so exposure times will vary in a regular manner from one step to another. The step factor is fixed for each disk, but various disks are available. The two most common step sector disks are those in which the angles α, β, and γ change by some fixed multiple, such as a ratio of 1, 2, 4, etc. The other common stepwise change is to make the steps logarithmically related to one another.

The third type of rotating sector disk, shown in Figure 5-5C, has the sector opening cut as a logarithmic spiral. This disk can be used to determine intensity ratios of spectral lines since the length of the line, as obtained on a photographic plate when using a logarithmic sector, is proportional to its intensity. Semiquantitative data on element concentrations also can be obtained using a logarithmic sector disk.

5. EXCITATION SOURCES

Any excitation source for analytical atomic emission spectroscopy must accomplish the following processes: (1) the analytical sample must be vaporized; (2) it must be dissociated into atoms; (3) the electrons in the atoms must be excited to energy levels above the ground state. The three steps

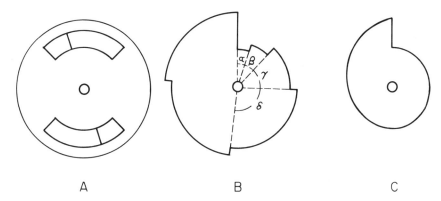

FIGURE 5-5. Types of rotating sector disks.

listed above are difficult to separate and in most excitation sources no attempt is made to separate them.

The energy required to produce spectral emission can be provided in several ways, including discharge tubes, flames, electric arcs, electric sparks, plasmas, and lasers. The first two, discharge tubes and flames, are not discussed here. Flames are treated in Chapter 9 (Flame Emission Spectroscopy) and discharge tubes are discussed in Chapter 10 (Atomic Absorption Spectroscopy).

5.1. The Direct Current Arc

The dc arc is a widely used excitation source for both qualitative and quantitative spectrochemical emission analysis. Usually electrodes are made of carbon or graphite, but occasionally metal electrodes are used. Equipment required for the dc arc is simple and requires only a dc voltage source of 100–150 V at 5–30 A and a ballast resistor. A circuit diagram such as shown in Figure 5-6 is all that is required. The ballast resistor is necessary to stabilize and adjust the arc current. A dc arc has negative resistance characteristics and is therefore unstable. The ballast resistor stabilizes the arc current and permits adjustment of the current to a predetermined level. The relatively high voltage permits use of the ballast resistor since the voltage drop across the arc is approximately 25–50 V, depending on the type of sample vaporized into the arc and the electrode material.

Frequently dc generators have been used as sources of dc energy; however, ac sources may be used and rectified by vacuum tube rectifier circuits or by use of high-current solid state diodes. A rectifier circuit for producing dc energy from an ac source is shown in Figure 5-7. The inductance in the primary circuit is a variable core reactor for current control. Numerous variations of this basic circuit are commercially available. A dc source such

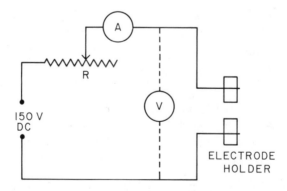

FIGURE 5-6. Circuit diagram of a dc arc power supply.

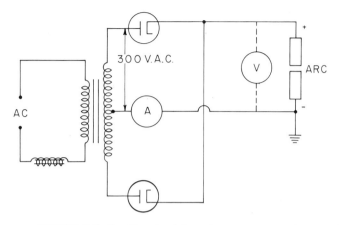

FIGURE 5-7. Rectifier circuit for a dc arc power supply.

as shown in Figure 5-7 produces a pulsating dc current. The current can be "smoothed" by placing an inductance in series with one output lead and a capacitor across the two output leads.

The dc arc excitation is primarily thermal in nature. The temperature in the arc varies across the arc gap and increases as the current increases. Temperatures of 4000–8000°C can be obtained using a dc arc. The dc arc is subject to considerable "wandering" and thus reproducibility is not as good as with some other excitation sources. Selective volatilization of the sample into the arc also is a problem. The dc arc is very sensitive and can be used to detect very low concentration levels. Spectra contain primarily atom lines although some ion lines are observed.

5.2. The Alternating Current Arc

Alternating currents can be used to sustain an arc, although higher voltages are required than for a dc arc. The discharge is more uniform than in the case of the dc arc since its polarity reverses 120 times per second if a 60-cycle source is used. Potentials of 2000–3000 V frequently are used with currents of 3–6 A. The ac arc provides better reproducibility than the dc arc but the sensitivity is decreased and more ion lines occur in the spectrum. Figure 5-8 is a diagram of a typical ac arc power source. A variable core reactor can be used to control the input voltage to the transformer primary. An ac ammeter is placed in the secondary to measure the arc current.

The voltage drop across the ac arc reaches an initial maximum each half-cycle, followed by a rapid decrease and a constant low voltage until the half-cycle is completed. The high-voltage maximum triggers the discharge. The operation of the ac arc is therefore primarily low-voltage with a peak

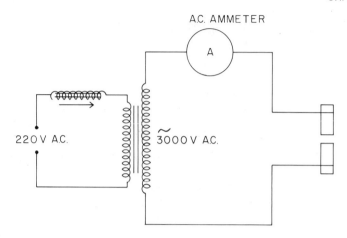

FIGURE 5-8. Circuit diagram of an ac arc power supply.

voltage to ignite the discharge each half-cycle and with a reversed polarity each half-cycle.

The ac arc is used primarily for liquid samples which can be evaporated on the flat ends of the electrodes. The selective volatilization that is a problem with the dc arc does not occur with the ac arc; however, it is important to keep the gap distance constant in the ac arc. Variations in line intensity ratios occur when the electrode separation changes, due probably to temperature changes in the arc gap.

5.3. The Electric Spark

An electric spark is an electrical discharge across a gap between electrical conductors caused by a high potential difference across the gap. The voltage requirement depends on the gap dimensions and the shapes of the electrodes, but potentials of 10,000–50,000 V are common.

A typical high-voltage ac spark source is shown schematically in Figure 5-9. The transformer raises the voltage to a high level and charges capacitor C until reaching a potential sufficiently high to discharge across the electrode gap G. The capacitor is charged and discharged on each half-cycle to the voltage that is necessary to ignite the spark. An oscillating current then flows across the gap with a frequency determined by the inductance L and capacitance C, and with a damping rate determined by the Ohmic resistance R. The frequency of the discharge, in cycles per second, is given by

$$f = 1/[2\pi(LC)^{1/2}] \qquad (5\text{-}1)$$

when L is measured in henrys and C in farads. The current flow, neglecting Ohmic resistance, is given by

$$I = V(C/L)^{1/2} \qquad (5\text{-}2)$$

where V is the capacitor voltage, C is the capacitance in farads, and L is the inductance in henrys. The instantaneous initial current can be very high, 100–200 A. The current on succeeding cycles decreases very rapidly. The result of this current behavior is the production of vaporized, excited atoms within the spark. Each half-cycle becomes nonconducting before the next half-cycle is initiated. Since this action is repeated rapidly, 120 times per second with a 60-cycle ac source, the result is the almost continuous production of atomic vapor between the spark gap electrodes.

Changes in inductance and capacitance change the frequency of oscillation in accordance with equation (5-1). An increase in either inductance or capacitance decreases the oscillation frequency. The current at the instant of discharge increases as the capacitance increases and decreases as the inductance increases. With large values of inductance and capacitance and the resulting decrease in oscillation frequency the spectra produced become more "arc-like". By controlling inductance, capacitance, and resistance, a wide variety of excitation conditions are available.

The condensed ac spark provides higher precision than does a dc arc, but with decreased sensitivity. Spectra produced by the ac spark contain a large number of ion lines as well as arc (atom) lines. Some doubly ionized species also are produced in a high-intensity ac spark.

"Controlled" spark systems have been devised to stabilize the spark. Feussner utilized a synchronous auxiliary spark gap arranged so the auxiliary

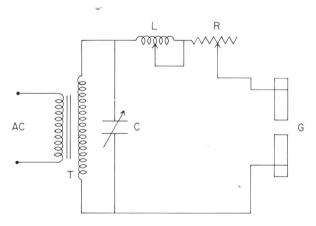

FIGURE 5-9. Circuit diagram of a high-voltage ac spark power supply.

gap is closed only at the moment of peak voltage of each half-cycle. The resulting discharge is more stable and steady than without the auxiliary gap and is frequently used for analytical spectrochemical purposes. The auxiliary gap is placed in series with one lead to the analytical gap.

Several other methods have been employed to accomplish the same goal as the Feussner system. Commercially available spectroscopic power sources make use of the mechanical auxiliary gap, but more recently electronic devices have been constructed to control the spark discharge time. These devices can control the moment of discharge to within 1 μsec and can be adjusted to produce discharge at almost any point on the time cycle. They result in highly reproducible, consistent spectra resulting in excellent analytical precision.

5.4. The Plasma Arc

A plasma may be defined as a gas containing a relatively large number of ions and free electrons. To produce a plasma, an energy source is required and for analytical atomic spectroscopy three different excitation methods have been used. They are (1) a dc arc, (2) radiofrequency energy coupled through a microwave cavity, and (3) radiofrequency energy inductively coupled to the plasma.

One design of a plasma using dc arc excitation is shown in Figure 5-10. A dc arc is ignited in the closed chamber between the anode and a ring-shaped cathode. A tangential stream of inert gas entering the chamber blows the arc

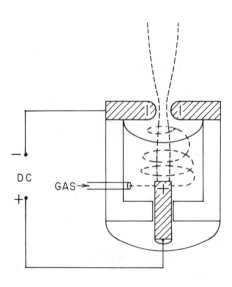

FIGURE 5-10. Basic diagram of a plasma jet excitation source. [Courtesy Spex Industries, Inc.]

EQUIPMENT FOR ARC/SPARK ANALYSIS

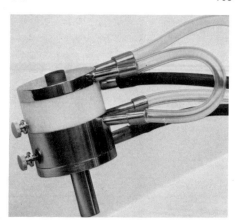

FIGURE 5-11. Photograph of a plasma jet solution analyzer. [Courtesy Spex Industries, Inc.]

through the opening in the cathode. The magnetic effect of the charged particles within the plasma constricts the plasma jet and increases its temperature. If an analytical sample is introduced into the plasma in liquid form, the high temperature of the plasma evaporates the solvent, dissociates the molecular species, and excites the atoms. Spectral analysis of the plasma thus offers an excellent means of determining element concentrations in the sample. The electronic temperature of the plasma may be as high as 8000–10,000° K.

Several variations of the dc-excited plama have been designed and some are commercially available. Figure 5-11 shows one such unit. A cathode "transfer" electrode is mounted vertically above the plasma unit. As the plasma emerges from the ring-shaped cathode, it is "transferred" to the vertical electrode. The result is a more stable plasma.

Another variation of the dc-excited plasma is shown in Figure 5-12. In this unit an arc is ignited between the two electrodes and the discharge appears as an inverted "V" between them. A liquid sample is injected through one electrode, is excited by the arc, and forms a plasma extending above the current-carrying portion of the inverted "V". Temperatures reach about 6000° K when using a dc arc source operating at 300 V and 10 A. The sensitivity is excellent, exceeding that of the dc arc in many cases, and working curves are linear over about four orders of magnitude if a photomultiplier tube read-out system is used. This particular plasma operates with much smaller gas flows than earlier models. Less than 2.5 liters of inert gas per minute is required and a conventional dc arc supply provides the energy to produce the plasma. Spectrometrics, Inc., has recently made a dc-excited plasma utilizing the inverted "V" configuration commercially available.

FIGURE 5-12. Photograph of a dc-arc-excited plasma excitation source. [From a Ph.D. dissertation by S. E. Valente, Kansas State University.]

The dc-excited plasmas are characterized by (1) high electronic temperatures, (2) good stability, (3) low detection limits, (4) linear calibration curves over a large concentration range, and (5) reduced interference effects from other elements in the plasma. They also have higher background.

Radiofrequency (microwave) energy can be coupled *capacitatively* to a gas flowing through a quartz tube by use of a microwave cavity. Energy is supplied to a metal cavity that is resonant to the microwave frequency. It is necessary that the cavity be properly tuned to the microwave frequency. Microwave power generators used for this purpose are most commonly of about 100 W output at 2450 MHz frequency. The plasma is maintained in a small quartz tube, about 2 mm inside diameter. If the microwave energy is efficiently coupled to the plasma and an ionization suppressing buffer is used, excellent analytical data can be obtained. Detection limits are comparable with those obtained with the dc arc plasma. The background is reduced as compared with the dc arc excited plasma, but the tuning of the cavity and the adjustments required for maximum emission signal are more critical.

Taylor, Gibson, and Skogerboe[1] used a capacitatively coupled plasma at atmospheric pressure to determine trace impurities in argon. Lichte and Skogerboe[2] used a similar system for the trace determination of arsenic. They report a detection limit of five parts per billion of arsenic when using a 1 ml sample.

A plasma also can be produced and maintained if radiofrequency energy is coupled *inductively* to the plasma. The frequency is not critical and radio frequencies from 2450 to 4.8 MHz have been used. A schematic diagram of a plasma tube with inductively coupled rf excitation is shown in Figure 5-13. This design was developed at the Ames Laboratories of the AEC by Dr. V. A. Fassel and his colleagues. The aerosol sample enters at the lower inlet, the plasma is supported by argon, and a two-turn inductance placed near the plasma orifice couples the rf energy to the plasma. Radiofrequency energy of 1-2 kW at 27 MHz provides excitation energy. The plasma is ignited with a Tesla coil to ionize the argon. The oscillating magnetic field associated with the high-frequency energy source interacts with the ionized argon and electrons, causing them to flow in closed annular paths inside the quartz tube. The accelerated electrons and ions meet resistance in this flow and heating results. These steps lead to the formation of a plasma with temperatures in the range of 9000–10,000° K.

The sample enters the plasma tube through the aerosol inlet and the usual vaporization, atomization, and excitation steps occur at an effective excitation temperature of approximately 7000° K. Excitation occurs in a chemically inert environment; thus some possible interference effects are reduced in magnitude. No electrodes are used, so contamination possibilities are reduced. Spectra produced do, however, include lines, some OH band emission at 2600–3250 Å, and weak band emission of NO, NH, CN, and C_2. The spectra are relatively free of general background radiation.

Fassel and Kniseley[3] reported experimentally determined detection limits for 61 elements that compare favorably with detection limits obtained by atomic absorption, atomic fluorescence, and flame emission methods. Interelement interferences are lower than by other methods and interferences due to PO_4^{3-} and Al^{3+} apparently are negligible. The technique seems well adapted to simultaneous multielement analyses through use of direct reading spectrometers.

Plasma excitation sources can generally be characterized by high stability, low detection limits, wide range concentration capability, and

---

[1] H. E. TAYLOR, J. H. GIBSON, and R. K. SKOGERBOE, *Anal. Chem.*, **42**, 876 (1970).
[2] F. E. LICHTE and R. K. SKOGERBOE, *Anal. Chem.*, **44**, 1480 (1972).
[3] V. A. FASSEL and R. V. KNISELEY, *Anal. Chem.*, **46** (November), 1110A (1974).

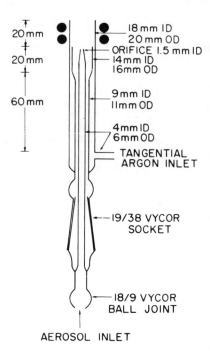

FIGURE 5-13. Diagram of an RF-excited plasma excitation tube. [From R. H. Scott, V. A. Fassel, R. V. Kniseley, and D. E. Dixon, Inductively Coupled Plasma-Optical Emission Analytical Spectrometry, *Anal. Chem.*, **46**, 75 (1974). Used by permission of the American Chemical Society.]

reduced interference effects. The required equipment is not expensive. For a dc plasma, existing dc arc sources may be used; for radiofrequency-excited plasmas an rf source of suitable power and frequency range is required.

5.5. The Laser Source

The mechanism of laser action was described in Chapter 2. Among its characteristics is its monochromatic, in-phase nature; a laser beam of energy is said to be coherent. Laser beams also produce very high-intensity pulses of energy concentrated in a very small region. A laser beam can therefore be used to vaporize and excite analytical samples for spectroscopic examination. The laser spectroscopic source can be used to sample areas as small as 25 μm in diameter. In the laser source designed by Jarrell-Ash the vaporized sample produced by the laser beam rises between two electrodes positioned above the sample and triggers a discharge from a capacitor charged to 2000 V. The capacitor discharge augments the laser energy to further energize the sample to produce spectral lines.

The laser source is especially useful for sampling small areas, where the amount of sample vaporized is very small and the bulk of the sample is not destroyed. It can be used on living tissue without destruction of the tissue. Thus the laser source is a valuable addition to the list of spectroscopic sources with special applications in the microsample area.

5.6. Multiple Source Units

Two approaches are commonly used to provide versatility and flexibility in power sources for analytical spectroscopy. One method is to use modular construction of units with a common set of start and stop switches, timers, power input circuit, and igniter. Units can be added to this system as needed. Modules are available for the dc arc, ac arc, low-voltage ac spark, and high-voltage spark. A common timing circuit is used to preset excitation timing, an igniter circuit initiates the discharge, and an oscilloscope allows observation of the discharge.

Another approach to making a variety of discharge conditions available in analytical spectroscopy is the development of the Multisource or Varisource power units. The basic circuit of the Multisource power unit as designed by Applied Research Laboratories is shown in Figure 5-14. The source includes a high-voltage and a low-voltage transformer, a variable inductor, capacitors and resistors, a synchronous gap, and an oscillograph

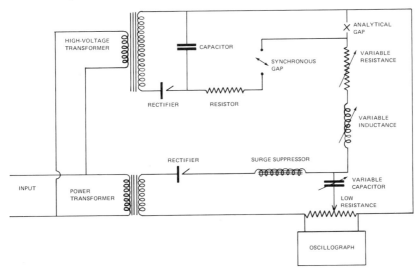

FIGURE 5-14. Basic circuit of a "Multisource" power supply. [Courtesy Applied Research Laboratories.]

for observation of the discharge signal. A variable capacitor is charged through a rectifier during one half-cycle and discharged through the analytical gap during the next half-cycle. The discharge is triggered by a high-voltage, low-power spark. With large values of inductance and capacitance, the frequency is low and the discharge possesses most of the characteristics of a dc arc. Increasing the resistance damps the oscillatory discharge to further add to the dc character of the discharge.

A circuit such as shown in Figure 5-14 provides a wide variety of excitation conditions but lacks the high-power spark so useful, in particular, for analysis of alloys and related substances. A high-voltage spark added to the Multisource power unit, based on the principle of charging and discharging a capacitor, provides controlled spark type excitation to meet this need. The result of this combination of sources is a high-precision unit capable of a wide range of excitation conditions with sufficient stability to be used for direct reading of spectral line intensities.

Other multiple source units are available for analytical spectroscopy. Usually they provide from dc arc to high-voltage spark excitation conditions by using several different circuits easily selected from a control panel. All of these include circuitry to ignite an arc and timing circuits to set accurately the time interval of excitation. Many of them also include oscilloscopes to provide visual display of the excitation cycle.

6. ARC AND SPARK STANDS

For analytical spectroscopy it is necessary to have the arc or spark excitation stand rigidly mounted on an optical bench. Care in optical alignment and care in spacing the electrodes are necessary and each adjustment must be reproducible. Early work in analytical emission spectroscopy made use of very simple electrode holders, open to the surroundings. With the development of improved, more complicated power sources and the need to further develop accuracy and precision, many changes in electrode holders and accessories occurred. The result has been the development of versatile enclosed units for excitation of analytical samples.

Modern arc and spark stands are enclosed systems constructed so fumes and volatile substances enter an exhaust system and are withdrawn from the laboratory. Electrical connections are shielded, doors have safety interlocks, a focusing lens usually is part of the arc–spark stand, electrodes can be positioned laterally and horizontally by external means, and electrical, gas, air, and water facilities usually are available within the excitation enclosure.

The enclosed systems also provide the capability of igniting the arc or spark in an inert atmosphere. Special electrodes or electrode assemblies can

EQUIPMENT FOR ARC/SPARK ANALYSIS

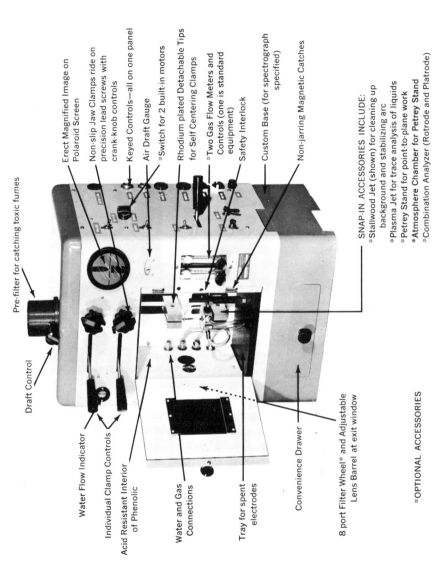

FIGURE 5-15. Photograph of an arc-spark stand for emission spectroscopy. [Courtesy Spex Industries, Inc.]

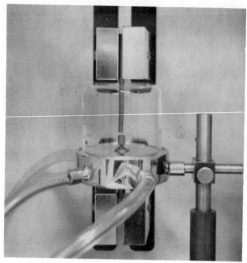

FIGURE 5-16. Photograph of an enclosed Stallwood jet excitation unit. [Courtesy Spex Industries, Inc.]

be mounted in the arc–spark enclosure. A modern arc–spark stand is shown in Figure 5-15 with parts identified in the figure.

6.1. Special Assemblies for the Arc–Spark Stand

Arc–spark stands are designed to accept carbon or metal rod electrodes. The special devices that have been developed to aid with some spectroscopic problems include the following.

6.1.1. The Stallwood Jet

Stallwood designed a device to allow a flow of gas to rise along the axis of the electrodes. Its original purpose was to stabilize the arc and it accomplished this objective in part. If the gas that surrounds the graphite or carbon electrodes does not include nitrogen, the troublesome cyanogen bands will not be formed. The Stallwood jet technique therefore can serve two purposes if an inert gas surrounds the electrodes. Several commercial Stallwood jet units are available that can be mounted in the electrode holder jaws of an arc–spark stand. One design is open and excludes air by flowing the inert gas vertically around the electrodes. Another design encloses the electrodes in a glass or quartz container for more complete protection from residual air. If important spectral lines lie in the cyanogen band region, the use of the Stallwood jet makes them available for analytical purposes. A commercial design of an enclosed Stallwood jet is shown in Figure 5-16.

EQUIPMENT FOR ARC/SPARK ANALYSIS

6.1.2. The Petry Stand

A Petry stand is designed to sample a flat surface spectroscopically. This design is especially useful to study metal or metal alloy surfaces. The counter electrode usually is a sharp, pointed graphite or carbon rod and spark excitation is most common. Some Petry stands have a motor-driven turntable so the flat sample can be rotated during excitation. The system also may include provision to flow a gas across the analytical gap. Petry stands are also constructed so they may be mounted in the electrode holder jaws of an arc–spark stand.

6.1.3. Rotating Disk Electrode Device

Direct spectroscopic analysis of liquids is possible using a rotating graphite disk as the lower electrode. The disk rotates slowly with its lower edge dipping into the solution to be analyzed. The liquid is carried upward

FIGURE 5-17. Photograph of a rotating disk electrode unit. [Courtesy Jarrell-Ash Division, Fisher Scientific Co.]

on the disk and enters the discharge between the upper edge of the disk and a pointed counter electrode. The disk is mounted to rotate in the same plane as the entrance slit. Devices to utilize this method of excitation are constructed to mount in the electrode jaws of an arc–spark stand. One such unit is shown in Figure 5-17.

Several other devices to use with arc–spark stands are available but are not so frequently used as the three that have been described. Some of these include controlled-atmosphere chambers, briquette holders, rotating platform assemblies, and water-cooled electrode holders. These devices are available and are very useful for special applications.

7. ORDER SORTERS

As described in Chapter 3, grating spectrometers will respond to harmonically related wavelengths; that is, if a grating spectrometer is adjusted to record spectral emission at 6000 Å in the first order, spectral emission at 3000 Å in the second order also will be recorded at 6000 Å and emission at 2000 Å will be recorded as third-order emission at the same 6000 Å position. Overlapping spectral orders therefore may unnecessarily complicate the observed spectrum and, in some cases, interfere with spectral lines used for analytical purposes.

Overlapping spectral orders can be separated by the use of a device to prevent the shorter wavelengths from reaching the photoreceptor. Filters that pass longer wavelengths but cut off shorter wavelengths can be used but are not highly efficient in the ultraviolet region. Another method is to use a "predispersing" prism in the optical path just in front of the entrance slit of the monochromator. The prism and associated focusing devices are oriented to disperse the entering beam vertically. Proper adjustment of the "order sorting" prism will prevent higher order spectra from entering the monochromator. Alternatively, the predispersing element can be adjusted so, for example, the first and second orders enter the monochromator vertically displaced to be recorded simultaneously as separate spectra on a photographic plate or film. Order sorters are commercially available from several manufacturers of spectrographic equipment.

8. DENSITOMETERS AND COMPARATORS

A densitometer is an instrument that can be used to determine the density of a spectral line on a photographic plate or film. A comparator is a device that allows easy comparison of two or more spectra recorded on separate photographic plates or films. In modern instruments these two functions usually are combined in one instrument.

EQUIPMENT FOR ARC/SPARK ANALYSIS

For quantitative spectrochemical analysis, using photographic recording, a densitometer is essential since it permits the measurement of relative spectral line intensities. A densitometer for measuring the opacity of a spectral line should have the following characteristics:

(1) the ability to measure the opacity from very low to very high intensity spectral lines;
(2) the ability to identify clearly the spectral region and the particular line being measured;
(3) methods of aligning the spectral line and the slit opening to the detector circuit;
(4) a method to move the plate rapidly from one spectral region to another; and
(5) methods to eliminate errors due to stray light entering the slit opening to the detector circuit.

Densitometers use a variety of optical systems. The optics utilizes either a slit at a fixed position with a mechanism to slowly move the spectral plate across the slit, or a slit that moves across the spectral line with the photographic plate in a fixed position. Either technique permits scanning the spectral line. Slits must be sufficiently narrow so that, when centered on the image of the spectral line, no light enters the slit from alongside the line. Some densitometer slits are adjustable for length and width. Others are of fixed width, but a series of fixed slits is available for interchange.

The light that passes through the spectral line strikes a photoreceptor and is converted into an electrical signal for some type of read-out device. Most instruments are arranged so that light passing through the clear emulsion of the photographic plate can be used as a reference. The electrical system usually is adjustable to provide for 100% transmission through the clear photographic emulsion. This arrangement is convenient for further processing of the data.

A variety of read-out devices are used. Frequently a galvanometer with a 0–100% T scale is used. Digital read-out also is used and arrangements for chart recording of line intensities are available with most densitometers. Recently instrumental design has switched to solid state devices to serve as amplifiers of the signal from the photoreceptor.

Comparators are commonly available as an integral part of a densitometer–comparator combination. Comparators are constructed so a reference spectrum from one photographic plate and another spectrum from a second plate may be projected onto the same viewing screen. Alignment of the two spectra provides a convenient method for comparing them. The technique is especially useful for qualitative analysis of an unknown, and in some cases semiquantitative information can be obtained. It also can be used to determine if a constituent of a sample is above or below a specified concentration.

The optics of one type densitometer–comparator is shown in Figure 5-18. One lamp is used and light paths from it pass through a comparison spectral plate and a sample plate; then both are projected on a viewing screen. Magnification is the same for both plates. In addition, the light passing through the sample plate is split by a rhomb mirror into two beams. The second beam strikes a narrow mirror and passes through an adjustable slit to a photoreceptor and then to an amplifier and a read-out system. The narrow mirror M and the adjustable photometer slit form a very narrow light beam that permits the scanning of a spectral line to determine its density. A photograph of a modern densitometer–comparator is shown in Figure 5-19.

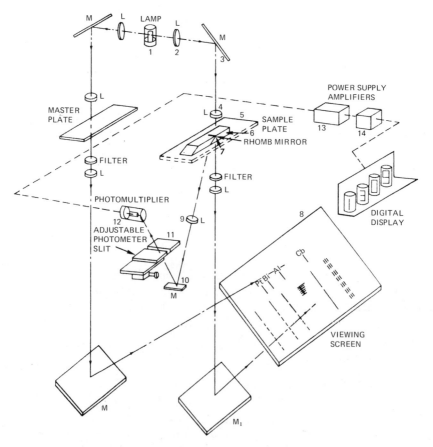

FIGURE 5-18. Optical arrangement of a comparator–densitometer. [Courtesy Jarrell–Ash Division, Fisher Scientific Co.]

FIGURE 5-19. Photograph of a comparator–densitometer. [Courtesy Jarrell–Ash Division, Fisher Scientific Co.]

9. MISCELLANEOUS ACCESSORY EQUIPMENT

Operation of a spectrometer for spectrochemical analysis utilizes, in addition to the accessory equipment already discussed, a variety of items. An optical bench is needed on which the items external to the spectrometer can be rigidly mounted. An assortment of lenses and lens holders is useful for different optical arrangements. Lenses preferably should be quartz, and a cylindrical lens is useful to focus radiation along the entrance slit of the spectrometer. Colored filters frequently are used in the external optical path to absorb radiation in certain wavelength regions and to separate overlapping spectral orders. An order sorter can be used for the same purpose.

9.1. Electrodes

Electrodes for analytical spectroscopy should properly be considered as supplies but will be discussed in this chapter since they are an essential part of the excitation process. Photographic plates, emulsions, and developing processes will be included in the following chapter, dealing with methods of recording and reading spectra.

Metal alloys frequently are formed into electrodes for spectrochemical analysis and usually are referred to as "self-electrodes." Since metals are electrically conducting, this method is quite satisfactory. Spark analysis of self-electrodes is common since good precision can be attained and concentrations of metals in the alloy range are easily determined. The Petry stand can be used for flat metal surfaces if desired.

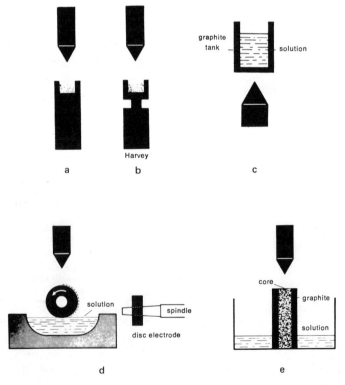

FIGURE 5-20. Some basic types of carbon electrodes. [Courtesy Atomergic-Chemetals Co.]

EQUIPMENT FOR ARC/SPARK ANALYSIS

Nonmetallic substances must be introduced into the spark or arc by being placed in or on some electrically conducting material. Carbon or graphite is used almost universally for this purpose. Both conduct electricity, although graphite is a better electrical conductor than carbon. Both will withstand temperatures of about 3500°C. Graphite is softer than carbon and more easily worked and machined than carbon. Graphite is the better heat conductor; as a result, the tips of graphite electrodes are cooler than those of carbon. This results in different rates of sample volatilization from the electrodes.

Other desirable properties of carbon or graphite as electrode material include a high excitation potential (about 10 eV), ease of preparation with a high degree of purity, ease of machining, resistance to corrosion, resistance to thermal shock, the possession of a simple spectrum of its own, and the fact that it sublimes at arc temperatures rather than melting.

Electrodes of many shapes and sizes have been used and almost any shape of electrode can be made if desired. Figure 5-20 shows some of the basic electrode types and shapes. A simple cup, such as shown in Figure 5-20a, is often used for powder samples. The sample is placed in the cup and an arc struck between the cup and a counter electrode. The sample is vaporized and excited in the arc. In Figure 5-20b the cup is undercut. The temperature of the cup rises more rapidly in this case. Often the sample and cup are both volatilized into the arc. A porous cup electrode is shown in Figure 5-20c. The cup is filled with liquid and passes slowly through the porous bottom end of the electrode. An arc or spark is struck to a lower counter electrode to excite the sample. Another technique useful for liquids uses a disk electrode, as shown in Figure 5-20d. The disk rotates and feeds liquid into the arc or spark gap for excitation. Another method to introduce a liquid sample into the arc or spark is to use a cored electrode as shown in Figure 5-20e. The solution enters the electrode gap by capillary action in the core.

Selected Reading

Ahrens, L. H. and Taylor, S. R., *Spectrochemical Analysis*, 2nd ed., Addison-Wesley, Reading, Massachusetts (1961).
Boumans, P. W. J. M., *Theory of Spectrochemical Excitation*, Plenum Press, New York (1966).
Brode, Wallace R., *Chemical Spectroscopy*, 2nd ed., Wiley, New York (1949).
Clark, George L., Editor, *The Encyclopedia of Spectroscopy*, Reinhold Publishing Corp. (1960).
Dodd, R. E., *Chemical Spectroscopy*, Elsevier Press (1962).
Harrison, George R., Lord, Richard C., and Loofbourow, John R., *Practical Spectroscopy*, Prentice-Hall (1948).
Nachtrieb, Norman H., *Principles and Practice of Spectrochemical Analysis*, McGraw-Hill, New York (1950).
Sawyer, Ralph A., *Experimental Spectroscopy*, Prentice-Hall (1944).

Chapter 6

Recording and Reading Spectra

The interpretation of spectra requires accurate information on spectral line intensities; this is essential for quantitative analytical data. Three general procedures are used to obtain this information: (1) visual inspection of spectral lines, (2) photographic recording of spectra, and (3) the use of a photocell and associated amplifiers with some type of read-out device. Visual inspection of spectral lines is possible but inconvenient and not very accurate. Photographic recording of spectra is a very common and useful technique since a spectral region may be photographed that includes many spectral lines of many elements. The photoplate also becomes a permanent record of the spectra.

Phototubes can be used in two different ways. A single phototube may be mounted in a fixed position and a scan of a spectral region obtained by slowly rotating a grating or prism. The spectra may then be recorded on a strip chart recorder. A series of phototubes can be mounted at exit slits so radiation of only one wavelength will strike each phototube. In this case, the signal from the phototube usually is stored in a capacitor and, after the excitation period is completed, its voltage read-out is a measure of the integrated signal at that wavelength. This chapter deals with both methods of obtaining spectral data, although the detailed amplifier and read-out system electronics are not included.

1. THE PHOTOGRAPHIC PROCESS

The photographic plate (or film) offers several advantages over the phototube as a receptor of radiant energy. It can be used to record a spectral region; thus the plate includes many spectral lines in one exposure. It also provides a permanent record of the spectra and acts as an integrating device since the spectral intensities recorded are time-averaged over the total exposure time. The resolving power of the photographic emulsion is good and closely lying lines usually can be separated. The photographic plate also is useful for measurement of spectral line wavelengths if that is necessary.

The following discussion of the photographic process is limited to those factors important in photographing spectra and in using the result for interpretation of spectral data.

1.1. Characteristics and Properties of the Photographic Emulsion

A photographic emulsion is a coating, about 0.03–0.04 mm thick, of small crystals of a silver halide scattered through a gelatin support. The grains vary in size, but generally are smaller than 6–8 μm in diameter. When the silver halide emulsion is exposed to radiant energy of suitable wavelength it is more easily reduced to metallic silver than if it has not been so exposed. Controlled action of a reducing agent will convert the exposed silver halide to metallic silver, while unexposed portions of the emulsion are not affected. After reduction, the remaining unreducèd silver halide is removed from the photographic plate. The result is a photographic negative image of the relative light intensities to which the emulsion had been exposed. Important properties of a photographic emulsion include speed, contrast, latitude, graininess, and spectral sensitivity. Some other properties of photographic emulsions that are somewhat interdependent are resolving power, halation, intermittency effect, and reciprocity law features.

1.2. The Characteristic Curve

The photographic emulsion properties of speed, contrast, and latitude may be obtained from the characteristic curve of the emulsion, which shows the relation that exists between exposure and the optical density of the photographic image. A typical characteristic curve is shown in Figure 6-1. If the exposure is plotted as the logarithm of exposure, the characteristic curve has a toe (AB) where the slope increases, followed by a linear portion (BC), and, finally, a region where the slope decreases (CDE) and becomes negative. The density is defined as

$$\text{density} = \log_{10} \frac{I_0}{I} \qquad (6\text{-}1)$$

where I_0 is the incident light intensity of the metering beam and I is the intensity of the light beam passing through the spectral line. In practice I_0 is usually measured through the clear emulsion adjacent to the spectral line. Usually I_0 is adjusted to read 100% transmittance, with 0% transmission when no light enters the densitometer slit. The value of I then is obtained as

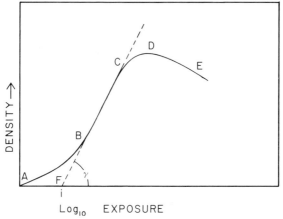

FIGURE 6-1. Characteristic curve for a photographic emulsion.

a percentage transmittance and equation (6-1) can then be written

$$\text{density} = \log_{10} I_0 - \log_{10} I = 2 - \log_{10}(\%T) \tag{6-2}$$

Exposure is the product of light intensity I and time t, or

$$E = It \tag{6-3}$$

Reference to Figure 6-1 indicates that the photographic image is just visible with an exposure equal to A, called the threshold exposure. After passing through the curved region AB, a relatively long region (BC) occurs which has a response where the density of the image is linearly related to the logarithm of the exposure; then the curve CDE occurs. The decrease in density beyond D is referred to as solarization or reversal. The intercept of the linear portion of the characteristic curve extending from B to F defines emulsion inertia. The slope of the linear portion of the characteristic curve (γ) is called the development factor or emulsion gamma. The linear portion of the characteristic curve fits the equation

$$D = \gamma \log_{10} E - \gamma \log_{10} i$$

where i is the emulsion inertia as shown in Figure 6-1, and E is the exposure. The equation may be rewritten as

$$D = \gamma \log_{10} \frac{E}{i} \tag{6-4}$$

The contrast constant γ varies with developing time and also varies for different emulsions and different wavelengths. Figure 6-2 shows the effect of developing time on gamma and Figure 6-3 shows the same effect on the characteristic curve. The value of i does not change with developing time.

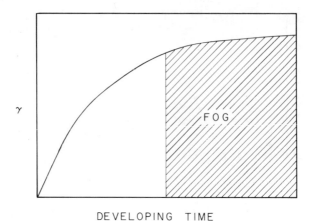

FIGURE 6-2. Effect of developing time on photographic emulsion gamma.

Relative speeds of different emulsions will differ, as illustrated in Figure 6-4. Emulsion II has a higher contrast constant γ than emulsion I, and its inertia is greater, as indicated by i_I and i_{II}. The choice between emulsion I or II depends on the purpose for which it is to be used. If very low intensities are to be recorded, emulsion I would be chosen because of its lower emulsion inertia. If high contrast is desired, emulsion II would be preferred. Emulsion latitude is a measure of the minimum to maximum exposure that will satisfactorily reproduce in the emulsion. It may be considered to be the exposure range of the linear portion of the characteristic curve of the emul-

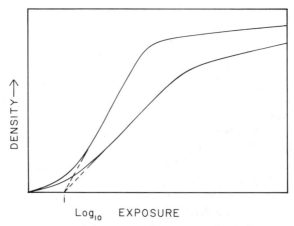

FIGURE 6-3. Effect of developing time on the characteristic curve of a photographic emulsion.

RECORDING AND READING SPECTRA

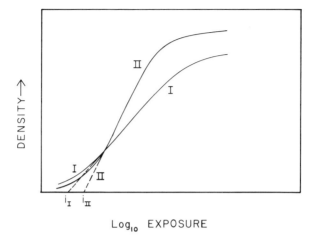

FIGURE 6-4. Characteristic curves of two different emulsions showing different relative speeds.

sion. In general, the higher the contrast (γ), the smaller the latitude. This is illustrated in Figure 6-4, where emulsion I has a greater latitude than emulsion II, but emulsion I has a smaller contrast constant.

1.3. The Reciprocity Law

The reciprocity law states that the extent of a photochemical reaction is proportional to the total energy employed. Equation (6-3) ($E = It$) is a mathematical statement of this law, which implies that I and t are independent of one another. Over a large portion of the exposure range the reciprocity law accurately describes the behavior of a photographic emulsion. However, photographic emulsions show a loss of sensitivity at very low or very high exposure levels. This loss in sensitivity is called the reciprocity effect or is referred to as the reciprocity law failure. A typical graph of the reciprocity law effect is shown in Figure 6-5. The 45° lines are lines of constant time and vertical lines represent constant intensity. If the curve is parallel to the ordinate, it would represent a behavior as predicted by the reciprocity law. For most emulsions the curves decrease at the left, become relatively parallel to the base, and then increase at the right. This corresponds to a reciprocity law failure at low exposure and higher exposure levels. For most analytical spectroscopy, it is desirable to obtain exposures in the midpoint of the exposure region so reciprocity law failure does not affect analytical results. If this is not possible, the data still may be valuable if proper interpretation of results is made. This can be done, especially at low exposure levels, by proper use of emulsion calibration techniques that will be described later.

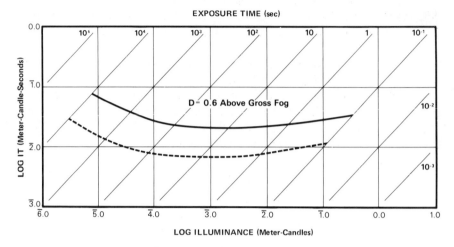

FIGURE 6-5. Typical curves to illustrate the reciprocity effect. [From a copyrighted Kodak publication. Courtesy Eastman Kodak Co.]

The range of concentrations satisfactorily covered by one working curve will not exceed about 20. For a larger concentration range a second and even a third working curve can be constructed using different exposure times.

1.4. The Intermittency Effect

Two exposures, each equal to one-half of a single given exposure, will not produce the same optical density as the single given exposure. This is known as the intermittency effect. Intermittent exposures frequently are used in analytical spectroscopy, and thus the effect is important. In arc and spark emission it is common to use a rotating sector to reduce exposure time. It has been shown, however, that if the intermittent exposures include from 50 to 100 segments, they will produce the same average intensity as a single exposure of the same effective time duration, and the intermittency effect can be neglected. The critical frequency varies with intensity but proper use of the rotating sector usually provides many more interruptions than necessary to avoid problems with the intermittency effect.

1.5. The Eberhard Effect

The Eberhard effect, sometimes called the edge effect, occurs during the developing process when a dense image is being produced. During developing a concentration of reaction products and an exhaustion of developer can occur at sites of dense image formation. Further development is inhibited by this situation and localized high density areas may not develop as normally expected. This process is of particular importance in quantitative spectrochemical analysis.

This effect is shown graphically in Figure 6-6 and results in a reduced density at the center of a dense spectral line. The spectral line profile is thus distorted and an inaccurate reading of line density will result. The Eberhard effect can be minimized by slow agitation of the developer or by carefully brushing of the emulsion surface during developing. Brushing is the more effective method since agitation of the developer does not replace the very thin layer of developer that is in contact with the emulsion surface.

1.6. Graininess and Granularity

A photographic image consists of very small silver particles and is not homogeneous. The particles are referred to as grains. Graininess is the nonuniformity observed in the photographic image and granularity is the variation in the transmitting or reflecting properties of the photographic image. Different photographic emulsions have differing degrees of granularity. Microdensitometer tracings across a photographic image can be used to measure the relative granularity of the emulsion. Figure 6-7 illustrates this property of photographic emulsions. Granularity also is affected by the kind of developer used and the length of time used in the developing process.

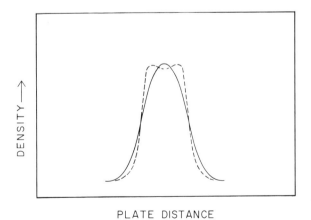

FIGURE 6-6. Eberhard effect on a spectral line density profile.

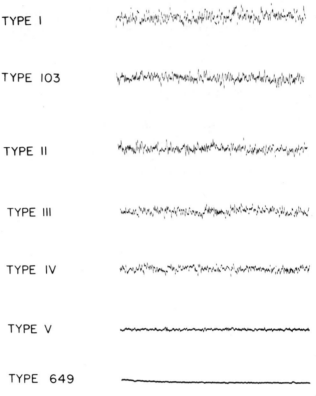

FIGURE 6-7. Microdensitometer tracings of the granularity of photographic emulsions. [From a copyrighted Kodak publication. Courtesy Eastman Kodak Co.]

1.7. Resolving Power

Resolving power is related to granularity and is defined as the ability of an emulsion to record separate adjacent lines. The resolving power of an emulsion usually is given in terms of the maximum number of lines that can be resolved per millimeter. For example, a resolution of 50 means that 50 equally spaced lines is the maximum that can be observed per millimeter using suitable magnification for observation.

The resolving power of an emulsion depends in part on the development process. It is affected by low- or high-exposure levels, with resolving power decreasing at both high and low levels of exposure. Resolving power usually is higher at short wavelengths. The spectral sensitivity of the emulsion plays no important role in the resolving power. The optical system employed must be of high quality to obtain maximum resolution.

RECORDING AND READING SPECTRA

Photographic emulsions are available which provide resolving powers from 50 to 225 lines/mm. If extremely high resolution is required, special emulsions are available with resolving powers of up to 1500 lines/mm. In general, photographic emulsions of high contrast show high resolving power but respond at a low-intensity level only to moderately intense exposures.

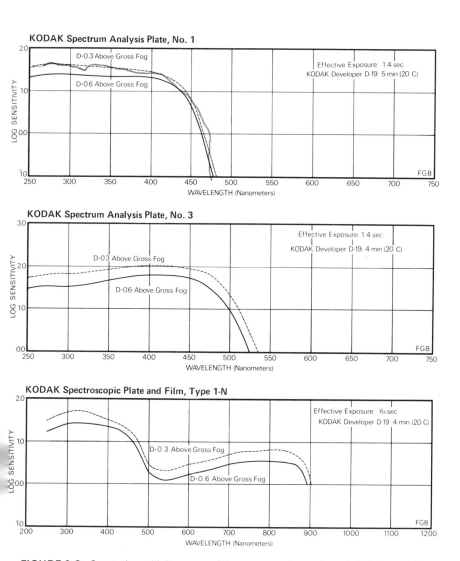

FIGURE 6-8. Spectral sensitivity curves of three commonly used spectral plates and films. [From a copyrighted Kodak publication. Courtesy Eastman Kodak Co.]

1.8. Spectral Sensitivity

Photographic emulsions exhibit wide variations in sensitivity at different wavelengths. Sensitivity generally decreases rapidly above about 5000 Å but the longer wavelength regions can be successfully photographed by special sensitization techniques. The lower limit of detection without special techniques is about 2000 Å.

For longer wavelength sensitivities special sensitizing dyes are incorporated into the emulsion to extend the useful range to even infrared regions of up to about 13,000 Å. For increased ultraviolet sensitivity the emulsion frequently is coated with a fluorescent substance. The ultraviolet radiation thus is transformed to a longer wavelength, which can expose the emulsion. Short wavelengths do not penetrate the gelatin support for the emulsion so very thin emulsions of low gelatin content frequently also are used in the ultraviolet region.

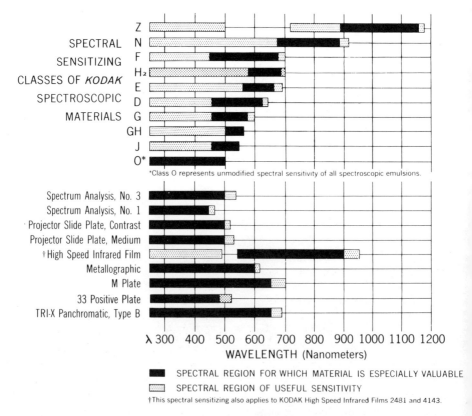

FIGURE 6-9. Spectral regions covered by different sensitization classes. [From a copyrighted Kodak publication. Courtesy Eastman Kodak Co.]

The contrast (γ) of a photographic emulsion also is dependent on wavelength, where the contrast generally increases as the wavelength increases. A calibration curve for an emulsion, therefore, is only correct at the particular wavelength at which it was obtained. Spectral sensitivity curves for some commonly used spectral plates and films are shown in Figure 6-8.

The Eastman Kodak Company provides a wide variety of spectral films and plates sensitized for maximum sensitivity in various spectral regions. The sensitizing types and their most useful spectral regions are shown in Figure 6-9.

2. PROCESSING OF SPECTROSCOPIC FILMS AND PLATES

Spectroscopic plates and films require special handling between the time of purchase and use. A latent image can be produced by proximity of the plates to x-rays, radioactive substances, heat, pressure, and exposure to some chemical gases such as ammonia. Emulsions deteriorate with time and deterioration is accelerated at higher temperatures. High relative humidity also can damage photographic emulsions. Photographic plates and films should be stored in light-tight, humidity-tight containers at reduced temperatures. They should never be stored near chemical fumes or exposed to heat. A good storage method is to use a second light-tight, humidity-tight container in addition to the box provided with the plates and store in a mechanical refrigerator. When plates are to be used they should be removed from the refrigerator about 4 hr prior to use and the container allowed to warm to room temperature before opening. This will prevent moisture from condensing on the plates. Plates are available to fit almost any standard size plate holder, so cutting of plates normally is unnecessary. Film also is supplied in sizes to fit almost all spectroscopic film holders, including rolls of 35 mm film.

2.1. The Developing Process

A darkroom is essential for the loading and unloading of plates and films and for their processing after exposure. The photographic materials must be handled in total darkness until processing and fixing is complete or only under the proper safelight. Different classes of photographic materials require different types of safelights and the instructions furnished with each particular emulsion should be followed. Safelights are available that will accommodate different safelight filters and if a variety of emulsions are to be used the darkroom should be equipped with a safelight and a series of safelight filters.

After exposure the spectroscopic plate or film must be developed, fixed, and dried. The degree of development is a function of the particular emulsion, the time of development, the temperature of development, the amount of agitation, and the concentration of the developer solution.

Kodak D-19 solution is recommended for most spectroscopic plates and films. It can be obtained in package form or can be prepared from the data given in Table 6-1. It provides good contrast, has a low tendency to fogging, and stores well. It is best to always use fresh developer. Old or used developer requires more developing time and the possibility of fogging is increased. Recommended development time is available for various combinations of developers and spectral emulsions, and for uniformity of development, the time and temperature of development should be carefully controlled. A development temperature of 68°F usually is recommended. A time–temperature curve for D-19 developer, centered on 68°F, is given in

TABLE 6-1
Formulas for Processing Spectroscopic Plates and Films[a,b]

Kodak developer D-19	
Water, about 125°F (50°C)	500 cc
ELON developing agent	2.0 g
Sodium sulfate, desiccated	90.0 g
Hydroquinone	8.0 g
Sodium carbonate, monohydrated	52.5 g
Potassium bromide	5.0 g
Cold water to make	1.0 liter
Kodak stop bath SB-5	
Water	500 cc
Acetic acid, 28%	32.0 cc
Sodium sulfate, desiccated[c]	45.0 g
Water to make	1.0 liter
Kodak fixing bath F-5	
Water, about 125°F (50°C)	600 cc
Sodium thiosulfate (Hypo)	240.0 g
Sodium sulfite, desiccated	15.0 g
Acetic acid, 28%[d]	48.0 cc
Boric acid, crystals[e]	7.5 g
Potassium alum	15.0 g
Cold water to make	1.0 liter

[a] Reproduced with permission from copyrighted Eastman Kodak publication.
[b] Dissolve the chemicals in the order given in each formula.
[c] If crystalline sodium sulfate is preferred to the desiccated form, use $2\frac{1}{4}$ times the quantities listed.
[d] To make approximately 28% acetic acid from glacial acetic acid, dilute three parts of glacial acetic acid with eight parts of water.
[e] Crystalline boric acid should be used as specified. Powdered boric acid dissolves only with great difficulty and its use should be avoided.

RECORDING AND READING SPECTRA 137

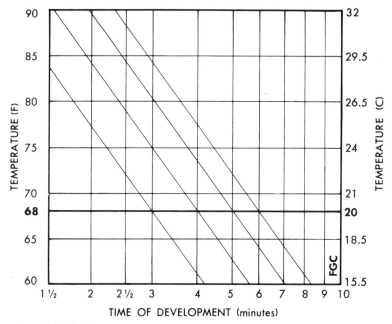

FIGURE 6-10. Time–temperature developing curves for D-19 developer. [From a copyrighted Kodak publication. Courtesy Eastman Kodak Co.]

Figure 6-10. To use the curves, reference is made to the recommended development time, as given in Table 6-2 at 68°F, and the development time is adjusted for other temperatures from Figure 6-10. For example, it is recommended that SA-1 emulsion be developed for 5 min at 68°F. If the temperature of the developer is 60°F, the time then should be increased to 7 min as determined from line three of Figure 6-10.

Continuous agitation of the developer or the plate is recommended. If tray development is used, the emulsion may be carefully brushed with a soft brush. This will bring fresh developer into contact with the emulsion and reduce the Eberhard effect. Some manufacturers of spectroscopic equipment have available photoprocessing equipment especially designed for the processing of spectroscopic plates and films. These processing units include provision for developer, stop baths, and fixers, and usually also include a device to dry the plates rapidly. A plate or film holder is provided to simplify the problem of transferring the plate or film from one solution to another. A thermostat controls the solution temperatures. Arrangements also include provision for continuous agitation of the plate or film to produce greater uniformity of developing. One such unit is shown in Figure 6-11. The photographic solutions are in containers along the lower part of the unit, the controls are on the vertical panel, and the plate drier is across the top of the unit.

TABLE 6-2
Recommended Developer and Developing Times for Various Spectroscopic Emulsions[a]

Kodak sensitized material	Kodak developer	Development time, (min)
Spectroscopic plates and films		
Type 103	D-19	4
Type 103a	D-19	4
Type I, Classes O, D, F	D-19	5
Type I, Classes N and Z	D-19	4
Type II	D-19	3
Type III	D-19	3
Type IV	D-19	2
Type V	D-19	3
Type 649	D-19	5
	D-8 (2:1)	2
Spectrum analysis plates and films		
#1	D-19	5
#3	D-19	4

[a]Reproduced with permission from copyrighted Eastman Kodak publication.

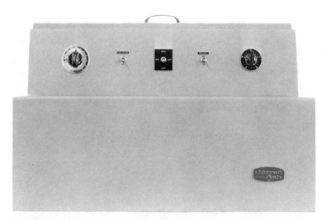

FIGURE 6-11. Photograph of a developing unit for spectrographic films and plates. [Courtesy Jarrell-Ash Division, Fisher Scientific Co.]

After expiration of the developing time the plate or film is removed from the developer and dipped into a stop bath solution. The stop bath immediately halts development. The plate or film is then placed in the fixing solution, to remove all unreduced silver, and then washed thoroughly and dried. It is important that all solutions be at the same temperature since the emulsion can be adversely affected by sudden temperature changes. Cleanliness in the darkroom and its surroundings is essential to prevent damage to the emulsion due to chemicals and dust.

3. HADAMARD TRANSFORM AND FOURIER TRANSFORM SPECTROSCOPY

Two methods of recovering spectra by use of complex mathematical methods are known as Hadamard transform and Fourier transform spectroscopy. Both have been applied successfully to infrared spectroscopy.

Plankey et al.[1] have reported on the application of Hadamard transform spectrometry to the ultraviolet and visible spectral regions. The Hadamard technique utilizes the dispersed radiation from a conventional spectrometer. The radiation is passed through a coded mask, recombined, and recorded. The mask is composed of N slits, and N measurements are made with the mask in different positions. After measurement, N simultaneous equations must be solved with a Hadamard matrix to obtain the spectrum. The mask must be moved mechanically and reproducibly and a computer is used to solve the simultaneous equations.

Plankey et al. reported that spectra in the ultraviolet and visible regions can be obtained but that sensitivity and detection limits are poor because multiplexing decreases the signal-to-noise ratio for any particular spectral line.

Fourier transform spectroscopy has had wide application in the infrared region but not as yet in the ultraviolet and visible regions. This procedure utilizes an interferometer to obtain an interferogram. By using an inverse Fourier transform, the interferogram can be converted into a conventional spectrum. The mathematical manipulations are unwieldy and time consuming and the transformation of the interferogram into a conventional spectrum in any reasonable time requires the use of a small digital computer. If some of the problems associated with signal-to-noise ratio in the ultraviolet and visible regions and the rigid mechanical conditions required of the interferometer can be overcome, this technique may have application in atomic spectroscopy.

---

[1] F. W. PLANKEY, T. H. GLENN, L. P. HART, and J. D. WINEFORDNER, *Anal. Chem.*, **46**, 1000 (1974).

4. LIGHT-SENSITIVE PHOTOTUBES

Light-sensitive phototubes also can be used to determine relative spectral line intensities. Two approaches are used for this purpose. The large, direct reading spectrometers use a battery of phototubes, one for each spectral line desired, located at the individual focal points. Usually the output of the phototube is collected over a specified time interval and stored in a capacitor. After exposure the capacitor is discharged into some type of read-out device. This method integrates the total energy over a time interval to provide a measure of spectral energy.

A second method is to use one phototube positioned at a single fixed slit at the focal position of the spectrometer. The dispersing element is then mechanically rotated so the spectrum can be scanned by the single phototube. The Beckman DU spectrophotometer functions in this manner, as do numerous other spectrometers. Most of the smaller monochromators used for flame emission and atomic absorption spectroscopy also use the scanning method. Scanning can be used with instantaneous read-out to record the spectrum on a strip chart recorder. The scanning spectrometers also can be positioned on a single spectral line for intensity measurement. In this application the spectral line intensity usually is determined with an instantaneous reading, although integration over a time interval also is possible.

Different types of photosensitive detectors have been used for spectral intensity measurements, including barrier layer photocells, vacuum and gas photodiodes, and multiplier phototubes. By far the most commonly used device is the multiplier phototube because of its extremely high sensitivity and precision when powered by a voltage-regulated power supply. A variety of multiplier phototubes are available that have maximum response in different wavelength regions.

4.1. Spectral Response Designation

For maximum sensitivity it is essential to choose the proper multiplier phototube. The Electronic Industries Association (EIA) has developed a spectral response designation system to facilitate designation of spectral response. A series of "S" numbers indicates the total response of the tube, including a combination of the transmission characteristics of the envelope as well as the sensitivity of the photosensitive material. Table 6-3 gives the spectral response number and the associated wavelength region of maximum response. Multiplier phototubes are useful beyond the range designated for maximum response but at decreased sensitivity. For example, the RCA 1P28 multiplier phototube has an S-5 spectral response (wavelength of maximum response is 3400 ± 500 Å) but is useful over a range of 2100–7000 Å. The EMI-6256 (EMI Electronics Ltd, England), with an S-13 spectral response,

RECORDING AND READING SPECTRA

TABLE 6-3
Spectral Response Numbers and
Wavelength Regions of Maximum
Response

Spectral response number	Wavelength of maximum response, Å
S-1	8000 ± 1000
S-3	4200 ± 1000
S-4	4000 ± 500
S-5	3400 ± 500
S-8	3650 ± 500
S-9	4800 ± 500
S-10	4500 ± 300
S-11	4400 ± 500
S-13	4400 ± 500
S-17	4900 ± 500
S-19	3300 ± 500
S-20	4200 ± 500

has a useful range of 2000–6500 Å. Table 6-4 lists some of the more commonly used multiplier phototubes in analytical spectroscopy.

4.2. General Characteristics of Multiplier Phototubes

A multiplier phototube consists of a light-sensitive photocathode followed by a series of dynodes arranged to multiply the electron emission from the photocathode. The electrical arrangement and basic circuitry used are described in Chapter 9.

Dynode configuration refers to the shaping and positioning of the dynodes so all stages are efficiently utilized. A number of different configurations are

TABLE 6-4
Commonly Used Multiplier Phototubes for
Analytical Spectroscopy

Multiplier phototube	Spectral response	Wavelength range, Å
FW-118	S-1	3200–10,000
RCA-1P28	S-5	2000–7000
RCA-6217	S-10	3000–8000
EMI-6256	S-13	2000–6500
R-106	S-19	1800–6500
EMI-9558AQ	S-20	2000–8500

used. One of the more common is to use a circular arrangement, which permits a compact layout in a small envelope. Linear configurations are also used, which permit a larger number of dynodes. Box-type configurations and venetian blind configurations also are used, but the circular and linear arrangements are the most popular.

Transit time is the time interval between the arrival of the light pulse at the entrance window of the tube and the time at which the output pulse at the anode terminal reaches a peak. The transit time is usually not a factor in analytical spectroscopy, although it is important in scintillation counting. The transit time is a function of the geometry of the multiplier phototube as well as the voltage between the anode and cathode. Transit time will be of the order of 20–100 nsec in most multiplier phototubes at normal operating voltages.

Dark current is the current that flows through a multiplier phototube when the tube is in complete darkness and electrical energy is applied to the tube. Any electron component of the tube, regardless of origin, can initiate a dark current. Only the dark current in the anode circuit of the tube is of importance. The dark current results in *noise* in a multiplier phototube and thus becomes a critical factor in limiting the lower level of light detection.

Three types of dark current effects are possible: (1) Ohmic leakage, (2) thermionic emission, and (3) regenerative effects. Ohmic leakage results from imperfect insulation. This type of leakage usually is very small; however, condensation of water vapor and dirt or grease on the tube envelope can cause an erratic increase in dark current. Precautions therefore should be taken to handle tubes carefully. The glass envelope of multiplier phototubes should never be handled with the hands. Tube containers often include a noncorrosive desiccant to reduce humidity. Ohmic leakage is proportional to the voltage applied to the tube and can be recognized by this fact.

Thermionic dark current emission occurs whenever thermally produced electrons are accelerated in the dynode train. The detection of low light energy levels is limited by the thermionic dark current. Thermionic noise cannot be balanced out of the measurement. Thermionic noise can be decreased by operating the multiplier phototube at reduced temperatures and in some applications this is done. For most analytical spectroscopy applications, however, the lower thermionic noise obtained at lower temperatures is not worth the considerable inconvenience of operating at low temperatures.

Regenerative ionization effects can be observed at high dynode voltages. The dark current, under these conditions, becomes highly unstable and may increase to dangerously high levels. Continued operation under these conditions will damage the tube. All multiplier phototubes eventually become unstable as the dynode voltage is increased and best operation is where the voltage is limited to the region where regeneration effects do not occur.

Noise effects include all those effects that produce a random response at the output of the multiplier phototube. Noise effects in the tube itself have been described; however, noise also may be a part of the incoming light signal. This type of noise, external to the multiplier phototube, cannot be controlled by adjustment of the tube operating parameters. External noise also limits the precision of the analytical determination.

Linearity of response of a multiplier phototube is proportional to light input over a wide range of values and may range over approximately eight orders of magnitude. The limit of linearity occurs when the space charge in the tube reaches a significant level. (In this respect multiplier phototubes cover a much wider range of light intensities than does a photographic plate.) Nonlinear phototube response to light intensity may increase with tube age, and tube sensitivity may decrease with age, especially if the multiplier phototube has been subjected to current overload.

Multiplier phototubes are sensitive to magnetic fields, especially if relatively long path lengths exist between elements in the tube. If magnetic fields are near the phototube, the tube will require a magnetic shield.

Phototubes also show "fatigue" characteristics, that is, the relative intensity of emission may decrease with time, especially if the light intensity to the tube is high. A rest period usually will restore the tube to its original response. When a multiplier phototube is to be placed in operation, it is best to turn on the power supply and open the entrance slit to the tube for 10–15 min prior to collecting analytical data. The power supply and the tube become more stable and better data can be obtained. When a phototube is not in use, it should be stored in the dark in a dry container. It is important that the photocathodes not be exposed to intense light.

Multiplier phototube life is very long when operated according to recommended conditions. The lifetime will be decreased by too high voltages and very high incident light intensities. High temperatures also will shorten tube life. Over an extended use period the sensitivity of a multiplier phototube decreases and the change becomes permanent. The type of dynode material also affects tube life since some surfaces exhibit longer lifetimes than others. For maximum life the tubes should be operated within recommended ratings and handled with due care.

4.3. Solar Blind Phototubes

Phototubes with restricted spectral response curves frequently are referred to as "solar blind" phototubes. Typical of such tubes is the Hamamatsu R166, which has a sensitivity curve as shown in Figure 6-12, where its spectral response may be compared with the spectral response curve of the R106 (S-5) phototube. It is apparent that the R166 has little response at

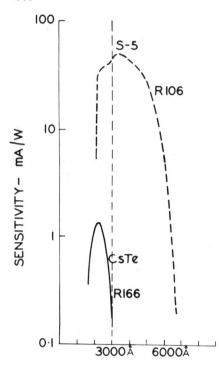

FIGURE 6-12. Spectral sensitivities of the R106 and R166 photomultiplier tubes. [From A. Walsh, *Physical Aspects of Atomic Absorption*, ASTM–STP 433 (1968). Used by permission of the American Society for Testing Materials.]

wavelengths longer than 3200 Å. By restricting photoresponse to wavelengths below 3200 Å, it is possible to eliminate the noise generated at wavelengths above 3200 Å. Thus for spectral lines that lie below 3200 Å a more desirable signal-to-noise ratio can be obtained. The result is a lower detection limit for elements whose spectral lines lie within the spectral band pass of the solar blind phototube.

5. RESONANCE DETECTORS

Walsh[2] has described a system of isolation and detection of radiation by a resonance technique. The system as used for atomic absorption is shown in Figure 6-13. Radiation from a hollow cathode source is passed through the flame sample cell into a resonance detector. The resonance detector contains an atomic vapor of the specific element under analysis. The atomic vapor in the resonance detector may be produced by cathodic sputtering or thermally. The atomic vapor in the resonance detector absorbs a portion of the

---

[2] A. WALSH, *Physical Aspects of Atomic Absorption*, ASTM Publication STP-443 (1968).

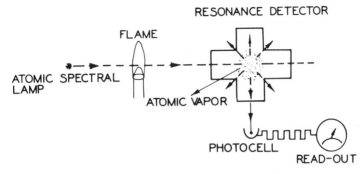

FIGURE 6-13. Resonance detector for atomic absorption spectroscopy. [From A. Walsh, *Physical Aspects of Atomic Absorption*, ASTM STP 443 (1968). Used by permission of the American Society for Testing Materials.]

energy from the light beam entering the detector, and some of the absorbed energy is emitted as resonance radiation. The reemitted radiation can be detected by the phototube and its intensity determined through a suitable read-out system.

A resonance detector is a simple device that requires no adjustment as does a conventional monochromator, and thus can be used under quite rigorous conditions. If a multielement hollow cathode source is used, resonance detectors may be aligned in series in the optical path emerging from the sample cell to provide simultaneous determination of several elements.

6. VIDICON DETECTORS

A vidicon detector is a photosensitive device composed of a two-dimensional array of several thousand very small detectors, each capable of responding to incident radiant energy. The most useful vidicon detector for analytical spectroscopy is the silicon vidicon detector.

The target of a silicon vidicon tube is shown in Figure 6-14. It is composed of an array of p-type semiconductor areas insulated from one another and formed over an n-type silicon base. The spaces between the p-type areas are coated with silicon dioxide to shield the silicon n-type base from the electron beam.

An electron beam is focused on the surface of the mosaic of p-type semiconductors and, by proper electronic control, scans the vidicon tube target. The electron beam charges each p-type cell to a negative potential. A positive voltage is applied to the silicon n-type base to create a depletion zone. Each diode thus acts as a capacitor, storing electrical energy.

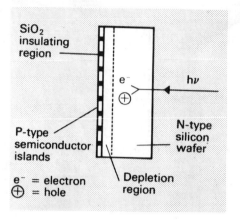

FIGURE 6-14. Target of a silicon vidicon tube. [From K. W. Busch and G. H. Morrison, Multielement Flame Spectroscopy, *Anal. Chem.*, **45**, 719A (1973). Used by permission of the American Chemical Society.]

When a photon is absorbed by the *n*-type base, electron–hole pairs are produced. The holes migrate through the depletion layer to the *p*-type cell, discharging the capacitor. An output signal results when the scanning beam passes over a cell which has lost its charge. The current produced during the recharging of the partially discharged capacitor forms the output signal.

The vidicon detector has a range of 2000–10,000 Å and a linear response over a wide intensity range. It is less sensitive than a photomultiplier and has a higher signal-to-noise ratio. The signal from a vidicon detector can be displayed on an oscilloscope or a recorder. If a wavelength range is focused on the target of a vidicon tube, the spectrum produced over that wavelength range can be displayed as an output signal. The vidicon tube thus provides an excellent method to observe a spectral region where the wavelength range covered will depend on the dispersion of the monochromator.

SELECTED READING

Ahrens, L. H., and Taylor, S. R., *Spectrochemical Analysis*, 2nd ed., Addison-Wesley, Reading, Massachusetts (1961).
Brode, Wallace B., *Chemical Spectroscopy*, 2nd ed., Wiley, New York (1943).
Kodak Materials for Spectrum Analysis, Eastman Kodak Company (1950).
Kodak Plates and Films for Science and Industry, Kodak Publication No. P-9 (1962).
Kodak Plates and Films for Scientific Photography, Kodak Publication No. P-315 (1973).
Mees, C. E. K., *The Theory of the Photographic Process*, Macmillan (1942).
Processing Chemicals and Formulas for Black-and-White Photography, Kodak Publication J-1 (1973).
RCA Phototubes and Photocells, Technical Manual PT-60, RCA, Lancaster, Pennsylvania (1963).

Chapter 7

Qualitative and Semiquantitative Arc–Spark Emission Spectrochemical Analysis

Emission spectroscopy provides an ideal method for qualitative analysis, since each atomic species has its own unique line spectrum. Spectral lines have two characteristics useful for qualitative analysis: (1) their wavelengths and (2) their intensities. It is the pattern of wavelength distribution that is primarily used for qualitative analysis, although the relative intensity distribution also can be helpful to verify spectral lines to identify an element. About 70 elements are easily identified by spectral methods. Those that are more difficult to identify include the gases and a few nonmetals, primarily because sensitive lines lie in the short ultraviolet portion of the spectrum that is difficult to observe.

For qualitative spectrochemical analysis it is desirable to identify from three to five or more spectral lines of the element. This is necessary since the spectrum of a complicated sample may contain many spectral lines and the possibility of line overlap or of misidentification of a spectral line exists. Photographic recording of spectra for qualitative analysis is essential since identification depends on identifying several lines for each element rather than just one spectral line. Photorecording also provides a permanent record of the sample that can be referred to later if necessary.

Qualitative spectrochemical analysis requires only a very small sample. Frequently a complete qualitative analysis can be obtained from a 1–5 mg sample and a single exposure. All readily detectable elements can be observed from one spectrum of the sample. Qualitative analysis also is possible with samples difficult to handle by more traditional chemical methods; for example, glasses, refractory materials, slags, minerals, etc., can be handled by reducing the sample to a fine powder. No chemical treatment or chemical separation is required.

Spectroscopic detection limits differ for different elements; many elements can be detected at very low concentration levels, some as low as 10^{-8} g. The spectroscopist should become familiar with detection limits of elements of most concern in his particular field and under his excitation conditions. The sensitivity of qualitative spectral analysis is dependent on the type and size of the sample, the excitation conditions, and the sensitivity of the photographic emulsion and the optical system used with the spectrograph. For best results excitation conditions should be maintained as uniform as possible.

1. SAMPLE EXCITATION

Direct current arc excitation is preferred for qualitative analysis since it is simple to use and is more sensitive than flame or spark excitation. It is common practice to use 1–5 mg of sample placed in a cup electrode that serves as the lower electrode and is the anode (plus electrode). The counter electrode is mounted immediately above the anode and a dc voltage of 200–250 V is applied. A current of 10–15 A is desirable. Electrode spacing should be maintained constant at about 3–5 mm. Current and voltage also should be maintained constant.

It is important in qualitative analysis that the sample be completely volatilized into the arc. With dc arc excitation different elements are vaporized at different times, with the low-melting elements appearing first and the high-melting elements appearing last. The time required to volatilize completely the sample may cause overexposure problems. The exposure can be reduced, if necessary, by using either a rotating sector or a neutral filter in the optical path just before entrance into the spectrometer slit. A combination of light intensity reduction and exposure time should be selected to provide a spectrum of suitable intensity.

The qualitative sample can be placed in the electrode cup in several different forms. If the sample is a powder, it can be conveniently placed in the cup electrode without weighing. Small metal filings also can be placed directly in the cup. If the sample is a solution, it can be placed in the cup and the solvent evaporated. If this procedure is followed, it is best to pretreat the electrode cup with carnauba wax dissolved in carbon tetrachloride. The cup is filled with the carnauba wax solution and dried at about 100°C to prevent the sample solution from seeping into the electrode. Another technique is to fill the cup loosely with graphite powder, place the solution on the powder, allow it to soak into the graphite, and then dry the sample. The powdered graphite provides a smooth burning process in the dc arc. Liquid samples also can be introduced into the arc by using a rotating disk or a porous cup electrode.

QUALITATIVE AND SEMIQUANTITATIVE ANALYSIS

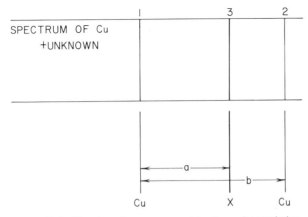

FIGURE 7-1. Wavelength measurement by linear interpolation.

2. WAVELENGTH MEASUREMENTS

Precise wavelength measurements are not required for qualitative spectrochemical analysis. Usually measurements to ± 0.05 to ± 0.10 Å will suffice, since the spectroscopist usually relies on the identification of three or four spectral lines to prove the presence of an element in the analytical sample. Use also is made of unknown spectra of elements to compare with the unknown sample. This technique does not require measurements of wavelengths of spectral lines.

If it is necessary to measure line wavelengths to identify spectral lines, reference should be made to wavelength standards. The international standard of wavelength, adopted in 1960, is the red line of krypton-84, with a wavelength assignment of 6057.802106 Å. This standard is not especially useful for qualitative analysis but is the basis for wavelength tables for iron lines, which can serve as secondary standards. Stanley and Meggers[1] have reported more precise iron wavelengths than the original data. They obtained these results by using an iron halide in a radiofrequency-excited electrodeless discharge lamp.

2.1. Line Identification by Wavelength Measurement

Occasionally it is necessary to identify spectral lines by wavelength measurement. The most common procedure in the analytical spectrochemical laboratory is to use a linear interpolation method based on known wavelengths of nearby lines. Figure 7-1 illustrates this technique, utilizing the known wavelengths of the two copper lines. The upper part of Figure 7-1 is a portion of a spectrum of a sample containing copper (spectral lines 1 and 2)

---

[1] R. W. STANLEY, and W. F. MEGGERS, *J. Res. Nat. Bur. Std.*, **58**, 41 1957.

and an unknown element (spectral line 3). The element producing spectral line 3 may be at least tentatively identified if its wavelength can be calculated. Inspection of a wavelength table of copper lines indicates that line 1 has a wavelength of 3247.5 Å and line 2 of 3274.0 Å. Distances a and b are measured carefully using a traveling microscope or a small magnifying calibrated scale graduated to 0.1 mm. With these data the wavelength of the unknown element line can be obtained from the formula

$$\lambda_x = \lambda_1 + \frac{a}{b}(\lambda_2 - \lambda_1) \qquad (7\text{-}1)$$

Measurement of a and b in Figure 7-1 gives $a = 6.1$ mm and $b = 3.2$ mm. Substituting these data into equation (7-1) gives

$$x = 3247.5 + \frac{3.2}{6.1}(3274.0 - 3247.5) = 3261.3 \text{ Å}$$

A search of wavelength tables, arranged by wavelength, reveals a sensitive cadmium line at 3261.1 Å; thus, cadmium may be the element producing the observed spectral line. To verify this tentative identification, a table of spectral lines of cadmium should be examined and the spectrum should be inspected to determine if other cadmium lines are present. Three or four lines in the spectrum coinciding with the wavelengths and relative intensities of other cadmium lines is considered positive proof that cadmium is present in the sample.

Linear interpolation can be used with grating spectrometers operated in the normal position. It also can be used to obtain approximate wavelengths with prism instruments if the wavelength difference between the two known spectral lines is small. For prism instruments interpolation is possible using the Hartmann formula:

$$\lambda = \lambda_0 + \frac{C}{(n - n_0)^{1/a}} \qquad (7\text{-}2)$$

where λ is the unknown wavelength, λ_0 is a constant, C is a constant valid for wavelengths close to λ, n is the index of refraction at λ, n_0 is a constant, and a is a constant and equal to unity over a spectral interval of 100–200 Å. Over a limited spectral range this equation can be simplified to

$$\lambda = \lambda_0 + \frac{C}{d_0 - d} \qquad (7\text{-}3)$$

The three constants λ_0, C, and d_0 may be evaluated from three known wavelengths if the separations of the three lines are obtained as d_1, d_2, and d_3. The three simultaneous equations can then be solved for λ_0, C, and d_0 and the resulting equation used to determine unknown spectral lines.

3. COMPARISON SPECTRA

A very common and useful technique to identify elements in a spectrum is by comparison with known spectra. It is useful for the spectroscopist to prepare spectra of known elements on his own spectrograph for comparison purposes. Figure 7-2 is a typical spectral plate, prepared with high-purity substances. The more useful spectral lines for each element are identified in each spectrum. The spectral plate also includes an iron spectrum, which is convenient for orienting the known and unknown spectra. When an unknown spectrum is obtained, an iron spectrum also should be placed on the spectral plate to aid in "lining up" the known and unknown spectra.

It is desirable to view the known and unknown spectra simultaneously. This can be done with a viewing box or with a projection-type comparator. Commercially constructed viewing boxes are available or one can be made with opal glass and a light source mounted back of the glass (Figures 7-3 and 7-4). A projection-type comparator usually is part of a densitometer–comparator instrument. They usually are arranged to provide a magnified image on a translucent screen with an optical system that will project the image of two plates on one screen. The images can be brought into coincidence for viewing and comparing spectra recorded on two different photographic plates or films.

Some commercially available mixtures of elements are available to produce comparison spectra containing more than one element and are especially useful for qualitative analysis. The Jarrell-Ash Division, Fisher Scientific Co., produces a powder standard that contains 1.30% of 45 different elements and another containing 1.20% of 48 elements. The one powder standard includes Al, Sb, As, B, Ba, Be, Bi, Ca, Cd, Ce, Cs, Cr, Co, Cb, Cu, Ga, Ge, Hf, In, Fe, Pb, Li, Mg, Mn, Hg, Mo, Ni, P, K, Rb, Si, Ag, Na, Sr, Ta, Tl, Th, Sn, Ti, W, U, V, Zn, and Zr. The second powder standard includes all the above elements plus Au, Pd, and Pt. The powders can be used to produce reference spectra for qualitative analysis and also can serve for semiquantitative procedures by proper dilutions with other matrices.

Johnson-Mathey Chemicals, Ltd., produces several standard mixtures. One contains 53 elements at 1.18 wt % of each individual reference element. A second mixture of eight elements, (Au, Ir, Os, Pd, Pt, Rh, Ru, and Ag) is available for identifying precious metals and a third powder is available for the rare earth elements. Johnson-Mathey recommend use of lithium fluoride as a spectroscopic buffer if one is needed to suppress matrix effects. If the recommended line is observed in the unknown spectrum, the presence of the element should be confirmed by identifying two or three other spectral lines of the same element.

Spex Industries, Inc., also has a series of qualitative analysis standards available. One such mixture includes 49 elements so blended that several lines

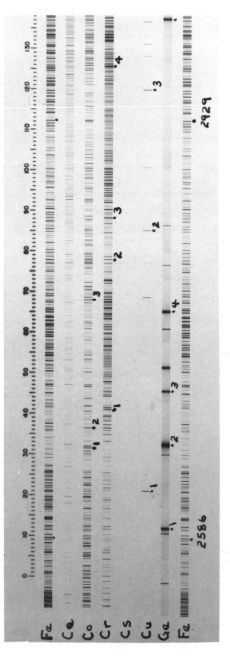

FIGURE 7-2. Spectra of some elements for qualitative analysis.

QUALITATIVE AND SEMIQUANTITATIVE ANALYSIS

FIGURE 7-3. Spectrum measuring magnifier. [Courtesy Baird-Atomic, Inc.]

FIGURE 7-4. Spectrum viewing box. [Courtesy Baird-Atomic, Inc.]

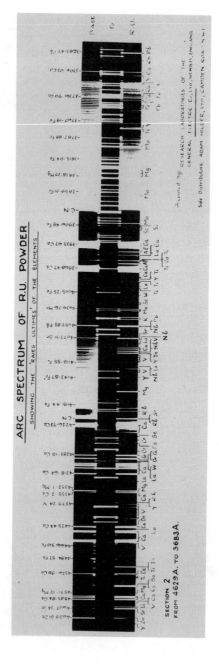

FIGURE 7-5. Section of the spectrum of RU powder with an iron spectrum. [Courtesy A. Hilger, Ltd.]

of each metal will be recorded on a spectrographic plate between 2000 and 4700 Å. They also have available a rare earth element mix and a noble metal element mix. Spex also provides other mixtures to be used for semiquantitative analysis in which each element is present as a fixed percentage by weight of the mix. For example, they have a mix of 49 elements, all present at 1.27% by weight.

Adam Hilger, Ltd., of London has an RU (*raies ultimes*) powder mixture of 50 elements. The concentrations are adjusted so only the persistent spectral lines (*raies ultimes*) will appear when the sample is excited in a dc arc. The base material for the RU powder is a mixture of zinc, magnesium, and calcium oxides. A spectral plate of RU powder with the iron spectrum included is shown in Figure 7-5. The plate was obtained using a prism spectrograph, so the dispersion is not linear.

4. SPECTRAL CHARTS

When master plates taken with the spectrograph being used are not available it is possible to identify spectral lines by comparison with spectral charts. If charts are used, it is often necessary to make proper allowance for differences in dispersion and for differences in changes in dispersion with wavelength if the charts have been prepared using a prism spectrograph.

Some charts are obtained by photographic processes. The spectrum of the standard and another of iron are obtained and a photographic enlargement is prepared. Spectral lines and wavelengths are then identified alongside the spectrum. The iron spectrum serves to orient the standard chart with the unknown spectrum if an iron spectrum also is included with the unknown.

Adam Hilger, Ltd., has such charts prepared for its RU powder standard. The charts were made with a prism spectrograph and thus the dispersion is nonlinear. Six separate enlargements, each 10 in. long, cover the spectral region from 2440 to 6717 Å. Each chart includes an iron spectrum, RU powder spectrum, and a spectrum of the base material. All persistent lines of the 50 elements in the RU powder are identified and frequent identification of wavelength is included. The spectra were obtained using dc arc excitation.

Brode[2] gives a set of charts covering the spectral range 2310–5090 Å. The charts are of the iron spectrum with iron lines identified and the positions of principal lines of other elements indicated by position and wavelength. The charts are arranged so an unknown spectrum can be aligned alongside the iron spectrum. The charts are very useful but visual alignment of an unknown spectrum requires a dispersion exactly the same as that of the charts.

[2]W. R. BRODE, *Chemical Spectroscopy*, 2nd ed., Wiley, New York (1943).

Spex Industries also has charts available of their Qual Mix. The charts are very similar to the RU powder charts but were obtained using a grating spectrograph. The charts therefore present the data with a linear dispersion.

5. WAVELENGTH TABLES

A number of wavelength tables with spectral descriptions and relative line intensities are available to the analytical spectroscopist. The most complete and most widely used tables are the *MIT Wavelength Tables*,[3] which include about 110,000 wavelengths for 87 elements from 2000 to 10,000 Å. Spectral line intensities are estimated from 1 to 9000, for excitation in a dc arc and a high-voltage ac spark. The atom lines are designated with the Roman numeral I and the singly ionized lines by the Roman numeral II.

Several other tables are available that are condensations of the MIT table. Earlier editions of Lange's *Handbook of Chemistry*, have a table of lines from 2000 to 9200 Å by wavelength and another of sensitive lines by element. Shorter lists of 500 sensitive lines for 85 elements arranged by wavelength and also by element have been prepared from the longer MIT tables by Harrison.[4]

Other tables based on the MIT compilation are given by Brode (see footnote 2), Harrison, Lord, and Loofbourow,[5] and Ahrens and Taylor.[6] Ahrens and Taylor also identify possible interfering spectral lines in their table.

An extensive study of spectral line intensities has been made at the National Bureau of Standards by Meggers, Corliss, and Scribner. They used copper electrodes and a dc arc to obtain their data on 70 elements. About 39,000 wavelengths are listed. Tables are available by wavelength and also by element. The data are included in the following four monographs available from the U. S. Superintendent of Documents:

> NBS Monograph 53, *Experimental Transition Probabilities of Spectral Lines of Seventy Elements* (1962).
> NSRDS-NBS, 22, Vol. II. *Atomic Transition Probabilities, Sodium through Calcium* (1969).
> NBS Monograph 32—Part I. *Tables of Spectral-Line Intensities Arranged by Elements* (1961).
> NBS Monograph 32—Part II. *Tables of Spectral-Line Intensities Arranged by Wavelengths* (1961).

---

[3]G. R. HARRISON, Editor, *MIT Wavelength Tables,* Wiley, New York (1939).
[4]G. R. HARRISON, Editor, *MIT Wavelength Tables, Revised,* MIT Press, Cambridge, Massachusetts (1969).
[5]G. R. HARRISON, R. C. LORD, and J. R. LOOFBOUROW, *Practical Spectroscopy,* Prentice-Hall (1948).
[6]L. H. AHRENS, and S. R. TAYLOR, *Spectrochemical Analysis,* 2nd ed., Addison-Wesley, Reading, Massachusetts (1961).

In NBS Monograph 32, Parts I and II, line intensities are reported on a scale of one to 6500. Intensities are related to the concentration required to produce a faint but unmistakable line at a given wavelength, which is then assigned an intensity of one when the copper electrode contains 0.1 % of the element. Any similar line appearing at unit intensity when the energy or concentration is reduced to one-fifth is assigned an intensity of five. Each line therefore is assigned a relative intensity proportional to its limiting detectability. Proper use of these data can lead to semiquantitative results. The data in NBS Monograph 32 were obtained using copper electrodes with the element under test added to the copper. A dc arc operating at 220 V was used. The spectra were photographed using a step sector in front of the entrance slit.

Appendices II and III of this text are abstracted from the National Bureau of Standards Monographs and include intensity data as well as gf values from NBS Monograph 53.

6. SOME SPECIAL PROBLEMS AND TECHNIQUES OF SPECTROCHEMICAL QUALITATIVE ANALYSIS

6.1. Spectral Line Interferences

Qualitative spectral analysis requires positive identification of one or more spectral lines of an element. If the sample is an unknown material, the identification may be complicated by the presence of lines of other elements that coincide or overlap the spectral lines of the suspected element. For this purpose it is essential to have available a table of spectral lines arranged by wavelength. After the spectral line of an element has been tentatively identified, reference should be made to a wavelength table to determine what interfering lines may occur. Standard reference tables may be used for this purpose. Ahrens and Taylor (see footnote 6) give tables that list sensitive lines of possible interfering elements and Harvey[7] gives an interference table that is useful. It is useful for the spectroscopist to learn as much about the unknown sample as possible since this information frequently can limit the search for interfering elements. A short table of sensitive lines is not always adequate for interference identification since weak lines of major constituents of the sample also may cause spectral line interference.

6.2. Spectral Band Interferences

The most common spectral band interference in qualitative spectral analysis is that produced by cyanogen. Cyanogen produces a number of bands, and three of them, with band heads at 4216.0, 3883.4, and 3590.4 Å,

[7] C. E. HARVEY, *A Method of Semiquantitative Spectrochemistry*, Applied Research Laboratories (1964).

are especially troublesome since they are in a spectral region containing many sensitive and useful spectral emission lines. If the desired spectral lines lie within the wavelength region of the cyanogen bands it may be necessary to reduce the band intensities. One method to reduce the cyanogen band intensity is to turn off the arc as soon as the sample has been completely volatilized. Continued arcing of the electrodes after the sample is volatilized only increases the intensity of the cyanogen bands. The entrance slit width to the spectrograph should be a minimum; a 10–20 μm slit width is sufficient. The background is increased with increased slit width but line intensity is not. Small, thin-walled electrode craters consume less carbon during arcing and should be used whenever possible.

Since cyanogen bands are produced by reaction of carbon with atmospheric nitrogen, it is possible to eliminate the bands by arcing the sample in the absence of nitrogen. Various devices are available for this purpose, with the most common being some variation of the Stallwood jet which is arranged to permit the operation of the arc in an inert gas atmosphere.

Copper electrodes are sometimes used if copper is not an interferent. Cyanogen bands are eliminated but spectral line sensitivity is also decreased. Graphite electrodes reach higher temperatures than copper electrodes during arcing and thus provide a higher excitation energy.

6.3. Arc Continuum Interference

A continuous background is usually the result of scattered light from optical surfaces and the incomplete volatilization of the sample and the carbon or graphite electrodes. Small solid particles in the arc are heated to incandescence to produce a continuum. A continuum can be reduced by use of thin-walled electrodes to reduce the amount of carbon consumed by the arc and by careful control of exposure time. It also is possible to reduce background by the use of a sector wheel or a light intensity filter. A proper adjustment of exposure time and neutral filter can reduce the background to below the threshold intensity of the photographic emulsion without affecting the photography of the spectral lines since they have an intensity greater than that of the continuum.

7. INCREASING SPECTRAL LINE INTENSITIES

As with any analytical technique, detection limits limit the ability of spectrochemical methods to detect the presence of minute traces of elements. Several methods are available to produce lower detection limits and can be used, especially if information is needed concerning one or more specific elements.

A *moving plate* technique can be used to increase the sensitivity of detection. A dc arc produces a fractional distillation of sample into the arc, with highly volatile elements appearing in the arc column early and those of lower volatility appearing later. A moving plate technique also reduces background effects. Two techniques may be used to produce moving plate spectra: (1) The plate may be moved quickly at regular time intervals, or (2) a Hartmann diaphragm may be adjusted to a different opening at regular intervals. This technique will concentrate different elements into different exposure segments, thus simplifying the search for elements, reducing the background and increasing the sensitivity.

Repetitive exposures also can be used to increase sensitivity. If sufficient sample is available, a series of repetitive exposures without moving the spectral plate can be useful. Care must be exercised to not build up the background when using this technique.

Concentration techniques can be used, especially if the search is to determine the presence or absence of certain specific elements. Chemical separations by any one or a variety of methods may be employed. Reagents that precipitate a group of elements are particularly useful. Standard references on chemical separations should be consulted if this method appears desirable.

The following is a list of techniques that can be used to increase the sensitivity of spectrochemical qualitative analysis.

1. Use of spectral plates of high sensitivity.
2. Optimum adjustment of the spectrometer entrance slit width.
3. Higher current for dc arc spectra.
4. Use of moving plate technique.
5. Use of fractional distillation methods.
6. Chemical removal of major constituents of the sample.
7. Use of special excitation conditions to reduce background effects.
8. Use of buffers to suppress band spectra.
9. Use of carrier distillation methods.
10. Observation of spectral lines in the most sensitive portion of the arc column.

8. SEMIQUANTITATIVE SPECTROCHEMICAL ANALYSIS

Spectrochemical methods are useful for what has become known as semiquantitative analysis. Frequently it is not necessary to have high-precision results and often it is only necessary to determine if an element has a concentration above or below some specified concentration. Several different techniques are available to determine the concentration level of an element in a sample and also to obtain semiquantitative data. In spectrochemical analysis semiquantitative results usually imply an accuracy of about $\pm 50\%$ of the amount that exists in the sample. Another type of semiquantitative method simply divides the constituents into "major" and "minor" elements. These techniques provide a great deal of information from a single spectrogram and thus often are used prior to a more precise determination of element concentrations.

8.1. Determination of a Concentration Level

One of the simplest semiquantitative methods is to determine if an element is above or below a specified concentration. These data are especially useful if impurity concentration specifications are given in terms of "the concentration does not exceed," as is done in many instances.

One approach to this problem is to devise a set of operating parameters, i.e., sample size, current and voltage values, exposure time, slit width, etc., so the maximum allowed concentration is the detection limit for the spectral plates used. If the spectral line is observed on the plate, the concentration exceeds specifications and if it does not appear, the concentration is below that specified.

An improvement to the above method is to photograph the spectrum of a sample containing the maximum concentration of the element being considered under conditions that will produce a spectral line of reasonable intensity, but to utilize either a step sector or a step filter in the process. If a standard is obtained by this method, to which the unknown may be compared, it is simple to determine visually if the concentration exceeds or meets specifications. The stepped spectrogram also provides a semiquantitative estimate of how much the concentration is above or below specifications.

The above methods of analysis are very useful for routine quality control; hence usually the samples are similar in terms of major constituents. Since matrix effects are considerable in spectrochemical analysis, it is important to realize that a spectral line intensity established for one matrix is not valid in another. If the same contamination element is to be determined in a different matrix, another comparison standard must be prepared.

8.2. The Harvey Method of Semiquantitative Spectrochemical Analysis

Harvey[8] developed a comprehensive method of semiquantitative spectrochemical analysis involving careful control of a large number of

---

[8]C. E. HARVEY, *A Method of Semiquantitative Spectrographic Analysis,* Applied Research Laboratories (1947).

QUALITATIVE AND SEMIQUANTITATIVE ANALYSIS   161

factors. The method involves the following:

1. A 10-mg sample is excited in a 10-A dc arc.
2. Excitation continues until the sample is completely vaporized as well as the cup part of the electrode.
3. Line densities and background densities adjacent to the spectral lines are obtained using a photometric-type densitometer.

Harvey's method requires care and attention to detail by the spectroscopist. Among the variables that must be controlled are the following.

(a) An electrode design as shown in Figure 7-6 must be used, where the electrode is undercut to raise the average temperature of the cup during arcing. The sample and electrode are arced until the sample is completely consumed and the cup walls are vaporized. The cup electrode dimensions should be as given in Figure 7-6.

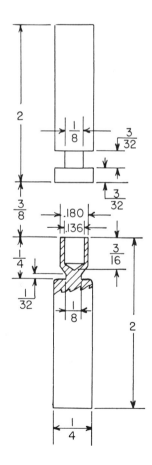

FIGURE 7-6. Electrode design for semiquantitative spectrochemical analysis by the Harvey method; standard cup electrode, 10 mg size. Dimensions in inches. [Courtesy Applied Research Laboratories.]

(b) The sample should be weighed and normally is 10 mg. Ten mg of powdered graphite is added to cover the sample.

(c) A 10-A dc arc should be used for sample excitation and the electrode gap should be set for 9 mm. The cupped electrode is the positive electrode. As the electrodes are burned the 9 mm gap spacing should be maintained.

(d) The slit width should be fairly wide and maintained constant. The optimum slit width will vary with the choice of spectrometer, but a 30–40 μm slit is commonly used.

(e) The photographic emulsion recommended is the SA-3; however, the SA-2 emulsion was used in Harvey's original work published in 1947.

(f) The center 4 mm of the arc should be selected to enter the spectrometer slit. This region is removed sufficiently from the cathode layer region to ensure uniform spectral lines and includes about half of the arc column.

(g) Matrix effects, caused by the major constituents of the sample, can be circumvented, in part, by use of lithium carbonate as a buffer.

Detection limits by the Harvey method are basic to use of the method and Harvey defines detectability as "the ability to see a spectrum line on a film preparatory to making density measurements." Harvey's method requires that a background appear in the spectrum, since spectral line intensities are compared to background intensity. Harvey has observed that a spectral line intensity about one and one-half times greater than background is necessary for the line to meet his definition of detectability. Background intensity thus serves the function of an internal standard. To obtain spectral line intensities from line densities also requires the preparation of an emulsion calibration curve relating intensity to density.

Since reading the density of a spectral line gives the intensity of the line plus the background, it is necessary to correct for background to obtain the net line intensity, where

$$I_{line} = I_{line + background} - I_{background} \qquad (7\text{-}4)$$

Since the background is used as an internal standard, equation (7-4) should be divided by $I_{background}$, giving

$$\frac{I_{line}}{I_{background}} = \frac{I_{line + background} - I_{background}}{I_{background}} \qquad (7\text{-}5)$$

or

$$\frac{I_{line}}{I_{background}} = \frac{I_{line + background}}{I_{background}} - 1 \qquad (7\text{-}6)$$

When I_{line} is at the limit of detectability, we have

$$\frac{I_{line\ min}}{I_{background}} = 1.5 - 1 = 0.5 \qquad (7\text{-}7)$$

QUALITATIVE AND SEMIQUANTITATIVE ANALYSIS

A linear relation exists between spectral line intensity and concentration from the minimum detectability level to higher line intensities. The percentage concentration of an element therefore can be calculated from the expression

$$\frac{\text{concn}(\%)}{k} = \frac{I_{\text{line}}}{I_{\text{line min}}} \quad (7\text{-}8)$$

where k is the minimum detectable percentage concentration of the element being considered. Equation (7-8) can be rewritten

$$\text{concn}(\%) = k\frac{I_{\text{line}}}{I_{\text{line min}}} \quad (7\text{-}9)$$

Using equations (7-6) and (7-7), we can rewrite equation (7-9) as

$$\text{concn}(\%) = 2k\left(\frac{I_{\text{line + background}}}{I_{\text{background}}} - 1\right) \quad (7\text{-}10)$$

Harvey has determined values of k for many elements and in different matrices. Use of these data can give semiquantitative results on a wide variety of samples containing many different elements. More recently Harvey (see footnote 7) has extended his original work to include data obtained using lithium carbonate as a buffer. These data permit buffering samples to a common matrix and make possible more uniform results.

To illustrate a typical calculation based on equation (7-10), an emulsion calibration curve is needed. Figure 7-7 is a typical calibration curve using percentage transmission as the ordinate and relative intensity as the abscissa,

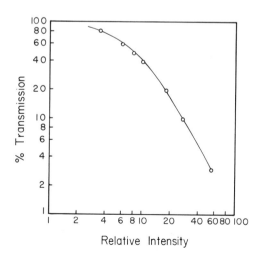

FIGURE 7-7. Typical photographic emulsion calibration curve.

plotted on a log–log scale. Assume the background near the 3415-Å nickel line in an aluminum alloy reads 80% transmission and the line plus background reads 20% transmission. Reference to the proper table lists the k value for the 3415-Å nickel line in aluminum as 0.0005. Reference to the emulsion calibration curve indicates a relative intensity value for 80% transmission to be 3.5 and that for 20% transmission to be 17. If these data are entered into equation (7-10), then we obtain

$$\text{concn}(\%) = 2(0.0005)\left(\frac{17}{3.5} - 1\right) \quad (7\text{-}11)$$

and

$$\text{concn}(\%) = 0.0039\%$$

If a calculating board is available, the calculation process can be simplified. Since an intensity ratio is needed, it can be obtained if the relative intensity scale is movable horizontally across the graph. If the scale is movable, an intensity of unity is set vertically below the intensity of the background (in this case 80% transmission) and the intensity then determined at the 20% transmission point. Using Figure 7-7, the reading at 20% transmission will then be 4.9. This is the same value as obtained by dividing the fraction 17/3.5 in equation (7-11) since, mathematically, the sliding of the intensity scale converts 17/3.5 to 4.9/1. Detailed information on the use of calculating boards is given in Chapter 8 in connection with calculations involved in quantitative spectrochemical analysis.

8.3. Matrix Effects

Harvey's original work (1947), in which he published k values for 42 elements in 36 different matrices, illustrated the significance of matrix effects on spectrochemical analytical data. If all samples could be reduced to a common matrix, semiquantitative and quantitative spectrochemical analysis could be simplified. A method to accomplish this, in part, involves the addition of another substance to the sample. The added substance is frequently referred to as a spectroscopic "buffer" and is added in sufficient quantity to become a major constituent of the sample mixture.

Germanium or germanium oxide has been used as a spectroscopic buffer. Strock[9] observed that germanium suppresses fractional distillation of the sample elements and he obtained excellent calibration curves for minor elements in synthetic standards when using powdered germanium metal as a buffer.

[9] L. W. STROCK, *Appl. Spectrosc.*, **7**, 64 (1953).

A semiquantitative method of spectrochemical analysis based on Strock's work was developed by Rudolph.[10]

The sample, as placed in the electrode cup, is one part sample, nine parts germanium, and 20 parts graphite. A thin-walled undercut electrode is used. Excitation is provided by an 11-A dc arc and the sample is arced until it is totally consumed.

Line intensities are determined by use of a photoelectric densitometer and the germanium line at 2417.4 Å is used as an internal standard. Intensity ratios are determined from the appropriate emulsion calibration curve. Standard comparison samples are prepared by mixing the elements as oxides or carbonates and are then buffered as described above.

Another commonly used buffer to suppress matrix effects is lithium carbonate. Lithium carbonate buffer provides a smoothly burning arc and compensates for many different matrices. A long arcing time is required with lithium carbonate buffer to totally consume the sample and this produces a strong background. A mixture of graphite and lithium carbonate overcomes some of these difficulties. Relative spectral line intensities of many elements are greater when using the lithium carbonate–graphite mixture than when using only lithium carbonate.

8.4. The Wang Method of Semiquantitative Spectrochemical Analysis

Wang et al.[11] have made use of the lithium carbonate–graphite buffer together with controlled-atmosphere excitation to devise another method of semiquantitative analysis. An intensity gain of 1.4–3.7 was observed when the sample was arced in an argon–oxygen atmosphere and the cyanogen bands were eliminated.

Wang prepared synthetic standards using Spex Mix, diluted with graphite. These dilutions are mixed with nine parts of a lithium carbonate–graphite (1:1) buffer. The sample is prepared by similarly mixing it with the buffer. A calibrated plate is made using a seven-step filter, which includes the standards at 0.1, 0.01, 0.001, and 0.0001 % concentrations and an iron spectrum. Comparison with the unknowns is made, using an iron spectrum to align the two plates. When a suitable line is selected it is compared with the intensity of the weakest exposure obtained with the seven-step filter to give a semiquantitiative estimate of the concentration range of the unknown.

[10] J. J. RUDOLPH, Materials Laboratories Report No. 6171-3030 A, Westinghouse Electric Corp., Pittsburgh, Pennsylvania.

[11] M. S. WANG et al., Qualitative and Semiquantitative Analysis, in *Analytical Emission Spectroscopy*, Part II, Edited by E. L. Grove, Marcel Dekker, New York, (1972), Chapter 8.

Operating conditions must be held constant and include a 13-A dc arc operating until the sample is entirely consumed. The analytical gap is 3 mm and is adjusted during arcing, with the center 2 mm of arc column being exposed. A seven-step neutral filter is used at the entrance slit. A controlled atmosphere of either argon or a mixture of argon and oxygen is used. Spectra are recorded on an SA-1 photographic emulsion.

9. SOME SPECIAL SPECTROCHEMICAL PROBLEMS

9.1. Microsamples

Occasionally the amount of sample available for analysis is very limited. To obtain information from a very small sample (10–100 μg) requires special handling. Small electrodes with small craters should be used for such samples. The optical alignment of the spectrograph should be arranged to collect the maximum of energy at the entrance slit and include the entire arc column, although care in alignment should be exercised to prevent light from entering the slit from the incandescent electrodes. Semiquantitative data are difficult with microsamples but qualitative data can be obtained quite readily by the usual comparison with previously prepared standards.

9.2. Microarea Sampling

Several methods for sampling small areas are available. The methods frequently are used to sample metal surfaces for nonuniformity but are not limited to this. One technique is to use a spark discharge from the metal to a sharply pointed counter electrode. A nonconducting substance with a small hole through it is used to shield the metal around the area to be sampled. A mica sheet with a small hole can be used as a shield. Plastic electrician's tape also can be used, as can Teflon.

Another method is to use a small wire in a capillary tube as the counter electrode with a spark discharge. The system does not possess high sensitivity but can reveal local differences in compositions that may be significant in quality control.

A third technique is to use a laser beam to vaporize a very small sample area. A laser can be used to direct very high energy to a very small area. A microscope can be used to assist in aligning the laser beam. The sample is vaporized by the laser beam and excited between auxiliary electrodes to provide the energy to record a spectrum of the sampled area. Multiple exposures frequently are used to increase the intensity of the spectrum. The technique has been applied to metals and alloys and also to biological systems.

QUALITATIVE AND SEMIQUANTITATIVE ANALYSIS

SELECTED READING

Books

Ahrens, L. H., and Taylor, S. R., *Spectrochemical Analysis*, 2nd ed., Addison-Wesley, Reading, Massachusetts (1961).
Brode, W. R., *Chemical Spectroscopy*, 2nd ed., Wiley, New York (1943).
Grove, E. L., Editor, *Analytical Emission Spectroscopy*, Vol. I, Part II, Marcel Dekker, New York (1972).
Harrison, G. R., Lord, R. C., and Loofbourow, J. R., *Practical Spectroscopy*, Prentice-Hall (1948).
Harvey, C. E., *A Method of Semi-Quantitative Spectrographic Analysis*, Applied Research Laboratories (1947).
Harvey, C. E., *A Method of Semi-Quantitative Spectrochemistry*, Applied Research Laboratories (1964).
Nachtrieb, Norman H., *Principles and Practice of Spectrochemical Analysis*, McGraw-Gill, New York (1950).
Slavin, Morris, *Emission Spectrochemical Analysis*, Wiley–Interscience (1971).

Spectral Line Wavelength and Intensity Tables

Ahrens, L. H., and Taylor, S. R., *Spectrochemical Analysis*, 2nd ed., Addison-Wesley, Reading, Massachusetts (1961).
Brode, W.R., *Chemical Spectroscopy*, 2nd ed., Wiley, New York (1943).
Corliss, C. H., and Bozman, W. R., *Experimental Transition Probabilities of Spectral Lines of Seventy Elements*, National Bureau of Standards, Monograph No. 53 (1962).
Harrison, G. R., *MIT Wavelength Tables*, Wiley, New York (1939).
Harrison, G. R., *MIT Wavelength Tables, Revised*, MIT Press (1969).
Meggers, W. F., Corliss, C. H., and Scribner, B. F., *Tables of Spectral Line Intensities, Arranged by Elements*, National Bureau of Standards, Monograph No. 32, Part I (1961).
Meggers, W. F., Corliss, C. H., and Scribner, B. F., *Tables of Spectral Line Intensities, Arranged by Wavelengths*, National Bureau of Standards, Monograph No. 32, Part II (1961).

Spectral Charts and Plates

Applied Research Laboratories, Sunland, California.
Arc Spectrum of RU Powder, A. Hilger, Ltd., London.
Brode, W. E., *Chemical Spectroscopy*, 2nd ed., Wiley, New York (1943).
Jarrell-Ash, Division of Fisher Scientific Co., Pittsburgh, Pennsylvania.
Spex Industries, Metuchen, New Jersey.

Chapter 8

Quantitative Spectrochemical Analysis

Quantitative spectrochemical analysis usually is considered to have originated about 1882 when W. N. Hartley, in Ireland, used a spark source to determine beryllium. In 1902, de Gramont, in France, used a spark to excite solid samples and used the method for metallurgical analysis.

Early estimates of concentrations were usually made by visual comparison of photographed spectra. Frequently the spectra were used only to obtain an order of magnitude of concentration. Since the early days of the "*raies ultimes*" and the "*letzen linien*" techniques, great changes have occurred in quantitative spectrochemical analysis. At the present time a precision of approximately 2% is possible using a spark excitation source with photomultiplier tube read-out. With a dc arc the precision that can be expected is about $\pm 4\%$ and with photographic methods of recording spectra a precision of about $\pm 5\%$ can be realized using dc arc excitation.

1. SOME GENERAL CONSIDERATIONS

Quantitative spectrochemical analysis has numerous advantages over other analytical techniques and also some disadvantages. The advantages of the method include the following.

1. A small sample (1–10 mg) is sufficient for analysis of up to 40 elemental constituents.

2. Usually no chemical separations or concentration steps are needed.

3. Very low detection limits are possible. Some elements can be detected at as low a concentration level as 0.0001%.

4. If suitable comparison standards have been prepared, the time required for analysis is short.

5. The method is suitable for a wide variety of samples, including

geological samples, metals and alloys, plants, forensic exhibits, soils, and biological and environmental samples.

6. The method is suitable for routine quality control work and, if automated, can provide analytical results on previously selected elements in 30 sec to 1 min.

The disadvantages to spectrochemical methods include the following:

1. Spectrochemical analysis is a comparison method. The precision of the analysis depends on the care taken in processing the sample and in all steps of the analytical process. Accuracy depends on the care exercised in developing the comparison standards.

2. "Cookbook" methods of spectrochemical analysis usually are not highly successful. Good technique, experience, and training, together with the adapting of sound spectrochemical principles to the existing problem, are essential.

3. Spectrochemical analysis destroys the sample. Although the amount of sample consumed is small, if a small sample should not be destroyed, another analytical technique will be required.

4. Analysis of major sample constituents cannot be readily accomplished. Spectrochemical methods are ideal for trace element concentrations. It is difficult to get high precision if the constituent desired is present in a concentration much above 30%.

It is important to emphasize that the present state of spectrochemical analysis is the result of careful, patient, and tedious developmental research over a period of 60–70 years. Precision such as indicated earlier in this section can only be attained by employing rigidly standardized control of all operational parameters.

2. THE INTERNAL STANDARD

Several authors indicate that modern spectrochemical analysis should be dated with the introduction of the concept of the internal standard by Gerlach in 1925. A variety of difficult to control operating parameters influence spectral line intensity. Among them are arc wandering, arc length, arc current, failure to time the exposure exactly, small errors in sample weight, and small losses of sample in transfer to the electrode.

In 1925 Gerlach introduced the concept of the internal standard and listed the criteria that should be used for the selection of an internal standard with his "homologous line pair" concept. An internal standard can be a weak line of a major constituent of the sample, but more likely will be a substance, not already a constituent of the sample, that is added to all samples and standards at a constant concentration level. In practice, an internal standard, whose concentration is constant, produces a spectral emission line whose

QUANTITATIVE SPECTROCHEMICAL ANALYSIS

intensity is compared with the intensity of a spectral line of the element whose concentration is desired. It is thus possible to obtain an intensity ratio rather than a single spectral line intensity. It is basic to the concept of an internal standard to consider that any factor that affects the intensity of the internal standard spectral line has a similar effect on the intensity of the unknown spectral line.

To function effectively as an internal standard, the element selected should meet certain criteria. These include the following:

1. The internal standard to be added should not be a constituent of the sample.
2. The internal standard and the unknown element should have similar rates of volatilization.
3. The internal standard line and the analysis line should have similar excitation energies.
4. The two lines should be in the same spectral region.
5. The internal standard substance should be of high purity.
6. The ionization potentials of both elements should be similar.

It is evident from the above statements that the intention is to provide an internal standard element that responds to excitation in a manner similar to the test element. Not all the criteria listed above can be met by a single element, but in practice one tries to match up as many similar characteristics as possible.

After selection of the internal standard element, its concentration in the unknown sample should be adjusted to provide an intensity level at about the middle of the calibration range for the analysis line. If the internal standard

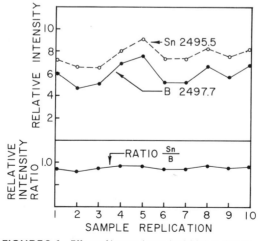

FIGURE 8-1. Effect of internal standard (tin) on boron.

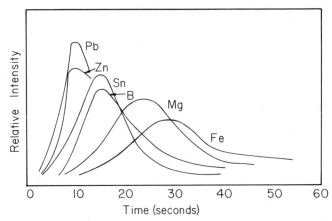

FIGURE 8-2. Some typical volatilization rates of slected elements in the dc arc.

concentration is too low (on the toe of the emulsion calibration) or too high, decreased accuracy will result.

The data presented in Figure 8-1 illustrate the value of the internal standard technique when applied to the determination of boron in plant tissue. The boron spectral line is at 2497.7 Å, and the tin line at 2495.5 Å is used as an internal standard. The data are replications of the same sample using dc arc excitation and graphite electrodes.

The choice of tin to serve as the internal standard for boron illustrates the application of the criteria for internal standards as listed earlier. For example, tin is not a constituent of plant tissue, its ionization potential is 7.30 eV and that of boron is 8.29 eV, the boiling points are 2260°C for tin and 2550°C for boron, and the two spectral lines used are only 2.2 Å apart.

It is difficult to decide which of the criteria listed earlier for selection of an internal standard element is most important; however, the rate of volatilization certainly is one of the more important. If the distillation rates of the internal standard and the analysis element differ greatly, the reproducibility of data will be adversely affected. If one element appears in the arc soon after the arc is ignited and the other appears later, the arc temperature will differ and variations in electrode spacing, arc current, and arc wandering will occur. Figure 8-2 illustrates volatilization rates of some selected elements.

3. SPECTROSCOPIC BUFFERS

Spectral line intensities are influenced by the chemical composition of the sample. Their effects, called matrix effects, are due primarily to the major

QUANTITATIVE SPECTROCHEMICAL ANALYSIS   173

constituents of the sample, and can cause serious errors in analysis for trace constituents of the sample if they are not recognized and compensated for in some way. The semiqualitative technique of Harvey, reported in Chapter 7, attempted to establish relative line sensitivities in each of a series of matrices. This approach leads to a large number of possibilities and is difficult to handle quantitatively, although is very useful for obtaining the semiquantitative results Harvey desired. A number of studies have been made that illustrate the effect of an extraneous substance on the spectral line intensity of another substance. Brode and Timma, in 1949, studied the effects of a series of elements on one another. They found that under their excitation conditions, in the sequence of elements Na, Sn, Fe, V, Al, Mn, Cd, Bi, Ca, Pb, Zn, and Cu, an element to the left in the sequence will depress the intensity of one to the right. One example of this effect is shown in Figure 8-3. In this case the effect of potassium on the spectral line intensity of phosphorus (2535.7 Å) was studied at three concentration levels of phosphorus. The data were normalized to a common starting density. The decrease in intensity for phosphorus emission as the concentration of potassium is increased is quite striking. Numerous other examples of matrix effects are well known.

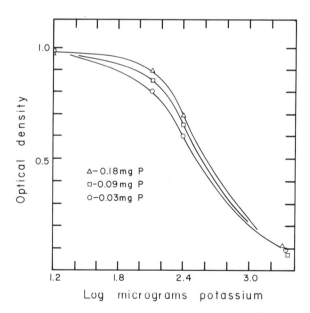

FIGURE 8-3. Effect of potassium on the spectral line density of phosphorus. [From W. G. Schrenk and H. E. Clements, Influence of Certain Elements on Line Intensity in the Direct Current Arc, *Anal. Chem.*, **23**, 1467 (1951). Used by permission of the American Chemical Society.]

The causes of matrix effects are not precisely known but probably include several factors, among them volatilization rate changes, changes in arc temperatures, atomic weights of matrix elements, kind of anions in the sample, and possible collisional deactivation processes.

The rate of volatilization of an element into an arc is affected, in part, by the matrix from which it escapes. For example, the volatilization rate is increased in the presence of powdered graphite or germanium oxide. If volatilization rates can be stabilized and made uniform from sample to sample, increased precision should result.

The arc temperature is affected by the composition of the arc column and its effective ionization potential. Alkali metals reduce the arc temperatures since they are easily ionized and thus produce an arc column that is highly conducting. As the arcing process continues, the arc temperature will increase since the composition of the arc column changes as the low-melting constituents of the sample are removed early in the arcing process by fractional distillation.

A collision of the "second kind," illustrated by the general reaction

$$A + B^* \rightarrow A + B + \text{kinetic energy} \qquad (8\text{-}1)$$

and

$$A + B^* \rightarrow A^* + B \qquad (8\text{-}2)$$

results in collisional deexcitation of B^* and thus removes B^* from the arc column. Spectroscopically, the intensities of spectral lines from B^* to some lower level are decreased, and from reaction (8-2) the intensities of spectral lines from A^* to lower levels are increased.

Some anions, especially the oxyanions, can affect the intensities of atomic lines of metallic elements. Some metals form stable oxides at high temperatures and thus, atomic lines of calcium and magnesium are decreased in intensity if oxygen is available from the sample to react with calcium and magnesium. Other refractory elements respond similarly to the presence of oxygen.

Matrix effects can be minimized by adding to the sample a substance that counteracts the effects described above. A substance that acts in this manner is called a spectroscopic buffer. In effect, its presence tends to reduce all samples to a common matrix. The properties required of a spectroscopic buffer have been described by Langstroth and McRae[1] and include the following: (1) The spectroscopic buffer should have a low ionization potential, (2) it should have a low boiling point, (3) it should have a simple spectrum, and (4) it should be of high purity.

---

[1] G. O. LANGSTROTH and D. R. MCRAE, *Can. J. Res.*, **16A**, 61 (1938).

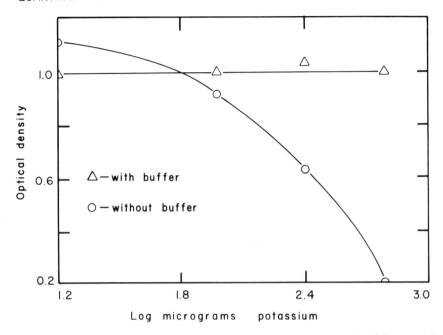

FIGURE 8-4. Effect of a spectroscopic buffer on the potassium repression of the spectral line intensity of phosphorus. [From W. G. Schrenk and H. E. Clements, Influence of Certain Elements on Line Intensity in the Direct Current Arc, *Anal. Chem.*, **23**, 1467 (1951). Used by permission of the American Chemical Society.]

A number of different substances have been used as spectroscopic buffers. Included are salts of the alkali metals as well as calcium carbonate and strontium and barium compounds.

The efficiency of lithium tartrate as a buffer to counteract the effect of potassium on phosphorus is shown in Figure 8-4. The quantity of phosphorus on the electrode was 0.10 mg for all points plotted and the amount of lithium tartrate on the electrode was 2.8 mg. The sample was in liquid form and was placed in the electrode cup and dried before dc arc excitation occurred. One of the most common spectroscopic buffers presently used for a variety of samples is lithium carbonate, which is available in a highly pure form from spectroscopic supply houses.

4. EXCITATION OF THE SAMPLE

The excitation source must provide the energy required to (1) vaporize the sample, (2) dissociate any molecular species present, and (3) excite the atoms. The excitation source may be a dc arc, an ac spark, a flame, or a plasma. Many variations in sources and operating conditions are used spectroscopically. Direct current arcs may be operated at low or high currents, and some are pulsed arcs operating from a pulsed dc power supply. The ac spark may be operated at high or low frequencies and some are operated at very high voltages. Flames and plasmas are being used more frequently, usually for samples in the liquid state. In flames various combinations of fuels and oxidants are in use to provide different temperatures and different flame environments for the sample. Plasmas, especially inductively coupled and radiofrequency powered, are used, usually with liquid samples. The basic circuitry and design of excitation sources is described in Chapter 5.

The choice of type of excitation is determined by several factors. Some of these are: type of sample, form of the sample (solid, liquid), sensitivity required, and time requirements of the analysis.

Since the dc arc is the most sensitive excitation source, it will be the excitation source of choice if extremely low detection limits are required. It also is the most difficult source to use in terms of reproducibility. The arc column wanders in the arc gap, the gap length changes during arcing, the background usually is high, and selective vaporization of the sample elements into the arc occurs. However, the dc arc is applicable to solid or liquid samples, sample preparation is a minimum, and the power source is inexpensive.

An ac spark source provides much better reproducibility than does the dc arc. It is ideal for metallurgical analytical problems, using "self-electrodes" or the Petry "point-to-plane" technique. Detection limits using the ac spark are not as low as with the dc arc, but the ac spark can be used satisfactorily to higher concentration levels. The ac spark is very noisy and more ion lines appear with ac spark discharge than appear with dc arc excitation. Fractional distillation is not a problem with the ac spark.

Flame sources usually are used for easily excited elements, with liquid samples, to produce simple spectra. Although they have been used with large spectrographs, flame sources are most commonly used with small spectrometers. A discussion of flame emission spectrometers is presented in Chapter 9.

Plasma sources, dc excited or excited with radiofrequency energy, provide temperatures from 6000 to 10,000°K and are very stable when compared with the dc arc. They are usually used with liquid samples, which are aspirated into the source continually during the excitation period. The problem of fractional distillation thus is avoided and the source is very stable. Dickinson and Fassel[2] devised a system to introduce a solid sample into a plasma by vaporizing the sample from an electrically heated tantalum boat.

[2] G. W. DICKINSON and V. A. FASSEL, *Anal. Chem.*, **41**, 1021 (1969).

5. SELECTION OF SPECTRAL LINES

If very low concentrations of trace constituents are to be determined, it is necessary to select a sensitive line of the element for analytical purposes. If somewhat higher concentrations are to be determined, more spectral lines are available from which to make the choice. Spectral line wavelength and intensity tables should be consulted in order to make the selection. These sources are listed in Chapter 7. Appendixes II and III also give information on some spectral line wavelengths and intensities.

In addition to selecting a spectral line (or lines) of suitable intensity in the desired concentration range, it also is essential to determine if any spectral line interferences may occur. This can be done by reference to a spectral line table arranged by wavelength. If the resolution of the spectrometer is known, it is simple to determine how closely an interfering line must be to the analyte element line to produce spectral line interference. Knowledge of the composition of the unknown sample also helps in determining if a particular spectral line of an interfering element will cause a problem. For example, a rubidium spectral line is not likely to interfere with the determination of copper in an alloy since rubidium is not present in the alloy.

6. COMPARISON STANDARDS

Since quantitative spectrochemical analysis is a comparison technique, it is important to exercise the greatest of care in selecting and preparing the comparison standards. Preparing useful spectroscopic standards is time-consuming and different standards usually are required for different matrices. It is important that standards resemble the unknown in chemical composition and physical properties. Matrix effects can be reduced by the use of a spectroscopic buffer, but it is best to approximate the matrix composition in the standard and, in addition, also use a spectroscopic buffer, since small concentration changes of some elements, notably the alkali and alkaline earth elements, may produce large changes in the light emission intensity of a minor constituent of the sample.

High-purity chemicals should be used to prepare comparison standards. If there is any doubt concerning their purity, they should be checked spectrographically. High-purity compounds are available from spectroscopic supply houses if reagent-grade chemicals are not satisfactory. Comparison standards usually are compounded with the matrix substances present in a fixed amount. The elements whose concentrations are to be determined are added to the matrix in varying known amounts, so their concentrations are below and above the range of concentrations expected. If the sample is a solid

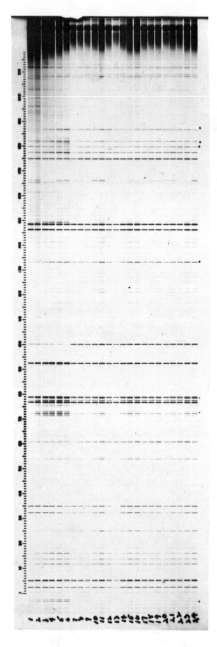

FIGURE 8-5. Spectrographic plate of comparison standards and samples of wheat grain.

powder, thorough mixing is essential. If the sample is liquid, the mixing process is simplified.

Metal alloy comparison standards can be made by mixing pure metals or metal alloys of known composition by heating in an induction furnace. The standards can then be formed into solid rods to serve as "self-electrodes" or flat disks if the "point-to-plane" technique is to be used. The National Bureau of Standards has available a number of metallurgical standards of certified composition.

Figure 8-5 is a spectral plate taken to determine a series of elements in wheat grain. The top six spectra are standards, the remaining spectra are wheat samples. Comparison of spectral lines that appear in the standards and the wheat indicates that the standards have a composition very similar to wheat. These standards and samples were prepared as solutions and dried on graphite electrodes before arcing. The buffer substance was lithium tartrate and cadmium was used as an internal standard. The wheat grain samples were ashed at 550°C and taken up in a solution containing the buffer and internal standard.

7. SAMPLE PREPARATION

Sample preparation for spectrochemical analysis is a most important process and requires extreme care. Samples should be handled as little as possible, to minimize the danger of accidental contamination. One of the special problems of spectrochemical analysis is obtaining a representative homogeneous sample since sample size is quite small.

Since a wide variety of sample types is likely, it follows that sample preparation will differ markedly from one type of sample to another. Metals can be formed into self-electrodes, as previously mentioned, or into flat disks. Metal alloys may need to be remelted to obtain homogeneity and also to be formed into a suitable shape. Metal powders can be pressed into pellets and placed in cupped graphite or carbon electrodes. Metals also may be converted into ionic solutions by proper treatment with acids or other chemical agents.

Solids frequently are converted into powders and thoroughly mixed to ensure homogeneity. If the solid cannot conveniently be put into liquid form, it can be excited from the solid state using carbon electrodes. Glasses, silicates, and many geological specimens fall into this category. If the sample is nonconducting, it can be mixed with powdered graphite to aid the excitation process. Buffer and internal standard also can be added in powdered form.

Samples of biological origin, plant or animal, are best handled by ashing to remove organic matter, followed by placing the sample in solution. The sample can then be excited as a solution by use of a porous cup electrode or a rotating disk electrode. Another approach is to place a known amount of the sample in the cup of a carbon electrode, dry it, and excite the solid residue. If this procedure is followed, it is best to render the electrode impervious to the solution by pretreating the electrode. One method is to fill the electrode cup with a saturated solution of carnauba wax in carbon tetrachloride and dry the electrode at about 100°C for 30 min. A liquid sample then will not penetrate the pores of the electrode.

Sometimes it is necessary to use separation and/or concentration procedures prior to spectrochemical analysis. Details of such procedures properly belong in texts on chemical separation and will not be dealt with here. However, some of the more common procedures that may be used include: ashing of organic substances; evaporation of solvent to concentrate the solute; chemical precipitation; ion exchange separations; liquid–liquid extractions; distillation; and electrolysis. The selection of the method to be used will be determined primarily by the type of sample and the elements to be separated.

8. EMULSION CALIBRATION AND ANALYTICAL WORKING CURVES

Quantitative spectrochemical analysis using photographic recording of spectra requires the accurate measurement of relative spectral line intensities. For this purpose it is necessary to use a densitometer, such as is described in Chapter 6. The spectral line density data obtained in this manner must then be converted into relative spectral line intensities. Since a photographic emulsion does not respond linearly to light intensity, it first is essential to prepare an emulsion calibration curve relating relative intensity to the response of the photographic emulsion as measured with the densitometer. Several techniques are used to obtain such data and several methods are used to plot the results. Some of the more common methods are described in the following section.

8.1. Emulsion Calibration

There are many methods that can be used to relate the response of a photographic emulsion to the intensity of the light striking the emulsion and there are several different graphical techniques that can be used to present the data. The problem of relating spectral line intensity to the concentration of the element producing the spectral line requires first an accurate measure of the photographic emulsion response. This section describes the four most commonly used methods of emulsion calibration and the three most commonly used methods of graphically presenting the data.

A step sector can be used to vary the exposure of the photographic emulsion to light in a regular manner. The sector must be used in such a way as to eliminate the intermittency effect. This can be done by rotating the sector at a rate that permits the interruptions in light to be more than 50–100 per exposure. It is best to rotate the sector disk at a rapid rate (up to 1000–1200 rpm) to avoid errors due to the intermittency effect and reciprocity failure. With the step sector factor, known from the geometry of the sector, it is simple to relate spectral line density changes to light intensity changes.

A stepped neutral filter can be used to vary the intensity of light striking the photographic emulsion. A neutral filter should reduce light intensity uniformly regardless of wavelength, but most of them do not. The difference is slight, however, so stepped, calibrated, neutral filters offer a suitable method to vary light intensity for the purpose of constructing an emulsion calibration curve. The filter method eliminates the possibility of errors due to intermittency effects and reciprocity law failures. It is important to illuminate the entrance slit of the spectrograph uniformly and since the filter may be 10–15 mm in length, an enlarged image of the source, using the central portion of the discharge, should be focused on the entrance slit.

Relative intensities of spectral lines, if they are accurately known, also can be used for photographic emulsion calibration. For this purpose, it is necessary that the relative spectral line intensities be well established and that they exhibit no self-absorption. The lines should have approximately the same excitation energies. The spectral lines should lie in the same general spectral region since emulsion contrast (emulsion gamma) varies with wavelength.

Dieke and Crosswhite[3] studied the feasibility of using selected iron lines for emulsion calibration, and later Crosswhite[4] extended the study. If the iron lines are selected from those with a common lower energy level, consistent results are possible. Table 8-1 includes a series of iron lines from 3157.0 to 3248.2 Å all having a common lower energy level (z^7D), selected from the more extensive tables of Crosswhite, which can be used for emulsion calibration. Ahrens and Taylor[5] give a shorter table of relative intensities of iron lines that also can be used.

---

[3]G. H. DIEKE and H. M. CROSSWHITE, *J. Opt. Soc. Am.*, **33**, 425 (1943).
[4]H. M. CROSSWHITE, *Spectrochim. Acta*, **4**, 122 (1950).
[5]L. H. AHRENS and S. R. TAYLOR, *Spectrochemical Analysis*, Addison-Wesley, Reading, Massachusetts (1961).

TABLE 8-1
Relative Intensities of Selected Iron Lines[a]

Line, Å	Intensity	Log intensity	Line, Å	Intensity	Log intensity
3157.0	100	2.00	3205.4	214	2.33
3157.9	73	1.86	3207.1	18	1.26
3161.9	64	1.81	3215.9	236	2.37
3165.0	25	1.40	3217.4	154	2.19
3165.9	40	1.60	3227.1	25	1.40
3168.9	19	1.28	3231.0	150	2.18
3175.4	125	2.10	3234.0	149	2.18
3178.0	95	1.98	3239.4	282	2.45
3194.4	39	1.59	3244.2	246	2.39
3196.9	505	2.70	3248.2	97	1.99
3200.5	270	2.43			

[a]Selected from Table 5, A. M. Crosswhite, Photoelectric Intensity Measurements in the Iron Arc, *Spectrochim Acta*, **4**, 122 (1950). Used by permission of Pergamon Press.

It is recommended that a metallic arc not be used for calibration purposes, because of the danger of self-absorption. One satisfactory method is to use iron oxide diluted with carbon to minimize the possible self-absorption effect of the metal electrode iron arc.

A two-line method of emulsion calibration has been described by Churchill[6]; it makes use of a pair of spectral lines whose intensity ratio is known. The spectral line pair selected must possess certain characteristics, as follows:

1. The intensity ratio of the two lines should lie between 1.2 and 2.0.
2. The intensity ratio must remain constant under usual excitation conditions. The lines should therefore be from the same element with similar excitation requirements.
3. The wavelengths of the two lines should be close (not more than 100 Å separation).
4. The lines should be sharp and free of spectral interference from other elements.

Table 8-2 lists three iron line pairs that Churchill reports are suitable in aluminum alloy analysis. Other pairs of lines that fit the requirements listed above and whose intensity ratios are known may be used.

[6]J. R. Churchill, Techniques of Quantitative Spectrographic Analysis, *Ind. Eng. Chem., Anal. Ed.*, **16**, 653 (1944).

QUANTITATIVE SPECTROCHEMICAL ANALYSIS

TABLE 8-2
Iron Line Pairs Suitable for the
Two-Line Method of Emulsion
Calibration[a]

Line pair	Intensity ratio
Fe(I) 3047.6/Fe(I) 3037.4	1.26
Fe(I) 2966.9/Fe(I) 3037.4	1.56
Fe(II) 2755.7/Fe(II) 2739.5	1.23

[a]From J. R. Churchill, Techniques of Quantitative Spectrographic Analysis, *Ind. Eng. Chem., Anal. Ed.*, **16**, 653 (1944). Used by permission of the American Chemical Society.

To calibrate an emulsion using the two-line method, a series of spectra are obtained that include the calibration line pair, using conditions that produce very light lines to very dense lines. The excitation conditions should remain constant for all exposures. One method of varying the spectral line intensities is to vary the distance from the source to the entrance slit of the spectrograph. Another technique that may be used is to use a neutral step filter in the optical path and still another is to use a rotating sector. These techniques permit the use of a constant exposure time. The series of exposures should cover the range of intensities that are encountered in spectrochemical analysis. Plate processing should be identical with that used for analytical samples.

TABLE 8-3
Relative Spectral Line Intensities of an
Iron Line Pair to Use for Emulsion Calibration by the Two-Line Method[a]

%T Fe 2739.5	%T Fe 2755.7
90.0	85.7
80.5	71.2
71.1	61.0
61.0	50.5
42.1	31.0
30.5	21.5
21.1	14.5
16.5	9.0
10.1	6.2

[a]Line pair Fe 2755.7/Fe 2739.5, intensity ratio 1.23 = r.

The use of the two-line method to obtain an emulsion calibration requires a preliminary curve prepared as follows. The percentage transmittance of both lines of the calibration pair is determined for all exposures. Percentage transmittance usually is a convenient measure of relative line density since most densitometers can be easily adjusted to provide a 100% transmittance reading through the clear plate and 0% when no light strikes the photoreceptor. Typical data obtained in this manner are shown in Table 8-3 for the iron line pair Fe 2755.7/Fe 2739.5 Å, which has an intensity ratio of 1.23. From these data a preliminary graph is prepared as shown in Figure 8-6. The percentage transmittance of one of the spectral lines is plotted against that of the other. The purpose of the preliminary curve is to determine the percentage transmittance produced by intensity I and that produced by intensity rI, where r is the intensity ratio of the calibration lines.

Data for preparation of the emulsion calibration curve are then obtained as follows. Select a starting point from the preliminary curve at a smaller percentage transmittance than is used for routine analysis and determine the percentage transmittance for r times this value. By reference to Figure 8-6, and selecting 93% transmittance for Fe 2739.5 Å, a reading of 90% is obtained for Fe 2755.7 Å. Continue alternating the abscissa and ordinate reading, recording each percentage transmittance until the reading is below that normally used for routine analysis. These data, obtained from Figure 8-6, are tabulated in Table 8-4. Along with the percentage transmittance, the relative intensity producing that reading is recorded after assigning units to the first reading. Each relative intensity value in the table is obtained by multiplying the preceding one by, in this case, 1.23. The data in Table 8-4 can then be used to prepare the emulsion calibration curve. If the value of r

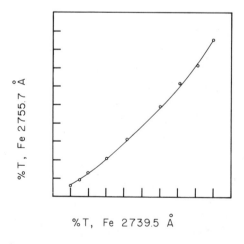

FIGURE 8-6. Preliminary curve for the two-line method of spectral plate calibration.

QUANTITATIVE SPECTROCHEMICAL ANALYSIS

TABLE 8-4
Data from Figure 8-6 to Be Used to Plot the Emulsion Calibration Curve

%T	Relative intensity	%T	Relative intensity
93	1.00	38	4.26
90	1.23	28	5.24
86	1.51	20	6.44
80	1.86	13	7.93
72	2.29	8	9.75
62	2.82	5	12.0
50	3.46	3	14.8

is precisely known, a typical H and D curve can be prepared. If the value of r is not precisely known, the data can still be used for preparation of the analytical working curve by use of an arbitrary constant. This will not affect analytical results since only relative intensity values are required rather than absolute intensity values. An H and D curve, however, will not provide data on plate contrast (plate gamma) unless the value of r is accurately known.

A variation of the two-line method of emulsion calibration is to use a neutral step filter with an accurately known intensity ratio. The percentage transmittances of the light and dense portions of a series of spectral lines covering a range of transmittances from very light to very dense is obtained. From these data a preliminary curve is obtained from which the calibration data can be determined in the same manner as in the two-line method. The lines selected for the *two-step* method should lie in the same spectral region as the analysis line.

8.2. The Emulsion Calibration Curve

Three different methods that are commonly used to present graphically emulsion calibrations will be described. These include the H and D curve, the percentage transmittance curve, and the Seidel function curve.

The H and D calibration curve, in addition to being useful to obtain intensity vs. line density data, also provides a quick measure of emulsion contrast since the slope of the straight line portion of an H and D curve is defined as the emulsion gamma.

The H and D curve is a plot of spectral line density vs. relative intensity. It is necessary therefore to convert the "%T" values of Tables 8-4 to density values. In the case of percentage transmittance, the relation

$$D = 2 - \log(\%T)$$

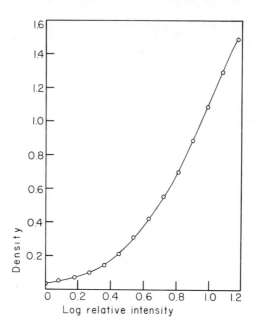

FIGURE 8-7. *H* and *D* photographic emulsion calibration curve (From data of Table 8-4).

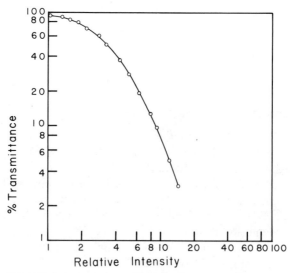

FIGURE 8-8. A photographic emulsion calibration curve plotting percentage transmittance versus relative intensity on a log–log scale.

QUANTITATIVE SPECTROCHEMICAL ANALYSIS

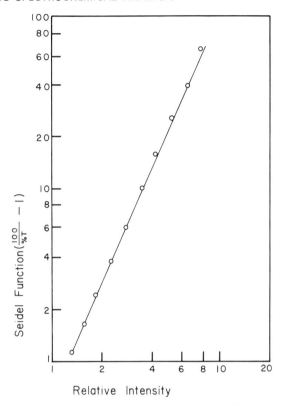

FIGURE 8-9. A photographic emulsion calibration curve using the Seidel function plotted versus relative intensity on a log–log scale.

can be used. Appendix V is a tabulation of these values and reference to it may be made to obtain the densities that correspond to percentage transmittances. In Figure 8-7, the percentage transmittances of Table 8-4 have been converted to densities and are plotted against the logarithms of the corresponding relative intensities to produce the conventional H and D curve.

Another method used to present graphically emulsion calibration data is to plot log percentage transmittance vs. log relative intensities. This type of plot is shown in Figure 8-8, using the data from Table 8-4. This method of presenting emulsion calibration data is widely used and convenient for the preparation of analytical working curves.

A third method of preparing an emulsion calibration curve is to plot the "Seidel function," $\log[(d_0/d) - 1]$, against log intensity, where d_0 is the galvanometer deflection through the clear plate and d is the deflection through the spectral line. If the galvanometer is adjusted to read 100 through the clear plate and 0 with no light, the Seidel function becomes $\log[(100/\%T) - 1]$. The term $(d_0/d) - 1$ may alternatively be plotted against intensity on log–log paper. The Seidel function is useful to extend the linear relationship of the usual H and D curve to lower levels of photographic response. Linearity of the curve results in increased accuracy of results at low intensity levels. Figure 8-9 is a plot of the Seidel function against log relative intensity. The curve should be compared with Figures 8-7 and 8-8, which were prepared using the same data (Table 8-4). Appendix VI gives the numerical values of $(d_0/d) - 1$ and of $\log[(d_0/d) - 1]$ when the densitometer is adjusted to read 100% transmittance through the clear emulsion.

9. THE WORKING CURVE

The working curve is the curve obtained when the intensity of the analysis line or the ratio of the intensity of the analysis line and internal standard line is plotted against the concentration of the analyte element in a series of standards. The emulsion calibration curve, regardless of the method of plotting, is used to convert spectral line response to relative intensity. Logarithmic coordinates commonly are used to plot the working curve. The construction and use of a typical analytical working curve are detailed in the following sections.

9.1. Construction of a Typical Analytical Working Curve

This section will describe, in step-by-step detail, the procedures involved in the establishment of a spectrographic working curve and its use in the determination of the concentration of the element in an unknown. Use will be made of the principles described in earlier sections of this chapter.

The problem is to determine germanium in a solution of unknown concentration. The first step is to obtain an emulsion calibration curve. For this purpose a seven-step neutral filter will be used, with each step having a relative intensity 1.5 times the preceding step. Since germanium is the element of interest, the germanium line at 3039 Å will be used for this purpose. Using a dc arc at 10 A and an exposure time of 30 sec with 50 ppm germanium on the electrode, the data in Table 8-5 were obtained. These data were used to construct the emulsion calibration curve shown in Figure 8-10.

To obtain the working curve, the following conditions and procedures were used. Standard solutions containing 2, 4, 8, 10, and 20 ppm germanium were prepared, all containing chromium at the same concentration, 50 ppm. The chromium serves as the internal standard.

High-purity, 3/16-in. diameter graphite electrodes were drilled to a depth of 1/4 in. to serve as anodes. Counter electrodes were blunt-nosed

QUANTITATIVE SPECTROCHEMICAL ANALYSIS

TABLE 8-5
Emulsion Calibration Data Using a Seven-Step Filter and the Germanium Emission Line at 3039 Å

Step	Relative intensity	%T
1	1.5	90.0
2	2.25	70.2
3	3.38	48.0
4	5.06	28.3
5	7.59	13.0
6	11.4	5.4
7	17.1	2.1

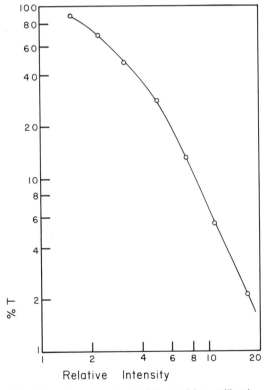

FIGURE 8-10. A photographic emulsion calibration curve using a seven-step neutral filter and the germanium spectral emission line at 3039 Å.

graphite electrodes 3/16 in. in diameter. The cupped electrodes were treated with a saturated solution of carnauba wax in carbon tetrachloride and dried at 100°C. A buffer, made of a mixture of nine parts powdered graphite and one part lithium carbonate, was prepared, and each electrode was filled with the mixture. The mixture was tamped slightly and additional buffer mixture added until the cup was filled. One drop of ethanol was added to the filled electrode, followed by 25 µl of standard. The alcohol aids in dispersing the aqueous standard solution through the buffer mixture. Electrodes were then dried at 110°C.

A dc arc, operating at 10 A for 25 sec, was used for excitation. Electrode spacing was 2 mm, entrance slit width was 10 µm and slit height was 2 mm. The germanium line at 3039 Å was used for the analysis and the chromium line at 3053 Å served as internal standard. Spectral plates were Eastman Kodak SA-1 and development was in D-19 developed for 4 min at 68°F. The unknown germanium solutions were prepared exactly as the standards and excitation and development conditions were the same.

After excitation and plate development the data in columns 2 and 3 of Table 8-6 were obtained from densitometer readings. The percentage transmittance readings of Table 8-6 must be converted to relative intensity readings by reference to the plate calibration curve previously prepared and

TABLE 8-6
Spectral Plate Data to Prepare a Working Curve for the Determination of Germanium

Concn. Ge, ppm	% T		Relative intensity		I_{Ge}/I_{Cr}
	Ge	Cr	Ge	Cr	
2	90.0	20.0	1.5	6.2	0.24
	95.3	21.5	1.3	6.0	0.22
4	54.3	19.5	3.0	6.3	0.48
	50.0	21.5	3.2	6.0	0.53
8	28.0	22.3	5.1	5.9	0.87
	25.5	21.3	5.4	6.0	0.90
10	21.3	23.0	6.0	5.9	1.02
	20.7	23.3	6.1	5.9	1.04
20	3.0	21.3	14.9	6.0	2.48
	3.7	18.5	14.1	6.4	2.20
Unknown	22.3	23.3	6.0	5.7	1.05
	25.7	25.0	5.4	5.8	0.93
	26.7	24.7	5.3	5.6	0.95

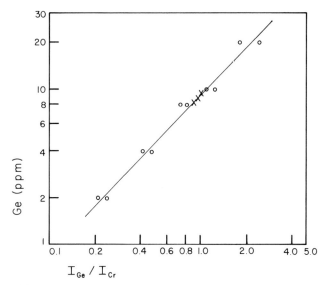

FIGURE 8-11. Analytical working curve using an internal standard and plotting I_{Ge}/I_{Cr} versus concentration of germanium.

shown in Figure 8-10. These data are given in columns 4 and 5 of Table 8-6. The last column of Table 8-6 gives the ratio I_{Ge}/I_{Cr} obtained by dividing the relative intensity data of columns 4 and 5.

The analytical working curve, shown in Figure 8-11, plots the concentrations of germanium versus the intensity ratios of I_{Ge}/I_{Cr} to provide the final working curve.

The solution of unknown germanium concentration to which had been added the internal standard was prepared in triplicate and the analytical data are shown at the bottom of Table 8-6. Reference to Figure 8-11 will show the analytical results. The circles represent the data used to construct the working curve and the crosses are the data for the three determinations of the unknown. The three determinations give, for the concentration of germanium in the unknown, 8.5, 8.9, and 9.6 ppm, respectively, for an average of 9.0 ppm.

Inspection of the working curve calibration data (the circles of Figure 8-11) gives some indication of the reproducibility of duplicate runs on the standards. Except for the data on the 8 ppm standard, the working curve lies between the two replications. The working curve is plotted using log–log paper and is linear over the concentration range studied. The slope is close to the 45° that is predicted theoretically, and the replication of results is close to 5%.

10. THE CALCULATING BOARD

If a large number of analytical determinations is to be made, a calculating board provides a method to obtain analytical results rapidly and conveniently. Calculating boards usually are arranged so a number of different elements can be determined rapidly. A typical calculating board is shown in Figure 8-12.

Included as parts of the calculating board are a movable vertical scale with a percentage T scale presented logarithmically. A Seidel function scale

FIGURE 8-12. Calculating board for processing spectrographic data from photographic recording of spectra. [Courtesy Jarrell-Ash Division, Fisher Scientific Co.]

QUANTITATIVE SPECTROCHEMICAL ANALYSIS

appears on the right side of the vertical scale. Below the plotting surface is a horizontal slide rule with five scales. Below the slide rule is a working strip board containing a number of colored plastic strips.

The movable vertical scale and the horizontal slide rule scales are convenient to plot emulsion calibration curves and working curves. The working strip board permits the preparation of intensity ratio scales for a series of different elements. Any calculating board uses the principles of emulsion calibration and working curve preparation that were presented in earlier sections of this chapter. The board makes the plotting of the data more rapid and convenient and is especially useful for routine repetitive analysis and multielement analysis, especially when one emulsion calibration curve is sufficient.

11. BACKGROUND CORRECTION

If a background of continuous radiation occurs at the wavelength of the analytical line, correction for its presence usually is necessary. If the background is small, for example, if the background reading is 95% as compared to a clear emulsion reading of 100%, correction is not necessary unless the spectral line is of very low intensity. Background intensity can be reduced by using a narrow slit width and by carefully adjusting exposure time. These techniques, however, may not always reduce the background to as low a level as desired. In fact, a light background may be useful to ensure that the exposure has been optimized.

Several methods of background correction can be used. One method is to adjust the read-out circuit to read 100% transmission through the background adjacent to the analytical line. This technique basically is a correction based on percentage transmission differences of the clear emulsion and the background. It is useful at low background intensity levels but mathematically is not correct.

A common method for correcting for background is based on subtraction of the background density from the density of the line plus the background. If the read-out system is adjusted to 100% T on the clear emulsion, then

$$D_{\text{background}} = \log \frac{100}{\%T_{\text{background}}} \tag{8-3}$$

and

$$D_{\text{line + background}} = \log \frac{100}{\%T_{\text{line + background}}} \tag{8-4}$$

The density of the spectral line is calculated by subtracting equation (8-3) from (8-4) to obtain

$$D_{line} = \log \frac{100}{\%T_{line + background}} - \log \frac{100}{\%T_{background}}$$

or

$$D_{line} = \log \frac{\%T_{background}}{\%T_{line + background}} \tag{8-5}$$

Equation (8-5) is almost equivalent to adjusting the galvanometer deflection to read $100\%T$ through the background.

Since common practice is to use an internal standard to prepare working curves, another type of approximate correction can be used. This involves a measurement of the percentage transmittances of the clear emulsion, of the background, of the spectral line plus background, and of the internal standard. These data are processed as shown in the following equation:

$$\log \frac{\%T_{line + background}}{\%T_{internal\,standard}} - \log \%T_{background} = \log \frac{\%T_{line}}{\%T_{internal\,standard}} \tag{8-6}$$

Equation (8-6) has been used successfully, although it is apparent from the equation that the internal standard line must be free of background. The equation also is not mathematically rigorous since two radiations are not always additive.

The most precise method to correct for background is to convert all $\%T$ values, or densities, to relative intensities based on the emulsion calibration curve. Background relative intensity can then be subtracted from the line plus background intensity to produce a relative intensity for the spectral line. The same calculation also is required on the internal standard line if background is present adjacent to the internal standard line.

12. MULTIELEMENT ANALYSIS WITH DIRECT READ-OUT

For routine analytical control the direct reading spectrometers described in Chapter 4 are ideal. These are special-purpose instruments that can be adjusted for specific purposes. Once adjusted they can provide a rapid, precise method for the analysis of the chosen elements in a single matrix. The instruments use photomultiplier tubes, one for each element to be determined, mounted along the focal plane of the spectrometer, immediately back of a carefully adjusted exit slit. The number of elements that can be determined simultaneously can be as many as 30–35. One channel usually is used for an internal standard and another may be used for a background measurement. It must be remembered, however, that the background measurement is strictly valid only at the wavelength at which it was determined. Most direct reading spectrometers integrate the energy of a spectral line over a specified exposure time and store it as electrical energy in a capacitor. After the exposure is completed the electronic system measures the stored energy of each capacitor and relates it to the element concentrations.

Direct reading spectrometers can provide analytical data on routine samples in 2–3 min with an average error of 1–2%. In the metallurgical industries the direct reader easily provides valuable data while a metal is in the molten state so composition can be adjusted as desired.

Direct reading has been applied successfully to carbon, sulfur, and phosphorus analyses in the iron and steel industries since concentrations of these elements are critical in the final product. Since the most sensitive lines of these elements lie in the vacuum ultraviolet, a vacuum spectrometer is required. These usually are separate units but can be constructed in combination with an air-path unit so a single sample can provide data on these three elements as well as those usually determined in the air-path spectrometer.

Since direct readers are not versatile or easily readjusted for different samples, most laboratories using them also have available another instrument using photographic recording. The photographic units usually are used for qualitative scanning, nonroutine quantitative analysis, and method development.

13. TYPES OF SAMPLES

One of the most important problems facing the analytical spectroscopist is suitable preparation of the sample for excitation. Although it is impossible to describe every situation that may arise, there are some general approaches to the problem that should be familiar to anyone using emission spectroscopy for analytical purposes.

13.1. Liquid Samples

One advantage of liquid samples is that they are homogeneous. They also can be excited by any one of several different methods to obtain their spectra. Standards can be easily synthesized and internal standards and buffers can be added conveniently.

Several different methods can be used to bring the solution into the analytical gap. The rotating electrode is a mechanically driven carbon disk with the lower edge of the electrode rotating through the sample. A conventional counter electrode is mounted above the rotating disk and a spark or interrupted arc is struck between the upper edge of the disk and the counter electrode. The sample continually enters the discharge as the disk rotates. The rotating disk electrode has been applied to petroleum-based samples as well as to aqueous solutions. To prevent possible ignition of inflammable samples, the sample can be bathed in an inert atmosphere.

The porous cup electrode is another device that can be used to bring a liquid into the analytical gap. The electrode is drilled axially to form a cylindrical hole with a thin base. Liquid sample is placed in the hole and the electrode is mounted in the upper jaws of the electrode holder. The counter electrode is placed in the lower electrode holder with an electrode gap of 3–4 mm and spark excitation is used. The liquid sample goes through the porous base of the electrode to enter the analytical gap.

Many different spray techniques have been used to introduce liquid samples into an analytical gap. One of the early successful uses of this technique was by Lundegårdh in 1928 with agriculturally important samples. Atomizers have been used to spray the sample into the analytical gap as a liquid aerosol and excellent analytical results can be obtained if care is exercised to keep operating parameters constant.

Another common method of utilizing liquid samples is to dry a known amount of liquid in or on an electrode. In all cases, when drying a liquid sample on a carbon or graphite electrode, the electrode should be made impervious to the sample solution. Electrodes can be coated with a resin, paraffin, or some other wax. Carnauba wax dissolved in carbon tetrachloride is convenient for this purpose. If ac arc excitation is used, it is possible to dry the liquid sample on flat electrodes utilizing both the upper and lower electrodes as sample carriers.

Another technique for using liquid samples is to drill an electrode cup, coat it with wax, fill the cup with powdered graphite or carbon, and allow the liquid sample to be dispersed through the powdered graphite or carbon. This technique was employed in the quantitative analysis example given in an earlier section of this chapter.

One result of utilizing the rotating disk, porous cup, or spray technique for introducing the liquid sample into the analytical gap is that fractional distillation of the sample is not a problem. If the sample is dried on an electrode, fractional distillation of the sample into the arc or spark may occur as with any solid sample. This effect is most evident with dc arc excitations.

13.2. Metallic Samples

Metal samples frequently can be shaped to fit electrode holders and serve as "self-electrodes." Spark excitation usually is used for metals and metal alloys since it is sufficiently sensitive and fractional distillation effects are very small. If additional sensitivity is required, an interrupted arc can be employed or an overdamped capacitor discharge can be used.

Metal samples also may be prepared as flat disks and used in the Petry stand designed for this purpose. The counter electrode usually is a pointed carbon or graphite electrode. Excitation conditions are similar to those used with the "self-electrodes."

Since some metals do not solidify into a homogeneous solid, it may be necessary to use the metal in the form of drillings, filings, or small chips. Such metals may be placed in solution and solution techniques utilized. Another method is to obtain as homogeneous a sample as possible from the metal particles and compress them into a solid pellet. The pellet may be placed in a graphite or carbon electrode crater for excitation.

13.3. Powder Samples

One of the most important problems associated with powder samples is that of obtaining representative samples. Powder samples should preferably pass a 300 mesh screen and they need to be carefully mixed and at the same time protected from contamination. Weights need to be obtained before and after heating (at about $110°C$) to determine if the sample has any appreciable moisture content.

Powder samples can be weighed and loaded directly into electrodes for excitation. More often, the powder sample is diluted with buffer and also frequently is mixed with graphite. Addition of graphite adds a conducting material to the sample and provides smoother burning with arc excitation. Sufficient graphite should be added to prevent bead formation of the sample during the arcing process. Powders subjected to dc arc excitation will undergo fractional distillation; thus it is important to consume totally the sample under these excitation conditions.

Powders can be made conducting by a pelleting process utilizing graphite with the powder to make the mixture electrically conducting. The pellet can then be analyzed utilizing spark excitation; thus the advantages of spark excitation can be obtained for use with a nonconducting powder sample.

13.4. Organic Samples

Preparation of all types of organic samples usually requires a preliminary ashing step. This includes a variety of substances, such as plant or animal tissue and fluids, soils, foods, grain, and organic compounds synthesized for some specific purpose. Spectroscopic analysis is applied to such samples to determine their inorganic constituents.

Usually such samples are ashed, either by wet ashing or dry ashing techniques, and the residue is handled as a dry powder. Frequently the ash is placed in solution, after which it can be handled by solution procedures. When wet ashing is used it is important to use high-purity chemicals to prevent contamination of the sample. If dry ashing is used the temperature of ashing should be low to prevent volatilization of metal elements in the sample. Care also must be exercised in the dry ashing technique to prevent the possible alloying of metal constituents of the sample into the crucible. This is particularly true when using platinum crucibles. Vycor crucibles are very useful for dry ashing of biological materials.

13.5. Special Samples

No single method can be recommended for handling special samples. These might include special research preparations, archeological samples, forensic materials, and other nonroutine substances. The size of the sample available, the desired data, and the value of the sample all affect the selection of analytical procedures. A qualitative analysis, obtained by totally vaporizing the sample, can often provide much useful information concerning major and minor constituents of the sample and the photographic plate provides a permanent record of the sample composition.

14. SOME SPECIAL TECHNIQUES

This section will deal with some special techniques that have been used to help solve some analytical spectroscopic problems. Some of the techniques described are used frequently and others occasionally. The section does not include every procedure that has been used but emphasizes some of the more important ones.

14.1. Fractional Distillation

Analytically, fractional distillation can be used to advantage in some cases. In a complex mixture, the lower boiling point elements enter the analytical gap before refractory elements. Therefore it is possible to excite the sample for a specified length of time and obtain complete vaporization and excitation of lower boiling elements while retaining higher boiling elements in the electrode cup. The spectra produced under these conditions will include only lines of the lower boiling elements. Spectra therefore will contain fewer lines than if the sample was totally consumed during excitation. Alkali metals vaporize into the arc early, as do low-melting metals, such as mercury, cadmium, and lead. The high-melting elements, such as rhenium, tungsten, and uranium, remain as residue in the electrode cup.

QUANTITATIVE SPECTROCHEMICAL ANALYSIS

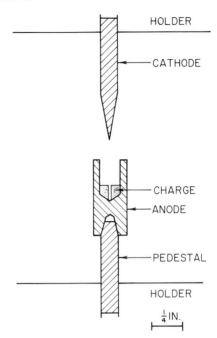

FIGURE 8-13. Electrode assembly for the carrier-distillation technique. [From M. Fred and B. F. Scribner, in *Analytical Chemistry of the Manhattan Project*, Editor-in-Chief, C. J. Rodden, McGraw-Hill, New York (1950), Chap. 26. Used by permission of the McGraw-Hill Book Co.]

For those metals that enter the arc gap later it is possible to use a moving plate technique. If the plate is moved rapidly at the proper time during excitation, some elements will appear in the first exposure, some will occur in both exposures, and some will appear only during the second exposure. The technique can be applied to the total consumption of the sample but the spectra obtained will be simpler and less complicated.

14.2. Carrier Distillation

Fred and Scribner[7] describe a variation of the fractional distillation technique. The analysis of uranium oxides or uranium ores is complicated by the complex spectrum of uranium. To obtain good temperature control of the sample during arcing, an electrode assembly as shown in Figure 8-13 was devised and constructed of graphite. An anode cup was prepared with fairly thick walls and mounted on a tapered pedestal. The sample was placed in the lower one-fourth of the electrode cavity. The arrangement of Figure

---

[7] M. FRED and B. F. SCRIBNER, in *Analytical Chemistry of the Manhattan Project*, Editor-in-Chief, C. J. Rodden, McGraw-Hill, New York (1950), Chapter 26.

8-13 provides quite uniform heating of the sample and efficiently separates the more volatile elements from those that are less volatile. To aid in the selective volatilization of the sample, a "carrier" substance is added. The carrier used with these uranium samples was gallium oxide, which was chosen since it was of moderate volatility and had an ionization potential sufficiently high so it would not suppress the excitation of impurities such as alkali metals. The carrier also should have a simple spectrum of its own and not have impurities present that would interfere with the desired analysis. By this technique the impurity elements were carried into the arc discharge before uranium entered the arc, making possible an analysis that could not have been carried out if uranium had been excited at the same time.

14.3. Transfer Methods

When sample excitation is accomplished using a dc arc, some of the volatilized sample condenses on the counter electrode. These deposits, representing the original material, can be analyzed in the dc arc if proper standardization occurs. Use has been made of this method to obtain metal samples that cannot be brought to the laboratory. A small arc welding unit can be used in the field to obtain such samples.

14.4. Laser Methods

A laser beam can be used to volatilize metal samples where the metal vapor thus produced can then be excited by a spark source. Laser vaporization can be applied to a very small area. A burst of laser energy can vaporize about 0.1 μg of metal. The process can be repeated to increase the intensity of the spectra produced. A diagram of a laser microprobe is shown in Figure 8-14. A microscope is used to focus on the area of interest; then the laser beam vaporizes the sample and spectral excitation occurs by use of the electrodes and power source.

Laser methods have been used to study the composition of occlusions in metals, variations in surface concentration of metals, and small areas of biological samples.

14.5. Controlled Atmospheres

The first use of a controlled atmosphere in emission spectroscopy was to prevent the formation of cyanogen bands. Removal of nitrogen from the atmosphere surrounding the dc arc effectively prevents cyanogen band formation. The Stallwood jet was designed to permit sample excitation in a controlled environment. A number of different gases or mixtures of gases

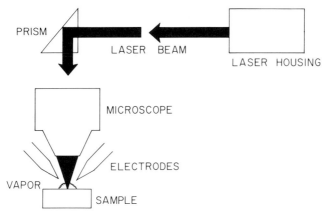

FIGURE 8-14. Schematic drawing of the laser microprobe. [Courtesy Jarrell-Ash Division, Fisher Scientific Co.]

have been used. These include carbon dioxide, argon, helium, and mixtures of argon or helium with oxygen.

The gases surrounding the excitation unit affect detection limits of certain elements in the analytical gap. One reason for improvement in detection limits is the absence of the cyanogen bands. A frequently used atmosphere of about 30% oxygen and 70% argon produces a smoothly burning arc and provides oxidizing conditions in the arc. Among elements that are much more easily detected in the argon–oxygen atmosphere are potassium, rubidium, calcium, strontium, chromium, and manganese. Wang and Cave[8] have demonstrated intensity gains ranging from 1.3 to 4.8 for a series of 18 elements when arcing occurs in an argon–oxygen atmosphere.

14.6. Cathode Layer Excitation

The usual procedure in dc arc excitation is to make the lower electrode, containing the sample, the anode and the counter (upper) electrode the cathode. This method facilitates the volatilization of the sample into the arc gap.

If, however, the lower sample holding electrode is made the cathode, metal vapors emerging from the sample are retained by the electrical field in a region just above the cathode for a longer period of time. The result is a region of high metal vapor concentration just above the cathode known as the "cathode layer." Spectroscopically, the result of the formation of the cathode layer is enhanced spectral line intensity in the cathode layer region.

[8] M. S. WANG, and W. T. CAVE, *Appl. Spectrosc.*, **18**, 189 (1964).

The cathode layer technique is especially useful for small samples since the electrical forces retaining the metal vapor in this region can have their maximum effect. Some increase in background also is noticed in the cathode layer so line-to-background measurements may not be noticeably improved when using this technique.

14.7. Gases

Emission spectroscopy has not been used extensively for analysis of gases. Problems of sampling and excitation are difficult to solve. Some gas analyses have been made by placing the gas in an enclosed glass or quartz tube and exciting the gas with radiofrequency energy. Tesla coil discharge also has been used to produce excitation of gases.

Gases have ionization potentials that are very high compared with those of metals; the values for many gases fall in the 10–15 eV range, while those for most metals are below 7 eV. When a mixture of metal atoms and nonmetal atoms is excited the metal atoms consume most of the excitation energy at the expense of the nonmetals.

Another important factor involves the wavelength locations of the sensitive spectral lines of gases. The most sensitive lines of most gases lie in the vacuum-ultraviolet region, below 2000 Å, and therefore require special vacuum spectrometers to provide suitable spectral line intensities.

In spite of the difficulties, some application of emission spectroscopy to gas analysis has been made. Halogens have been analyzed by use of an rf energy source and also by using a high intensity ac spark. Fluorine has been determined by the formation of calcium fluoride, utilizing the band spectrum of calcium fluoride.

14.8. Radioactive Samples

Radioactive samples can be analyzed by emission spectroscopy, although all the precautions usually associated with radioactive samples must be followed. Fortunately, sample sizes for spectrographic analysis are small, so total radioactivity in the laboratory should be small.

Of particular concern is the safe removal of the vaporized products of excitation. An enclosed arc–spark stand is mandatory, as is an efficient exhaust system. The exhaust system should be constructed so air can be removed but the particulate matter retained in a filter or some other collection device.

Sample preparation, prior to spectrographic analysis, should be done under approved safety conditions in a room separate from the spectrographic laboratory. Laboratories and personnel should be monitored on a regular basis to ensure that radioactivity remains below a safe level at all times. With

QUANTITATIVE SPECTROCHEMICAL ANALYSIS

suitable precautions, spectrographic analysis of radioactive samples can be performed safely, with the spectrographic conditions of sample size, preparation, buffering, and other procedures being the same as with nonradioactive samples. Grove et al.[9] quantitatively monitored gaseous impurities in inert gas shields surrounding welding operations. They used a 20-A dc arc for excitation.

One of the more interesting applications of gas analysis was that developed by Evens and Fassel[10] to determine oxygen in niobium. An argon atmosphere is used with dc arc excitation and concentrations of oxygen in the range of 0.004–0.60 wt % can be determined by their method.

Seeley and Skogerboe[11] have described a combined sampling–analysis technique for the determination of trace elements in particulate matter in the atmosphere. Porous cup graphite spectroscopic electrodes are used as filters to collect the particulates and then form the sample electrode for emission spectroscopic determination of element concentrations.

A 2.5-mm hole is drilled through the base of the electrode after sample collection and a technique utilizing argon saturated with HCl gas vaporizes the sample and carries it into a dc arc operating at 28 A. Indium is used as an internal standard. Fourteen elements are determined quantitatively in the particulate matter, including Al, Be, Cr, Co, Hg, Mg, Mn, Mo, Ni, Pb, Ti, V, W, and Zn.

15. TIME-RESOLVED SPECTROSCOPY

The technique of representing the intensities of spectral lines as a function of time is referred to as time-resolved spectroscopy. Time resolution of spectroscopic information has been applied to many problems, such as the kinetics of fast decay phosphorus, radiation from fast photolysis sources, and exploding wire phenomena. Of most importance to analytical spectroscopy is the use of time-resolved spectroscopy to study the characteristics of ac spark and ac arc discharges of the type normally used for analytical emission spectral analysis, since such information may be useful in optimizing operating conditions.

15.1. Time-Resolving Components

Time-resolved spectra may be produced with rotating mirrors, rotating disks, moving films, gated photocells, cathode ray tubes, and image ion

---

[9] E. L. Grove, V. Raziunas, W. A. Loseke, and B. K. Davis, *Welding J.*, **29**, 282 (1964).
[10] F. M. Evens, and V. A. Fassel, *Anal. Chem.*, **33**, 1056 (1961).
[11] J. L. Seeley, and R. K. Skogerboe, *Anal. Chem.*, **46**, 416 (1974).

vectors. Bardocz[12] has reviewed apparatus and techniques of time-resolved spectroscopy and suggests that the following criteria govern the selection of the time resolving component; (1) the periodicity of the discharge, (2) the duration of the discharge, and (3) the length of the spectrometer slit.

A rotating mirror is capable of very high time resolution (approximately 1–20 μsec) and consequently is very useful to resolve excitation from a high-voltage, high-frequency spark source. For lower voltage sources with considerable capacitance and inductance, a rotating disk can be shaped to resolve the excitation and is commonly used for discharges of the order of 2500–10,000 μsec.

Time resolution of spectroscopic sources also requires synchronization of the time resolving element with the event to be studied. Synchronization usually is accomplished either magnetically or with a photocell. A photocell converts a light impulse into an electrical impulse, which, when amplified, controls the spark discharge. Magnetic synchronization utilizes a soft iron armature fixed to the shaft of a synchronous motor driving the rotating mirror. When the armature passes near a coil mounted in a fixed position, an electrical current is generated in the coil which triggers the spark source.

Walters[13] has designed a complete system for time-resolved spectroscopy that is useful for transients from 0.2 to 90 μsec over a wavelength range of 1000 Å per exposure, using either photographic or photoelectric detection. The system utilizes a rotating mirror with photocell synchronization.

15.2. Some Characteristics of Time-Resolved Spectra

Some general observations made possible by time-resolved spectroscopy include the following.

For a high-voltage spark:

1. A continuum is produced at the beginning of the discharge. Its duration is directly proportional to the excitation energy.
2. The periodicity of the spectral intensity of spark lines of high excitation levels varies with the current until they cease to radiate.
3. Arc lines of lower excitation levels do not vary in intensity periodically. They may radiate after the spark discharge ceases.
4. An intense periodicity occurs in the vicinity of the cathode, probably due to ejection of electrode sample into the gap.
5. At the beginning of discharge, some spectral lines broaden and undergo a red shift caused by the high initial pressure and return to normal after 2–3 μsec.

---

[12] A. BARDOCZ, *Appl. Spectrosc.*, **21**, 100 (1967).
[13] J. P. WALTERS, *Anal. Chem.*, **39**, 770 (1967).

For a low-voltage spark the results are similar to those for the high voltage spark except for the following:

1. Emission is delayed after the ignition of the spark due to the time required to heat the plasma.
2. The radiation at the surface of the electrodes and in the middle of the gap is less than for the high-voltage spark.
3. The background is more intense.

Alternating current arc spectra exhibit the following differences when compared with spark sources.

1. Spectral line broadening is negligible.
2. The ignition spark spectra and the arc lines are easily distinguished.

15.3. Analytical Applications

In addition to providing fundamental information concerning the nature of spark excitation, several analytical applications of time-resolved emission spectroscopy are apparent. Proper timing of the observation of spectra can markedly reduce background radiation, and matrix effects also may be reduced. Analytical sensitivity also may be increased. Goto et al.,[14] utilizing time–intensity curves, were able to eliminate the interference of the chromium ion line at 2881.93 Å with the silicon atom line at 2881.58 Å in steel analysis. In another case, the calcium ion line at 3933.67 Å and the iron atom line at 3933.61 Å exhibited similar time–intensity curves in the high-voltage spark, but similar time–intensity curves with a low-voltage spark showed that the calcium line disappeared in 20 μsec. Goto et al. used a gated phototube as the time resolving element.

Walters[15] has reported on the formation and growth of the spark discharge as analyzed through time-resolved spectroscopy. Highly reproducible spectra were obtained and analyses of metal samples may be conducted in much the same manner as is now done with liquids using flame excitation.

16. CHEMICAL PREPARATION OF SAMPLES

Often it is necessary to use some chemical procedures to prepare a sample for spectroscopic analysis. This section will not detail such methods, since they are available in standard texts and references. Some of the more common techniques will be given to indicate some of the possibilities.

[14] H. Goto et al., Anal. Chem., **220**, 95 (1966).
[15] J. P. Walters, Appl. Spectrosc., **26**, 323 (1972).

A most common procedure with plant and animal tissue is to destroy and remove organic matter by ashing. Either a wet or dry ashing technique can be used. In addition to removing organic matter, this process also concentrates the metal ions and atoms into a more usable sample. The ashing procedure can be modified to produce either a solid or liquid sample.

Frequently a concentration step is desirable in the pretreatment of a sample. If the sample can be placed in solution, several concentration techniques are available, including precipitation, ion exchange, solvent extraction, and electrolysis. Concentration steps are more difficult if the sample must remain a solid; some possibilities for solid powder samples include high-temperature distillation, flotation, and magnetic separation.

If it is desired to remove considerable matrix material, it may be that quantitative removal is unnecessary. For example, if the matrix substances make up 95% of the sample and 90% of the matrix can be removed, a tenfold concentration of the desired constituents results.

Each sample concentration problem should be approached on the basis of type of sample and of the analytical data desired. A suitable separation method can then usually be devised.

17. APPLICATIONS OF SPECTROCHEMICAL ANALYSIS

Analytical emission spectroscopy has been used in many fields to provide useful information and a complete list of uses would be too lengthy to include in this book. Some of the more important fields of application are listed in this section.

17.1. Metals and Alloys

Metals and alloys as well as the raw materials from which they are made are analyzed primarily by spectroscopic methods. The iron and steel as well as the aluminum industries rely on spectroscopic analysis in all steps of their processes. Many of the analyses are needed in a very short time; thus these industries make wide use of multielement direct reading spectrometers to provide needed analytical data. It is important to control the composition of the molten metals before further processing and the direct reading spectrometers can supply routine analytical information in less than 2 min, something impossible with photographic recording or with chemical "wet" methods. Some control laboratories have direct readers capable of simultaneously determining 30 elements.

More recent analytical spectroscopic developments in the metals industries have been the successful determination of carbon, sulfur, and

QUANTITATIVE SPECTROCHEMICAL ANALYSIS

phosphorus. These elements are now determined routinely utilizing direct reading vacuum spectrometers with controlled-atmosphere excitation.

17.2. Geology

Geological samples frequently are analyzed by using emission spectroscopy and methods have been developed for determination of a large number of elements. Ahrens and Taylor (see footnote 5) give information to aid in the determination of some 48 elements in geological samples.

Much useful information on the composition of geological samples can be obtained by use of semiquantitative methods, such as those described in Chapter 7. Screening of a large number of elements in many samples can be accomplished in this manner. If concentrations of certain elements in such samples is sufficiently high, then a more precise quantitative technique may be used. The U. S. Geological Survey makes use of mobile spectroscopic laboratories in its study of geological materials and they have found this method very useful.

Spectroscopic methods are most useful to determine concentrations of minor elements in minerals; however, major element composition is also frequently desirable. Spectroscopy is not as well suited to determine the concentration of major elements as for minor constituents, yet it can yield valuable information and can serve especially to select samples that may require further study.

17.3. Oils and Water

Crude oil and products of the petroleum industry frequently are subjected to spectroscopic analysis to determine trace metal concentrations. It is important to the industry to know trace element concentrations in crude oil since some trace elements can poison the catalysts used in the cracking process. Some of the particularly critical trace elements are vanadium, copper, nickel, and iron.

Lubricating oils are routinely analyzed during use to determine the degree of wear of metal parts in motors and engines. If the metal element content of the lubricating oil increases with oil use, it may indicate excessive wear and the need for engine overhaul. The method is used routinely by railroads and airlines to aid their maintenance programs, and by large trucking firms.

Oil analysis may be performed by first ashing the sample, but a simpler technique is to use the rotating electrode. Sample preparation is a minimum and excellent precision is possible. Comparison standards are commercially available for routine studies.

Water analyses are performed to determine trace element concentrations and aid in judging water quality. Underground water samples can be analyzed to aid in mapping the flow of underground water. The petroleum industry frequently uses such methods as an aid to geological exploration.

17.4. Plants and Soils

It is generally agreed that in addition to major elements necessary to support plant growth, the elements boron, manganese, copper, zinc, molybdenum, and iron are required in trace quantities. In addition to these essential elements, some 40 other elements have been detected in plant tissue. Spectroscopic methods have been used to detect most of the above list of elements and quantitative data have been obtained in most cases.

If plants do not obtain sufficient quantities of the essential trace elements, growth is inhibited and crop yields are decreased. Spectroscopic methods are used to study trace element composition of plants and to diagnose deficiency problems in them. Recommendations can then be proposed for remedial treatment by application of deficient elements through soil applications or through sprays applied to the leaves. There are many examples of deficiencies that have been corrected through application of specific trace elements. Molybdenum frequently is added to fertilizers, as are boron and copper. In some areas iron and zinc are deficient and are added to fertilizers.

Some elements apparently are not essential to plant growth but are essential for the animals that feed on the plants. Cobalt is an example of such an element. Since cobalt is not essential to the plant, cobalt deficiency in animals can be corrected by adding it to the feed. Some elements found in plants can be toxic to animals if the concentration is too high. Selenium is found in certain plants in relatively high concentrations in a belt extending from western Texas to western North Dakota. The common name "locoweed" is attached to these plants; the botanical species involved are *Astragalis mollissimus* and *Oxygropis lamberts*. The importance of trace elements in agriculturally important crops has led many state agricultural experiment stations to establish laboratories to service the agricultural interests of their states.

Soil analyses frequently will reveal the lack of the abundance of trace elements in the soil. The important question is not the total amount of the element in the soil, but the amount present that is available to the plant. For this reason soil extracts are a better index of trace element availability than is a total trace element determination. Extraction solutions of various compositions have been proposed for this purpose. If research of this type is planned, reference should be made to the AOAC methods book.[16]

---

[16]WILLIAM HOROWITZ, Editor, *Official Methods of Analysis of the Association of Official Analytical Chemists*, 11th ed. (1970).

QUANTITATIVE SPECTROCHEMICAL ANALYSIS

17.5. Men and Animals

Many metal elements are found in animal tissue; some are essential, some are toxic, and some seem to have no noticeable effect. Considerable research has been done using emission spectroscopy, especially where information on a large number of metal elements is desired. Metal element concentrations in different body organs have been studied in an attempt to better understand body functions. Changes in trace metal concentrations also have been studied as related to the aging process.

Metal elements that are toxic also have received attention. Some elements, such as copper, are essential in small amounts and toxic at higher concentrations. Elements such as arsenic, bismuth, cadmium, mercury, lead, antimony, selenium, and thallium all are toxic and spectrographic methods have been used in their study.

Hospitals routinely determine blood calcium, potassium, and sodium to determine any changes from normal. These routine determinations are usually performed using flame emission or atomic absorption methods.

17.6. Environmental Studies

Spectrographic methods now are being used to provide data on environmental problems. These problems are varied in nature and include such items as the trace element composition of coal ash, the mercury content of water, the trace element content of effluent water from industrial process, atmospheric contamination from smelting operations, food contamination during processing and storage, exhaust emissions from automobiles, trace element composition of feedlot run-off, and atmospheric quality control in mines and other confined areas. Many other examples could be included; however, it should be clear that spectroscopic methods serve a vital role in providing useful information in attempting to solve environmental problems.

17.7. Some Other Applications

Miscellaneous other uses of emission spectroscopy should be mentioned. The cement and glass industries use spectroscopic methods for quality control. The food and beverage industries monitor trace element concentrations during processing. Spectroscopy is used for forensic purposes, usually to help identify samples as to source or origin. Meteorite composition also has been studied by spectroscopic methods, as have lunar samples returned to earth by astronauts. Emission spectroscopy also has served as a research tool in chemistry and physics by providing composition information on research samples.

The role of emission spectroscopy covers a broad spectrum of science and engineering and its uses are growing. New applications of emission spectroscopy will occur as new problems present themselves to be solved.

SELECTED READING

Ahrens, L. H., and Taylor, S. R., *Spectrochemical Analysis*, 2nd ed., Addison-Wesley, Reading, Massachusetts (1961).
Brode, Wallace, R., *Chemical Spectroscopy*, 2nd ed., Wiley, New York (1943).
Clark, George L., Editor, *The Encyclopedia of Spectroscopy*, Reinhold Publishing Corp. (1960).
Grove, E. L., Editor, *Analytical Spectroscopy Series*, Vol. 1, Parts 1 and 2, Marcel Dekker, New York (1971, 1972).
Harrison, George R., Lord, Richard C., and Loofbourow, John R., *Practical Spectroscopy*, Prentice-Hall (1948).
Nachtrieb, Norman H., *Principles and Practice of Spectrochemical Analysis*, McGraw-Hill, New York (1950).
Sawyer, Ralph A., *Experimental Spectroscopy*, Prentice-Hall (1944).
Slavin, Morris, *Emission Spectrochemical Analysis*, Wiley–Interscience (1971).

Chapter 9
Flame Emission Spectroscopy

Flame emission spectroscopy is so named because of the use of a flame to provide the energy of excitation to atoms introduced into the flame. Reference to Chapter 1 will provide some information concerning the historical development of this method of spectral excitation and its early use to detect the presence of metal elements in samples aspirated into a flame. Modern analytical flame emission spectroscopy can be considered to date from the work of Lundegårdh, reported in 1934, when he demonstrated its use for the determination of a variety of metal elements in samples of biological origin. Lundegårdh sprayed a solution of the sample material into a condensing chamber and then into an air–acetylene flame, where excitation occurred. He used a prism spectrograph to disperse the spectra excited by the flame with photographic recording of the spectra and a densitometer to determine spectral line intensities. His instrumentation requirements therefore were the same as for conventional arc–spark emission spectroscopy as described in Chapter 8, except for the replacement of the arc–spark excitation source with the flame.

The high stability of the flame source, when compared to arc or spark excitation, was recognized as the key to the construction of simple instruments for the determination of easily excited elements, such as the alkali metals. Thus the first "flame photometer" produced in the U.S. in 1945 by Barnes used filters rather than a prism or grating, and used a modified Meeker burner as the flame excitation source. The instrument was especially useful for sodium and potassium determinations and was also soon utilized for calcium and magnesium analyses despite the handicap of poor detection limits.

After 1945 the development of flame photometers or flame attachments for existing equipment was rapid. The Beckman Company introduced a flame attachment for their DU spectrophotometer in 1948 and subsequently introduced their very popular total-consumption burner. The signal was detected by either a phototube or photomultiplier with a dc amplifier coupled with a potentiometer circuit. Chart recording soon became available and other instrument manufacturers began to introduce their versions of flame photometers or spectrophotometers.

Flame excitation methods, coupled with simple read-out devices, provided high sensitivity and high reliability for the determination of the alkali metals in simple liquid systems. Further development of burners and aspirators, higher flame temperatures, better spectral isolation using gratings or prisms, and more sensitive detection and read-out devices has increased the list of elements that can be detected by flame excitation to between 50 and 60.

A table of elements, detection limits, and wavelengths used in flame emission and atomic absorption spectroscopy is presented in Appendix VIII. The table was compiled by Pickett and Koirtyohann.[1]

1. FLAME EMISSION INSTRUMENTATION REQUIREMENTS

A flame photometer requires (1) an atomizer–burner combination, (2) a device to isolate the desired spectral region, (3) a detector of radiant energy, and (4) a read-out device, or an amplifier plus read-out device. Such an arrangement, in block form, is shown in Figure 9-1. Each of the components of the flame photometer may be made of different parts, depending on the particular applications desired in the unit. For example, the spectral isolator can be a filter in certain applications, while a prism or grating may be required in other cases. The light-sensitive detector may be a barrier layer photocell, a phototube, or a photomultiplier, and the read-out system may be a galvanometer, a potentiometer, a strip chart recorder, or a digital voltmeter.

2. THE ANALYTICAL FLAME

The flame excitation source of a flame emission spectrometer must fulfill several requirements if it is to be satisfactory. These include the ability to (1) evaporate a liquid droplet sample, (2) vaporize the sample, (3) decompose the compound(s) in the evaporated sample, and (4) spectrally excite the ground state atoms. These processes must occur at a steady rate to achieve a steady emission signal.

---

[1]E. E. PICKETT and S. R. KOIRTYOHANN, Emission Flame Photometry—A New Look at an Old Method, *Anal. Chem.*, **41**, 28A (1969).

FLAME EMISSION SPECTROSCOPY

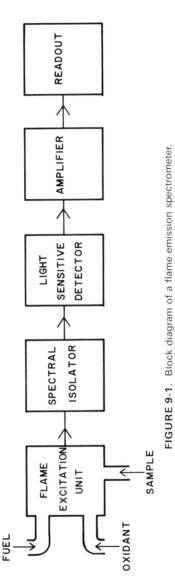

FIGURE 9-1. Block diagram of a flame emission spectrometer.

TABLE 9-1
Temperatures of Commonly Used
Fuel–Oxidant Combinations in
Analytical Flames

Fuel	Oxidant	Temperature, °C
Natural gas	Air	1700
Natural gas	Oxygen	2700
Acetylene	Air	2200
Acetylene	Oxygen	3200
Hydrogen	Oxygen	2800
Acetylene	Nitrous oxide	3400

The processes listed above occur in an environment that includes fuel and oxidant molecules and the products of the reactions between the fuel and the oxidant. The temperature of the flame, which is primarily responsible for the occurrence of these processes, is determined by several factors, including (1) type of fuel and oxidant, (2) the fuel-to-oxidant ratio, (3) type of solvent, (4) amount of solvent entering the flame, (5) type of burner, and (6) the region in the flame that is focused onto the entrance slit of the spectral isolation unit.

A variety of burners and fuel–oxidant combinations have been used to produce the analytical flame. Early usage was commonly natural gas and air. This combination, frequently used with a Meeker burner, produced relatively low temperatures and low excitation energies, leading to simple spectra of easily excited elements, so that its usefulness was limited primarily to the alkali metals. The successful development of flame methods for alkali

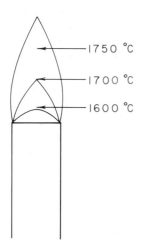

FIGURE 9-2. Temperature distribution in an air–natural gas flame.

FLAME EMISSION SPECTROSCOPY

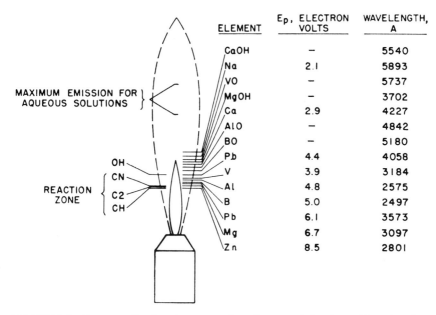

FIGURE 9-3. Flame profile showing areas of maximum emission intensities of various elements. [From B. E. Buell, Use of Organic Solvents in Limited Area Flame Spectrometry, *Anal. Chem.*, **34**, 636 (1962). Used by permission of the American Chemical Society.]

metals produced the desire to extend the technique to other elements, more difficult to excite; thus higher-temperature flames became desirable. Table 9-1 lists some of the commonly used fuel–oxidant combinations and the temperatures that can be attained with them.

Since flame temperatures are very important to the excitation process (refer to Chapter 2), it is important to realize that temperatures in flames are not uniform throughout; thus atomic excitation will not be uniform throughout the flame. Figure 9-2 is a typical temperature distribution of the Meeker burner operating with a stoichiometric ratio of natural gas and air.

The flame also is not homogeneous chemically. In the lower regions the reaction between fuel and oxidant is incomplete and in higher regions there will be higher concentrations of combustion products. Thus an "oxidizing" region and a "reducing" region appear in the flame. Since the processes leading to atomic excitation in the flame differ in the various regions in the flame, differing concentrations of excited atoms are observed in these regions. These are illustrated in Figure 9-3 for several elements as well as some oxides.

Flame temperatures and flame composition have an influence on interferences that may cause erroneous readings to occur. This aspect of the analytical flame will be detailed in a later section of this chapter.

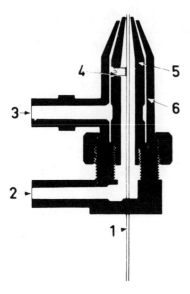

FIGURE 9-4. The Beckman total-consumption turbulent-flow burner–aspirator. [Courtesy Beckman Instruments.] 1. Solution capillary; 2. aspirating gas inlet; 3. fuel gas inlet; 4. centering screw; 5. gas inlet and jacket; 6. jacket.

2.1. Burners and Aspirators

The limitations of the Meeker burner to act as an excitation source soon led to the development of other burners and aspirators, or burner–aspiration combinations. One of the most successful was that designed by the Beckman Corporation. Their burner–aspirator is illustrated in Figure 9-4 and is known as a "total-consumption" burner since all the sample that enters the capillary tube enters the flame, regardless of droplet size. This burner–aspirator introduces the oxidant and fuel in separate parts of the burner, mixing them at the top exit of the burner. The liquid sample enters the flame

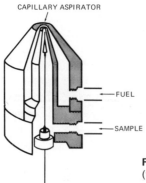

FIGURE 9-5. Jarrell-Ash burner for flame emission (turbulent flow). [Courtesy Jarrell-Ash Division, Fisher Scientific Co.]

FLAME EMISSION SPECTROSCOPY

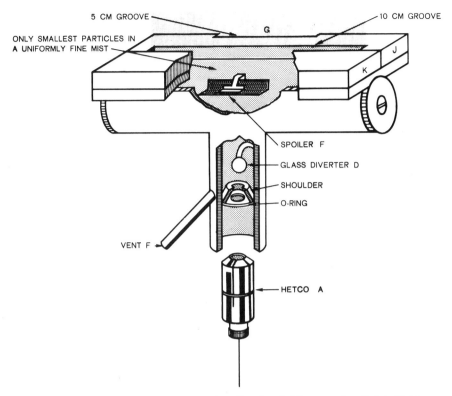

FIGURE 9-6. Jarrell-Ash burner with laminar flow head. [Courtesy Jarrell-Ash Division, Fisher Scientific Co.]

through the central capillary. The flame is noisy and turbulent, but can be adjusted to produce high temperatures by proper control of the fuel-to-oxidant ratio. The burner was widely used for a number of years but recently has been replaced by other types of burner.

The Jarrell-Ash Corporation has designed a similar burner, as shown in Figure 9-5, but have included a slot-type head to convert the burner to a laminar flow system. This arrangement is shown in Figure 9-6. The laminar flow head was intended primarily for atomic absorption measurements but also is useful for flame emission. The Perkin-Elmer laminar flow burner is shown in Figure 9-7.

Other laminar flow burners are available. Most of them premix the fuel and oxidant and have a separate nebulizer to break the liquid sample into small droplets. A drain, or chamber, is provided so the larger droplets do not enter the flame but are collected and removed from the system. This

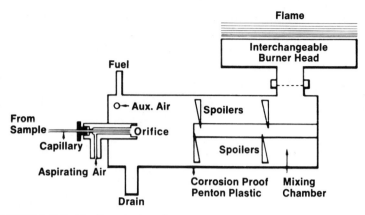

FIGURE 9-7 Perkin-Elmer laminar flow burner. [Courtesy the Perkin-Elmer Corp.]

results in more uniform sample droplet size in the flame with more uniform conversion into an atomic vapor.

Many burners and burner–aspirator combinations presently in use have been designed as dual-purpose devices. They can serve for either emission or atomic absorption processes. A burner that is a good unit for atomic absorption also is good for flame emission measurements.

2.2. Fuel–Oxidant Control

Careful adjustment of fuel-to-oxidant ratio, as well as total fuel–oxidant volume, is required for best response of a flame source. Some elements are best determined in a fuel-rich environment and others in a fuel-lean flame. Control of fuels and oxidants normally is made by use of pressure-reducing valves. Two sets of valves are required, one on the tank for pressure reduction, and the other for close control of the volume of gases entering the burner or mixing chamber. Control by pressure-reducing valves is usually sufficient to produce good, steady flames but does not provide close adjustment or monitoring of the quantities of fuel and oxidant consumed. In some cases it may be advisable to install flow meters in the gas supply systems so rates of gas consumption can be monitored and controlled.

Care needs to be exercised in the use of fuels and oxidants. If compressed air is used, it should be filtered before compression and as it leaves the high-pressure tank. Acetylene requires special care since it is supplied as a solution in acetone under pressure. As the pressure in the tank is reduced, considerable acetone may escape with the acetylene. A glass-wool filter will aid in restricting the flow of acetone into the burner.

Fuel–gas mixtures also may cause explosions. It is important that manufacturer's recommendations on gas and oxidant flow be followed,

3. THE EXCITATION PROCESS IN THE FLAME

There are several steps involved in the process of converting a liquid analytical sample into excited atoms capable of emitting energy. These steps are diagrammed in Figure 9-8.

The solution containing the elements of concern is converted into small droplets by a nebulization (step 1 to 2) process. The small droplets enter the flame and the solvent is evaporated to produce small, dry particles. This is followed by vaporization of the dry particles and their subsequent dissociation into atoms. The atoms are excited by the flame, usually to low lying energy states, followed by emission of the excitation energy as spectral lines.

Other processes also occur, as indicated in Figure 9-8. The early steps, 1–2, 2–3, 3–4, 4–5, occur rapidly and irreversibly; however, reversible reactions also can occur in the flame. The dissociated atoms (5) can react with other atoms or molecules in the flame to produce molecular species and/or radicals. These, in turn, may give rise to molecular spectra. Also, for elements of low ionization energy, ions and electrons may be produced. The ions may be excited and emission lines of the excited ions can be produced. Such lines usually are not useful analytically, but in certain cases cause interference with the desired process, that of producing emission lines of the excited atoms. Under certain conditions very stable molecular species are

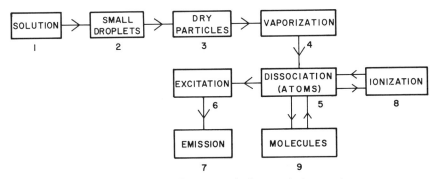

FIGURE 9-8. Sequence of processes in flame emission spectroscopy.

TABLE 9-2
Values of N^*/N^0 for Resonance Lines of Cesium, Sodium, Calcium, and Zinc[a]

Line, Å	P^*/P^0	N^*/N^0 2000°K	3000°K	4000°K
Cs 8521	2	4.44×10^{-4}	7.24×10^{-3}	2.98×10^{-2}
Na 5890	2	9.86×10^{-6}	5.88×10^{-4}	4.44×10^{-3}
Ca 4227	3	1.21×10^{-7}	3.69×10^{-5}	6.03×10^{-4}
Zn 2139	3	7.29×10^{-15}	5.58×10^{-10}	1.48×10^{-7}

[a]From A. Walsh, Application of Atomic Absorption to Chemical Analysis, *Spectrochim. Acta*, **7**, 108 (1955). Used by permission of Pergamon Press.

produced in the flame and resist dissociation. This process also interferes with the desired process. These special situations will be discussed later in this chapter under the topic of interferences.

The number of excited atoms in the flame in a specific excited state will determine the intensity of the emission line produced when the electrons return from that excited state to the lower state (ground state if a resonance line is considered). For any given emission line the number of atoms in the excited state N^* is related to the number in the ground state N^0 by the following expression:

$$N^* = N^0 \frac{P^*}{P^0} e^{-E^*/kT} \qquad (9\text{-}1)$$

Here P^* and P^0 are the statistical weights for the particular energy levels involved, k is the Boltzmann constant, and E^* is the excitation energy of the excited state. It can be observed from equation (9-1) that the number of atoms in the excited state is exponentially related to the absolute temperature. The equation also indicates why high-temperature flames have been successful in increasing the sensitivity of flame spectroscopy, especially for elements of high excitation energy.

In Table 9-2 values of N^*/N^0, as calculated by Walsh, illustrate the effect of temperature on the population of resonance excited states for cesium, sodium, calcium, and zinc.

3.1. Flame Emission Spectra

The emission spectrum of a flame into which a liquid sample has been aspirated includes a variety of components. The spectral lines of the elements

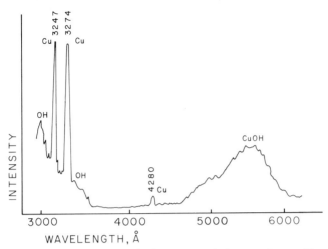

FIGURE 9-9. Flame-excited copper emission spectrum with background and OH band emission.

under investigation appear if excitation conditions are suitable and concentrations of the elements are sufficiently high. The origin of the emission line spectra of the elements has been described in Chapter 2. It should be recalled that since the energy available for the excitation process in a flame source is much less than with an arc or spark source, fewer emission lines will be produced. A flame source, however, will excite the lower energy levels of the spectrum of most elements and these are the lines most useful for analytical purposes. The absence of spectral lines originating from higher energy states results in a simple spectrum and reduces the possibility of spectral line interference. The emission spectrum produced by flame excitation produces relatively few lines of excited ions and this also results in fewer emission lines than produced by arc or spark excitation. The most commonly used lines in flame emission for elemental analysis are those produced from low-lying excited states as the electrons return to the ground state. Such lines are called "resonance lines."

Superimposed over the line spectra of the elements aspirated into the flame are other spectral components. These include line and band spectra of molecules and radicals contained in the flame. For example, excited OH radicals produce band spectra, as do CO, CH, C_2, O_2, CN, and H_2O. Most of these species are produced by the complex chemical reactions occurring in the flame among the fuel, oxidant, entrained air, sample solvent, and sample components.

Many band spectra, which arise from electron transitions of molecules and radicals as well as internal molecular vibrations and rotations, appear as a continuum since the monochromators commonly used for flame spectroscopy do not have the resolving power necessary to resolve the various components of the band spectra. Figure 9-9 is a spectrum showing the emission lines of copper superimposed over a continuum and band spectra of the OH radical. Such spectra require proper handling of line intensity measurements if data are to be used for quantitative analysis.

4. FLAME EMISSION INTERFERENCES

Flame emission spectroscopy is subject to interference processes that must be understood if adequate control of their effects is to be exercised. Properly used, flame emission can provide a simple, accurate technique of analysis, but it is only through adequate control of interferences that good analytical results can be obtained. The following section discusses some of the more commonly encountered interference processes.

4.1. Spectral

Direct spectral line interference occurs when the spectral line energy of two or more elements reaches the detector circuit. One type of spectral line interference involves spectral line overlap. This occurs because spectral lines have a finite linewidth. If the spectral energies of two lines overlap, the result is spectral interference regardless of the resolving power of the spectral isolation system of the spectrometer. At high flame temperatures, when

TABLE 9-3
Some Typical Spectral Interference Systems in Flame Emission Spectroscopy

Element	Wavelength, Å	Interferent	Wavelength, Å
Al	3961.5	Ca	3968.5
Ba	5535.6	CaOH	5540
Ca	4226.7	Sr	4215.5
Co	3453.5	Ni	3452.9
Cu	2274.0	OH	3274.2
Li	6707.8	SrOH	6720
Mg	2852.1	Na	2852.8
Ni	3524.5	Co	3526.8
Sn	3175.0	OH	3175
V	3185.4	OH	3185
Zn	4810.5	Sr	4811.9

FLAME EMISSION SPECTROSCOPY

numerous spectral lines are being produced, the possibility of such interference increases. For example, the iron line at 3247.28 Å overlaps the copper line at 3247.54 Å and the iron line at 2852.13 Å overlaps the magnesium line at 2852.12 Å. Other cases also occur; examples are silicon and vanadium, platinum and iron, and aluminum and vanadium.

Spectral line interferences also can occur when spectral lines of two or more elements are close but do not produce an actual overlap of their energy envelopes. This type of interference is especially troublesome when the spectral isolation device is a filter. With a filter, lines separated by as much as 50–100 Å may be passed through the filter to the detecting circuit, thus producing an incorrect read-out signal. As the resolution of the spectral isolation system increases, such interference possibilities decrease. They cannot be eliminated entirely, however, because of the finite width of the spectral isolation system and the finite slit widths required in such systems.

Another type of spectral interference can occur between the analytical spectral line and spectral band systems produced by molecules or radicals in the flame. This type of interference is encountered more often than atomic line interference. The band emission may be from various metal oxides or hydroxides, or may be produced by species generated from the fuel and oxidant system. If organic solvents are employed, these substances also can contribute species that produce band spectra. Examples of this type of interference include the copper line at 3274.0 Å and the OH band at 3274.2 Å, and the aluminum line at 3961.5 Å and the CH band series starting at 3872 Å.

A third type of spectral interference involves that occurring between a spectral line and a continuous background. This type of interference is most often caused by high concentration of salts in the sample. Of particular concern are salts of the alkali and alkaline earth elements. A continuous background also can be produced by some organic solvents. For example, methyl isobutyl ketone is particularly troublesome in flame emission. This substance is commonly used in extraction procedures when sample concentration is required prior to aspirating the sample into the flame.

If solid particles are present in the flame and are heated to incandescence, a continuum can be produced. Such a continuum is not common but could occur either from the production of a solid substance from the reactions of the fuel and oxidant, or from the production of very stable metal oxides in the flame, such as MoO, AlO, and SiO.

Table 9-3 presents information on a number of spectral interferences. The table is by no means a complete compilation of spectral interferences but will serve to illustrate the varied nature of such interferences.

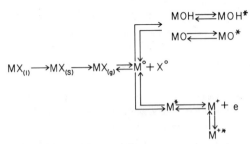

FIGURE 9-10. Some possible chemical processes during aspiration and excitation of the sample.

4.2. Ionization

Interferences caused by ionization of easily ionizable substances can be severe in some cases. To consider the origin of such interferences, it is important to consider certain reactions in the flame that involve the analyte element. Reference to the sequence of steps given in Figure 9-10 should make this clear.

In Figure 9-10, M is the metal element. The species whose concentration we wish to measure is M*, the excited metal atom. If M^0 or M* is

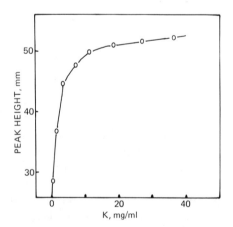

FIGURE 9-11. Effect of potassium on the flame emission intensity of rubidium. [From T. E. Shellenberger, R. E. Pyke, D. B. Parrish, and W. G. Schrenk, Some Factors Affecting the Flame Photometric Emission of Rubidium in an Oxy-Acetylene Flame, *Anal. Chem.*, **32**, 210 (1960). Used by permission of the American Chemical Society.]

TABLE 9-4
Percentage Ionization of the Alkali Metals as Calculated Using the Saha Equation at 2200°C

Element	Ionization, %
Lithium	<1
Sodium	4
Potassium	30
Rubidium	41
Cesium	65

provided sufficient energy, it is converted to $M^+ + e$. The following equilibrium can then occur in the flame:

$$M^0 \leftrightarrows M^* \leftrightarrows M^+ + e^- \quad (9\text{-}2)$$

As the temperature of the flame is increased the concentration of M^+ and e^- increases and M^0 decreases. An equilibrium as given above is subject to the usual stresses that may be placed on any such equilibrium. If a second, easily ionizable substance N is added to the flame, it also may ionize and produce electrons, thus:

$$N^0 \leftrightarrows N^* \leftrightarrows N^+ + e^- \quad (9\text{-}3)$$

thus increasing the electron concentration. The increased electron concentration will shift equilibrium in (9-2) to the left, increasing the concentration of M^*. The effect of N^0 then will be to enhance the concentration of M^*, producing an enhanced signal as compared with the signal produced in the absence of N^0. Figure 9-11 indicates the enhancement of the signal from rubidium as potassium is added to the sample solution. An enhancement of 80–100% occurs. The rubidium concentration was 1.0 mg/ml and the potassium concentration was varied from 0 to 40 mg/ml.

Saha has applied equilibrium principles to the reaction

$$M \leftrightarrows M^+ + e^- \quad (9\text{-}4)$$

and derived the following equation, which relates the ionization potential of the element under consideration E_i, the absolute temperature T, and the statistical weights g_{M^+}, g_{e^-}, and g_M to K, the equilibrium constant for equation (9-4):

$$\log K = \frac{-5050 E_i}{T} + \frac{5}{2} \log T - 6.50 + \log \frac{g_{M^+} g_{e^-}}{g_M} \quad (9\text{-}5)$$

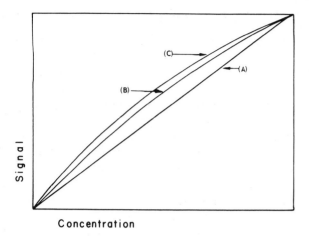

FIGURE 9-12. Flame emission calibration curves for magnesium (2852 Å): (A) 0–10, (B) 0–50, (C) 0–100 µg/ml. [From W. G. Schrenk, in *Flame Emission and Atomic Absorption Spectroscopy*, Vol. 2, Edited by J. A. Dean and T. C. Rains, Marcel Dekker, New York, (1971), Chapter 12. Used by permission of Marcel Dekker, Inc.]

The equation has been shown to be valid; thus it can be used to calculate the percentage ionization of metals in flames. Table 9-4 presents results of using the Saha equation to determine percentage ionization of several elements. The validity of the equation also indicates that the assumption that equilibria such as illustrated in equation (9-4) occur is correct.

Another type of spectral interference can occur in flames: self-absorption. It is especially noticeable at higher element concentration levels. If a metal element at relatively high concentration is excited and emits its energy near the center or hotter regions in a flame, that energy may be reduced in intensity as it passes through the cooler, outer regions of the flame. The mechanism responsible for the reduced intensity is the same as that which produces Fraunhofer lines in the spectrum of the sun. Neutral atoms of the element, in cooler flame regions, can absorb the energy emitted by the excited atoms in the hotter regions of the flame. This effect is largely responsible for the nonlinear calibration curves obtained at higher concentration levels.

The effect of self-absorption is evident in the calibration curves shown in Figure 9-12 for magnesium. The three curves are for magnesium concentration maxima of 10, 50, and 100 µg/ml, respectively, normalized so the shapes can be compared.

4.3. Cation–Anion Interferences

If the analyte element forms a stable molecule with an anion, or with some other substance present in the flame, the analyte element will not be available for the atomic excitation process. Several systems show this effect.

Calcium in the presence of the phosphate ion apparently forms a stable substance, so the calcium signal is depressed. Calcium in the presence of the borate ion also produces a smaller emission signal. The behavior of these two systems is illustrated in Figure 9-13, and although both curves show depression of the calcium signal, the curves are not similar. As the concentration of phosphorus increases, an almost linear decrease of calcium emission occurs to a 1:1 mole ratio of calcium to phosphorus. From this concentration of phosphorus to higher concentrations there appears to be no further decrease in the calcium signal. In contrast, the effect of boron on calcium is to produce a nonlinear decrease in the calcium signal. No sharp break occurs in the curve, although at higher boron concentrations the depressing effect of boron on calcium is small. Results with the Ca–P system suggest compound formation in the flame with a species produced that contains calcium and phosphorus atoms in a 1:1 ratio. The results with calcium and boron do not show this effect. The two curves of Figure 9-13 are typical of the curves obtained in many cases where interferences are caused by interactions between a cation and anions present in the flame.

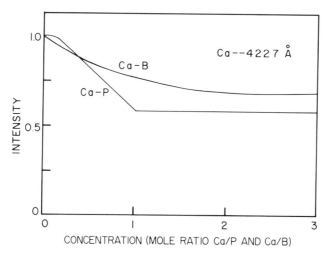

FIGURE 9-13. Effects of phosphorus and boron on the flame emission signal of calcium. [From a Kansas State University Ph.D. dissertation by Mitchell E. Doty.]

4.4. Cation–Cation Interferences

Mutual interferences of cations have been observed in many cases. Aluminum interferes with calcium and magnesium, as does silicon. Other examples include calcium and gallium, calcium and indium, and the interferences of beryllium, chromium, iron, molybdenum, silicon, titanium, tungsten, and the rare earth elements. These interferences are neither spectral nor ionic in nature and the mechanisms of their interactions are unknown. Cation–cation interferences, however, invariably decrease the signal intensity of the analyte element.

4.5. Oxide Formation

The sensitivity of any flame emission analytical procedure depends on the population of excited state metal atoms M* that can be produced in the flame. As metallic salts decompose in a flame, a variety of species are produced and a variety of reactions can occur. Some of the possibilities are shown in Figure 9-10. This section deals with the possible formation of oxides MO or hydroxides MOH in the flame. Formation of metal oxides in the flame reduces the population of M* and thus causes a decrease in the flame emission signal from M*.

If oxygen is present in a flame, the formation of stable oxides with free metal atoms can remove a large percentage of the metal atoms from the flame. For example, free magnesium atoms in a flame can readily form MgO if oxygen is present. The dissociation energy of MgO is approximately 7.9 eV and it has been calculated that MgO is only about 1.4% dissociated in an air–acetylene flame ($\sim 2200°C$). All of the alkaline earth elements readily form oxides and, as a family, are especially subject to this type of interference. Aluminum and boron, in family IIIA, also form stable oxides in most flames.

Hydroxides also are known to be formed in flames. Species such as LiOH, CsOH, CuOH, MnOH, InOH, and CaOH all have been observed. In some cases as much as 80% of the element can be present as the metal hydroxide, depending on flame temperatures and composition.

A few hydrides also have been reported in flames, such as CuH, AlH, TlH, and BH. These substances, however, are present in small concentrations and apparently are not an important factor in analytical applications of flame emission spectroscopy.

4.6 Chemiluminescence

The direct conversion of chemical energy to radiant energy is called chemiluminescence or chemiexcitation. This process can interfere with normal excitation processes. It can be used to provide better sensitivities and detection limits for certain elements. Chemical reactions in flames frequently release as much as 5–10 eV of energy. This is sufficient to excite and ionize many atoms. If theoretical calculations, based on thermal considerations, are made concerning expected intensities of spectral lines and are compared with experimental results, large discrepancies occur in some situations. For example, tin in isopropanol and aspirated into an air–hydrogen flame emits a rich line spectrum to 2100 Å, with some lines 1000 times more intense than expected on the basis of thermal calculations. Some other elements that emit abnormally intense line spectra include bismuth, cadmium, antimony, iron, lead, mercury, chromium, and magnesium. Rare earth element spectra also are reported to be enhanced in a fuel-rich oxyacetylene flame when ethanol is the solvent. These unusually high intensity line spectra are considered to be caused by chemiluminescent phenomena.

Some mechanisms proposed to explain these observations are the following. In hydrogen flames the recombination of hydrogen atoms may provide the energy, as

$$H + H + M \rightarrow H_2 + M^* \tag{9-6}$$

However, since hydrocarbon flames are especially effective in producing chemiluminescence phenomena, the following reactions also may provide excitation energy for the metal atoms:

$$CH + O + M \rightarrow CO + H + M^* \tag{9-7}$$

and

$$CH + OH + M \rightarrow CO + H_2 + M^* \tag{9-8}$$

Similar reactions also may be involved, such as

$$CH^* + O \rightarrow CHO^+ + e; \quad \Delta H^\circ \cong -3.5\,\text{eV} \tag{9-9}$$

and

$$H_3O^+ + e^- \rightarrow H_2O + H; \quad \Delta H^\circ \cong -8.7\,\text{eV} \tag{9-10}$$

Still other reactions of substances in the flame are possible sources of energy that may be transferred to metal atoms, producing non-thermally excited atoms. The above reactions are given to illustrate the possibilities.

4.7. Physical Interferences

Physical interferences are those caused by a physical property of the sample or produced by the physical properties involved in the process of converting the analyte element into excited atoms in the flame. Among the important physical properties of the sample solution are surface tension, vapor pressure, and viscosity. Some of the physical processes involved in conversion of the sample solution into an excited atom are droplet formation, solvent evaporation, and vaporization of the solid particle remaining after solvent evaporation.

Droplet formation (nebulization) occurs differently in a total-consumption burner than in a premixed flame–nebulizer combination. In the total-consumption burner the analyte solution is aspirated at high velocity into the flame, usually as the result of the Venturi effect created by the high velocity at which the fuel and oxidant are forced into the flame. Reference to the diagram of the Beckman type total-consumption burner in Figure 9-4 will make this action clear. The low pressure created at the tip of the capillary due to the rapid flow of fuel and oxidant gases past the tip brings liquid sample into the flame, with droplets of sample being formed at the orifice of the capillary. This system results in a noisy, turbulent flow of gases and sample into the flame. The rate of flow of the fuel and oxidant, the capillary tube diameter, and the relative spacings of the fuel, oxidant, and sample exits contribute to the efficiency of the process. In this system all the sample that exits from the capillary enters the flame. Droplet size is quite variable and some larger droplets pass through the flame without completing the processes involved in producing excited atoms in the flame. In fact, some larger droplets may pass through the flame without complete evaporation of the solvent.

Some burners premix the fuel and oxidant gases and have a separate device to nebulize the liquid sample. These burners usually are characterized as laminar flow systems, the burning proceeding in a silent fashion. In this type of system, sample droplets are formed in a spray chamber and are reduced in size by a "spoiler" which may be a dispersal ball, baffle, or similar device. A condensing chamber collects the larger droplets and only the smaller ones enter the flame. In some cases as much as 80% of the sample may be retained in the condensing chamber. Figures 9-6 and 9-7 show typical laminar flow premix burners with nebulizers, spoilers, and condensing chambers. Such nebulization systems produce small, uniform droplets that enter the flame and can respond quite uniformly to the processes of evaporation of solvent, vaporization of solid sample, and dissociation to atoms. In spite of the low efficiency of sample feed into the flame, the high efficiency of conversion of analyte element into excited atoms produces spectral intensities as good as or better than those obtained with the turbulent-flow total-consumption burner.

Solvent evaporation must occur as a first step in the flame. Most small droplets are evaporated quickly in a flame but larger ones may pass through the flame still not completely evaporated. The evaporation of the solvent lowers the flame temperature and therefore may affect subsequent steps of vaporization and dissociation. These facts help explain the advantages of reducing the sample to a very fine mist prior to entry into the flame. Use of organic solvents with their lower heats of vaporization also tends to maintain the flames at higher temperatures than when using water as a solvent. Organic solvents also may undergo combustion in the flame, contributing higher temperatures to the flame.

The solid analyte particle must undergo vaporization to the gaseous state prior to decomposition and excitation. The degree to which this is accomplished is related ultimately to the sensitivity of the determination. If the molecule containing the analyte element has a low dissociation energy, it will be almost completely dissociated. Flame temperature obviously is the principal controlling factor in this process.

If surface tension is low, improved nebulization of the analyte sample solution results. A larger percentage of small droplets is produced, and thus more sample enters the flame. The lower surface tension of organic solvents contributes to the signal enhancement observed when using these solvents. Since surface tension is temperature dependent, changes in the temperature of the analytical sample also may be reflected in changes in signal intensity.

Viscosity is a force that opposes the nebulization process; it is the tendency for a liquid to resist flow. Thus a highly viscous liquid would be almost impossible to force through a small capillary. Viscosity is inversely related to temperature; thus a highly viscous sample could be processed through a capillary if the temperature of the sample were raised. It is important that standard solutions and sample solutions have approximately the same viscosity since it is necessary that these solutions have the same sample uptake rate and produce the same particle size distribution.

5. CONTROL OF INTERFERENCES

5.1. Spectral

Spectral line interferences frequently can be minimized by using any one of a variety of procedures, depending on the particular situation. One important instrumental parameter that can be used to minimize spectral line interferences is the resolution of the spectral isolation device. Filter instruments are of little use in this respect except for very simple analytical samples since they may have a band pass of as much as 800–900 Å. Interference filters give better resolution (100–200 Å) and thus may be used in applications where colored filters are not acceptable.

Spectrometers, using prisms or gratings, have much better resolving power. The commonly used Beckman DU spectrophotometer, with its Littrow-mounted prism, can resolve spectral lines of 8–10 Å separation at the ultraviolet end of the spectrum. This instrument has a reciprocal linear dispersion of 37 Å/mm at 3000 Å. Larger prism spectrometers have correspondingly smaller reciprocal linear dispersions. Grating spectrometers offer similar reciprocal linear dispersions with the added advantage that the dispersion is constant as wavelength changes if the spectrum is viewed normal to the grating.

Most instruments constructed for atomic flame emission and atomic absorption spectroscopy have dispersions in the range of 15–35 Å/mm. For example, the Perkin-Elmer 290B uses a grating with a dispersion of 16 Å/mm, the Jarrell-Ash Dial-atom uses a grating with a dispersion of 33 Å/mm, the Hilger-Watts Atomspek uses a prism with a dispersion of 17 Å/mm at 2000 Å, and Aztec Instruments model AAA-3 uses a grating with a reciprocal linear dispersion of 16 Å/mm.

If the spectrometer has variable-width slits, it is possible to increase the resolving power by decreasing the slit width. The slit width does not affect the dispersion but does narrow the spectral band that is passed through to the readout system as slit width is decreased. It is important, therefore, if improved resolution is required, to use the narrowest slit width consistent with spectral line intensity and the ability of the read-out system to provide a satisfactory signal. However, no spectrometer can resolve two spectral lines that physically "overlap" one another. Since spectral lines have finite width, overlap is possible. An example of this situation is that of iron at 2719.03 Å and platinum at 2719.04 Å. Fortunately, not many systems fall into this category.

Spectral band interference also can be troublesome. At low dispersion a spectral band appears as a single, broad band. With high dispersion the fine structure of the band appears. Frequently encountered bands in flame emission spectroscopy include those produced by OH, C_2, CH, metal oxides, and metal hydroxides.

Several approaches are possible to minimize the effects of spectral band–spectral line overlap. One method is to use a high resolution spectrometer with narrow slit widths. This method frequently resolves the band into its separate components, thus permitting better separation of spectral line and spectral band components. Another method is to determine if another line is available for use in a different spectral region. For example, there is less OH band interference with the copper 3274.0 Å line than with the copper line at 3247.0 Å. Cobalt at 2873.1 Å is in a region of strong CH band interference, while the cobalt line at 3453.5 Å is not.

CH and C_2 bands can be eliminated by use of a hydrogen–oxygen flame, thus making these regions in the flame available for analytical purposes without the problems of the band interferences. A change of solvent system from water to an organic solvent also may help in certain instances. Organic solvents also often enhance the intensities of the line spectra of elements.

It is possible to subtract the effect of the band emission signal. This can be done by using a blank solution, measuring the signal it produces at the desired wavelength, then measuring the analyte solution at the same wavelength and subtracting the two to determine the signal due to the analyte. If the read-out system and spectrometer provide for spectral scanning, it is possible to scan the desired wavelength region and make the same correction to obtain the spectral energy produced by the analyte.

Background radiation usually can be corrected by the same scanning technique mentioned above for spectral band interference. It is important, however, to determine the background effects with a blank since small changes in the background intensity may be concealed by the spectral line emission. Since background radiation is different for different solvents, a solvent change may be effective in reducing the magnitude of the flame background signal.

5.2. Ionization Interference Control

Since ionization interferences in flames are equilibrium systems, the best method of control is to consider methods of dealing with the equilibrium. In the equilibrium

$$M^0 \leftrightarrows M^* \leftrightarrows M^+ + e^- \qquad (9\text{-}11)$$

it is apparent that increasing the electron concentration will shift the equilibrium to the left, thus increasing the concentrations of M^* and M^0. The effect of increasing the electron concentration in equation (9-11) is to enhance the emission signal from M^*. This effect is shown in Figure 9-11.

Ionization interferences are most serious with the alkali metals, although with higher temperature flames some of the alkaline earth elements also are significantly ionized. For example, in the nitrous oxide–acetylene flame (3000°C) magnesium is only about 6% ionized while strontium is 85% ionized.

The usual control of ionization interference is to add sufficient easily ionizable element to place the flame operating conditions on the plateau of Figure 9-11. CsCl is commonly used for this purpose and is a satisfactory control for all the other group IA elements as well as those of group IIA. Ionization interference for cesium is more difficult to control since no element more easily ionized than cesium can be added, although a rubidium or potassium salt will help.

5.3. Cation–Anion Interference Control

One of the most important factors involved in control of cation–anion interference effects is flame temperature. If compound formation occurs in the flame, as happens with calcium and the phosphate ion, a severe signal depression is observed. Use of high-temperature flames can minimize this effect. Flames that have been used successfully for this purpose are the nitrous oxide–acetylene flame and the premixed oxygen–acetylene flame.

The use of "releasing" and protective chelating agents also can minimize interferences due to formation of nonvolatile compounds. These agents are substances that will restore the emission intensity to the level that occurred when the interferent was not present. Lanthanum can be used as such an agent for calcium determinations in the presence of phosphate or sulfate. The mechanism usually proposed to explain this action is the preferential formation of $LaPO_4$ rather than a calcium–phosphorus compound. The $LaPO_4$ has greater thermal stability; thus the calcium is free to form excited atoms. Lanthanum is widely used as a releasing agent for alkaline earth elements since it is effective and its spectrum does not interfere with any of the alkaline earth spectra.

Other releasing agents have been used in other cases. Copper has been used as a releasing agent for noble metals and barium has been used for sodium and potassium determinations in the presence of aluminum and iron.

Two chelating agents, ethylenediaminetetraacetic acid (EDTA) and 8-hydroxyquinoline, also have been used for calcium and magnesium determinations. They are effective in controlling a number of interfering ions, including phosphate, sulfate, aluminum, silicon, boron, and selenium.

5.4. Control of Oxide Interference

Oxide formation can be a very troublesome interference, especially with group IIA and IIIA elements, since oxides of these elements are thermally very stable. Two approaches to minimizing this type of interference are available.

One method is to use very high temperature flames. Such flames tend to dissociate the oxide, thus producing free atoms for excitation. The oxygen–acetylene flame and the nitrous oxide–acetylene flame are the two most commonly used flames for this purpose.

The second method of control of oxide formation is to provide an oxygen-deficient (or fuel-rich) environment in which to produce the excited atoms. Such a flame minimizes oxide formation, thus aiding in the control of this type of interference. Most present-day, premixed, laminar flow burners can be adjusted to fuel-rich conditions. It is important, however, that the flow rate of the mixed gases be greater than the burning velocity of the gas mixture or the flame will be propagated back into the mixing chamber and an explosion will occur.

A few burners are available that can produce a "sheathed" flame. Such flames usually are surrounded by an inert gas that prevents the flame from entraining air from its surroundings. Thus oxygen in the air does not enter the flame to react with the metals atoms that are present.

5.5. Control of Physical Interference

Physical interferences cannot be controlled effectively except by proper preparation of reference standards. It is important that standard solutions have as near the same physical properties as the analyte solutions. If properties such as viscosity, surface tension, and vapor pressure are the same in the reference solution and the test solutions, the solutions should respond similarly in the burner. It is important that burner conditions, i.e., fuel and oxidant flow rates and pressures, be maintained constant so that sample uptake rate, droplet size, and transport into the flame will be constant.

6. SIMULTANEOUS MULTIELEMENT ANALYSIS

Multielement flame emission spectroscopy is a relatively new development, although multielement methods have been used in arc–spark emission spectroscopy for some years. Several multielement methods are available, including scanning, direct reading techniques similar to those employed in arc–spark emission spectroscopy and the more recently developed vidicon detector tubes. Vidicon detectors are described in Chapter 3.

Mitchell et al.[2] described the use of a vidicon detection system to permit the simultaneous determination of up to ten elements in solution. They used an 0.25-m monochromator and covered a wavelength range of 1680 Å displayed across the face of the vidicon detector. Spectral lines as close as 20 Å were readily resolved.

Busch et al.[3] used a silicon diode vidicon detector with an 0.5-m monochromator. They reported spectral lines 1.4 Å apart to be resolvable and obtained data on eight elements (Mo, Fe, Ca, Al, Ti, W, Mn, and K) simultaneously. A spectral range of 200 Å was possible with a single setting of the grating. Figure 9-14 is a multielement flame emission spectrum in the region 3886–4086 Å obtained with this technique.

[2] D. G. MITCHELL, K. W. JACKSON, and K. M. ALDOUS, Anal. Chem., **45**, 1215A (1973).
[3] K. W. BUSCH, V. G. HOWELL, and G. H. MORRISON, Anal. Chem., **46**, 575 (1974).

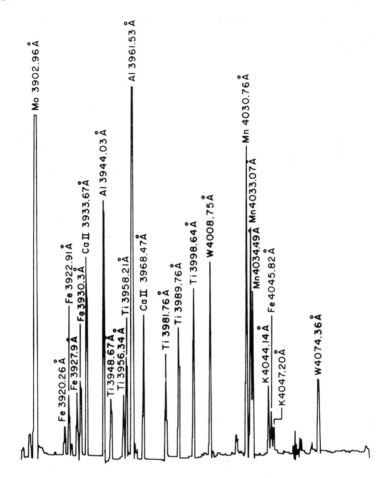

FIGURE 9-14. Multielement flame emission spectrum from 3886 to 4086 Å. [From K. W. Busch, N. G. Howell, and G. H. Morrison, The Vidicon Tube as a Detector for Multielement Flame Spectrometric Analysis, *Anal. Chem.*, **46**, 575 (1974). Used by permission of the American Chemical Society.]

An optical multichannel analyzer was used to obtain both peak height and peak area and was corrected for background. Results indicate either peak height or peak area can be used analytically except under electronic overload conditions. The method was used with a U. S. Geological Survey rock sample to determine Al, Fe, Ca, and Ti. The multielement technique compared favorably with concentrations of the four elements as reported by the U. S. Geological Survey.

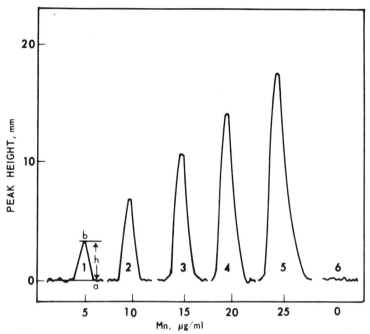

FIGURE 9-15. Flame emission signals of manganese standards (4032 Å). [From W. G. Schrenk, in *Flame Emission and Atomic Absorption Spectroscopy*, Vol. 2, Edited by J. A. Dean and T. C. Rains, Marcel Dekker, New York (1971), Chapter 12. Used by permission of Marcel Dekker Inc.]

7. ANALYTICAL TREATMENT OF DATA

The acquisition of quantitative analytical results is the ultimate goal of analytical emission spectroscopy; thus the procedures used to obtain these data are important. Read-out devices used with flame emission are usually a visual meter, a digital read-out (voltmeter), or a chart recording. A few instruments are equipped to provide internal adjustment to provide a direct read-out in concentration units, such as µg/ml. A few basic methods of treating data are presented here.

7.1. Establishment of a Working Curve

Any good analytical working curve requires first the establishment of optimum operating conditions, including spectral line selection, fuel and oxidant selection and control, sample uptake rate, and the positioning of the optical path and optical components. Once established, these conditions must be maintained throughout the acquisition of the analytical data.

The simplest analytical system to deal with is one in which the background is zero or very low and there are no spectral interferences of concern.

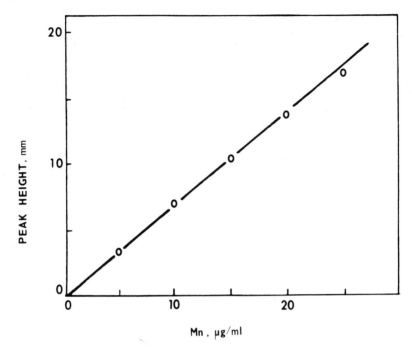

FIGURE 9-16. Concentration of manganese in μg/ml versus peak height in mm. [From W. G. Schrenk, in *Flame Emission and Atomic Absorption Spectroscopy*, Vol. 2, Edited by J. A. Dean and T. C. Rains, Marcel Dekker, New York (1971), Chapter 12. Used by permission of Marcel Dekker Inc.]

Recorded flame emission signals of such a system are shown in Figure 9-15, which shows tracings from strip chart recorded standards for manganese using the emission line at 4032 Å. These data could have been obtained with a visual meter, either by visually observing a scan of the spectral line or by carefully adjusting the wavelength setting of the monochromator to read the peak signal intensity.

These data may be plotted as intensity vs. concentration, as shown in Figure 9-16, and in this case a linear calibration results. Since the calibration is linear and passes through the origin, it is possible to reduce the curve to a simple algebraic equation:

$$\text{concn Mn} = \text{slope} \times \text{peak height} \qquad (9\text{-}12)$$

and dispense with the curve entirely. Regular use of such a relation, however, requires regular checks to determine that operating conditions have not changed. If minor changes do occur, a slight change in fuel or oxidant flow rate can be made to return the system to the original working curve.

7.2. Background Correction

If it is necessary to obtain the analytical data from a spectral region which includes background, another approach is necessary. From an analytical point of view the origin of the background is not important. Figure 9-17 illustrates the problem when a spectral line is superimposed on a background whose intensity is changing rapidly. In this case the emission signal is rubidium at 7800 Å superimposed over the base of a potassium emission line. The intensity of the emission of rubidium can be obtained as follows (refer to signal 4 of Figure 9-17). A tangent is drawn from a to b and the vertical distance from c to d is measured. These data can then be used to establish the analytical working curve for rubidium.

If scanning is not possible, signal intensities can be measured at three wavelengths, a, b, and c. The readings at a and b should be averaged and subtracted from the reading at c to give a measure of the signal intensity of rubidium. This technique is satisfactory if there are no significant "humps" or "valleys" in the background between a and b. Care must be exercised so all readings at a, b, and c are obtained at the same wavelengths for each separate analytical determination.

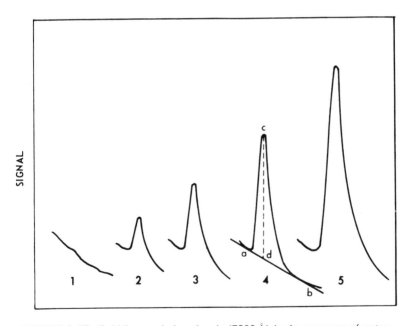

FIGURE 9-17. Rubidium emission signals (7800 Å) in the presence of potassium. [From W. G. Schrenk, in *Flame Emission and Atomic Absorption Spectroscopy*, Vol. 2, Edited by J. A. Dean and T. C. Rains, Marcel Dekker, New York (1971), Chapter 12. Used by permission of Marcel Dekker Inc.]

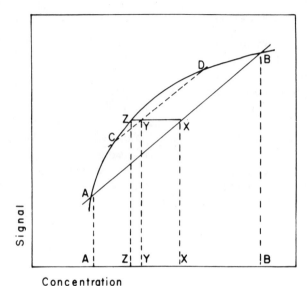

FIGURE 9-18. Sample bracketing technique with linear and nonlinear calibration curves. [From W. G. Schrenk, in *Flame Emission and Atomic Absorption Spectroscopy*, Vol. 2, Edited by J. A. Dean and T. C. Rains, Marcel Dekker, New York (1971), Chapter 12. Used by permission of Marcel Dekker Inc.]

7.3. Sample Bracketing

Sample bracketing is a technique especially useful if only a few samples are to be determined. After an initial run is made on the unknown, two standards, one of a concentration higher and one of a concentration lower than the unknown, are chosen and their signal intensities determined. These data are plotted as shown in Figure 9-18, where A and B are the standards and X is the unknown. If the calibration curve is linear, the concentration of X can be determined directly from the reading obtained from the concentration axis or from the equation

$$\text{concn } X = \text{concn } A + k(\text{concn } A \text{ concn } B)$$

where k is the slope of the calibration curve.

If the calibration curve is nonlinear, the same technique can be used but standards A and B must be chosen more carefully to produce a small concentration interval, with the unknown concentration being between the two standards. Figure 9-18 illustrates why this is necessary. If A and B are chosen as the bracketing standards and curve *ACZDB* is the correct calibration curve, a linear interpretation of the analytical results will produce a concentration value of X for the unknown when the true value corresponds to Z. If the bracketing standards are chosen as C and D, the unknown concentration would correspond to Y, which is closer to the true concentration of the unknown. If the interval between standards is still further reduced, the error will decrease. The method of sample bracketing assumes a linear calibration curve between the bracketing standards.

7.4. The Method of Standard Additions

The standard addition technique also is applicable to emission and atomic absorption spectroscopy. The technique involves the use of the unknown plus the addition of known amounts of standard. A signal intensity is obtained for the unknown X, then a series of solutions containing the unknown plus varying amounts of the standard are prepared and their signals obtained. The data are treated as shown in Figure 9-19. A known concentration of the element A is added to the unknown to produce solutions having concentrations of X, $X + A$, $X + 2A$, etc. These data are

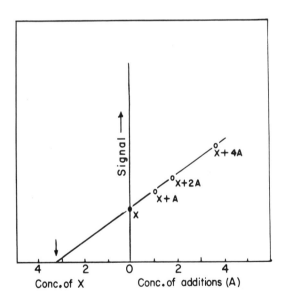

FIGURE 9-19. Standard addition method for the determination of the concentration of an unknown. [From W. G. Schrenk, in *Flame Emission and Atomic Absorption Spectroscopy*, Vol. 2, Edited by J. A. Dean and T. C. Rains, Marcel Dekker, New York (1971), Chapter 12. Used by permission of Marcel Dekker Inc.]

plotted as shown and the concentration of the unknown is determined from the intersection of the curve with the concentration axis. The standard addition procedure is especially useful where calibration data are doubtful and for very low concentrations. It is a useful technique only if the calibration curve is linear and passes through the origin.

SELECTED READING

American Society for Testing Materials, *Symposium on Flame Photometry*, ASTM Special Technical Publication No. 116, Philadelphia, Pennsylvania (1951).

Dean, J. A., *Flame Photometry*, McGraw-Hill, New York (1960).

Dean, J. A., and Rains, T. C., Editors, *Flame Emission and Atomic Absorption Spectrometry*, Vols. 1 and 2, Marcel Dekker, New York (1971).

Grove, E. L., Editor, *Analytical Emission Spectroscopy*, Vol. 1, Parts I and II, Marcel Dekker, New York (1972).

Hermann, R., and Alkemade, C. Th. J., *Flammenphotometrie*, Springer-Verlag, Berlin (1960).

Mavrodineanu, R., *Bibliography on Flame Spectroscopy, Analytical Applications*, National Bureau of Standards Miscellaneous Publication 281 (1967).

Mavrodineanu, R., Editor, *Analytical Flame Spectroscopy*, Springer-Verlag, New York (1969).

Parsons, M. L., and McElfresh, P. M., *Flame Spectroscopy, Atlas of Spectral Lines*, IFI/Plenum, New York (1971).

Poluektov, N. S., *Techniques in Flame Photometric Analysis*, Consultants Bureau, New York (1961).

Pungor, E., *Flame Photometry Theory*, Van Nostrand, London (1967).

Chapter 10
Analytical Atomic Absorption Spectroscopy

In 1955 Walsh established the foundations of modern analytical atomic absorption spectroscopy. In the same year Alkemade and Milatz also published a paper suggesting similar procedures. The work of Walsh, however, was much more detailed, since he examined the theory of the method, the basic principles involved, the instrumentation requirements, and its advantages over flame emission.

Since the work of Walsh, the growth of analytical atomic absorption spectroscopy has been phenomenal. About 70 different elements, including most of the common rare earth metals, have been determined by atomic absorption methods. Direct application of the technique is limited to metals, with the exceptions of B, Si, As, Se, and Te. Several of the nonmetals have been analytically determined by indirect methods. Recently detection limits have been lowered, new sample cells and sampling techniques have been developed, very high intensity sources have become available, and new and better systems developed to "read out" the analytical data. This chapter deals with the principles and techniques of modern analytical atomic absorption spectroscopy.

1. THE ATOMIC ABSORPTION PROCESS

The absorption of energy by ground state atoms in the gaseous state is the basis of atomic absorption spectroscopy. Understanding of this process of energy absorption had its foundations in the study of the origin of the Fraunhofer dark lines in the sun's spectrum. When radiation of proper wavelength passes through a vapor containing ground state atoms, some of the radiation can be absorbed by excitation of the atoms thus, $M^0 + h\nu \rightarrow M^*$, and the intensity of the radiation at a wavelength corresponding to the energy of the photon $h\nu$ is decreased. If the concentration of M^0 in the vapor is increased, the decrease in radiant energy will be greater. Since each species of atom can exist only in specific excited states, the photon energies required for each atomic species will be different, and thus will occur at different wavelengths. Only photons at the wavelengths corresponding to specific excitation states will be absorbed in each case. Thus, photons of wavelength 5890 Å can excite the sodium atom, while those of wavelength 2852 Å can excite magnesium. Each element therefore requires photons of specific energy to produce excited atoms of that element.

If a continuum is used as a radiant energy source and is passed through an atomic vapor, such as sodium, the spectrum obtained will have narrow absorption lines corresponding to several of the easily excited states of sodium. The absorption lines will be very sharp, indicating that only the energy corresponding to the particular electronic transition involved is absorbed. Two important interrelated considerations result from these observations: (1) The best emission source for atomic absorption measurements is a spectral line of the same wavelength as the analyte element and

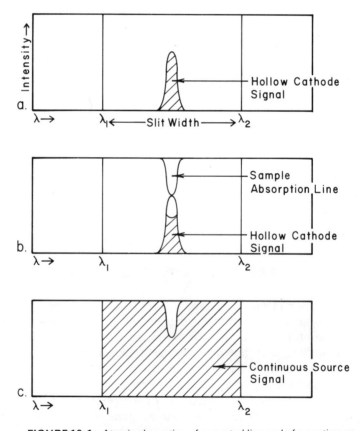

FIGURE 10-1. Atomic absorption of a spectral line and of a continuum.

ABSORPTION SPECTROSCOPY

(2) a continuum is a poor source for analytical atomic absorption spectroscopy. Reference to Figure 10-1 will help in understanding this. In Figure 10-1a, a spectral emission line from a hollow cathode source is shown centered in the band pass of the spectrometer slit. In Figure 10-1b, the same emission signal from the hollow cathode is shown with an absorption line of the same element at the same wavelength. The intensity of the emission line signal is reduced markedly, in this case by approximately 50%. In Figure 10-1c a continuous source signal is shown, filling the band pass of the spectral slit width, minus the energy absorbed due to the absorption line. The absorption in this case represents a very small fraction (perhaps 5%) of the total energy passing through the monochromator. It is obvious that the situation shown in Figure 10-1b is more efficient in detecting the effect of the absorption. Thus, a sharp line emission source is best for atomic absorption spectroscopy if no other spectral lines or spectral background emission enters the band pass of the spectrometer.

The magnitude of the atomic absorption signal is directly related to the number of ground state atoms in the optical path of the spectrometer. Ground state atoms are produced from the sample material, usually by evaporation of solvent and vaporization of the solid particle followed by decomposition of the molecular species into neutral atoms. Normally these steps are carried out using an aspirator and flame. These are the same processes that are involved in flame emission spectroscopy as described in Chapter 9. When ground state atoms are produced, some excited state atoms also occur and, for easily ionizable elements, some ions and electrons are produced.

The relative number of atoms in a particular energy state can be determined by use of the Boltzmann equation [refer to equation (2-23)]. Walsh has calculated these ratios for the lowest excited states of several typical elements and several flame temperatures. Table 9-2 indicates that the number of atoms in the ground state is much greater than the number in the lowest excited state at temperatures commonly used in atomic absorption spectroscopy.

Since atomic absorption spectroscopy utilizes the ground state atom population for its measurements, it would appear that atomic absorption has a great advantage over flame emission in terms of detection limits and sensitivities of detection. An inspection of Appendix VIII, where detection limits are given for a number of elements for flame emission and atomic absorption, indicates this is not true. The reason for this apparent discrepancy lies in the relative stabilities of ground state and excited state atoms. An excited atom has a lifetime of the order of 10^{-9}–10^{-10} sec, and thus emits its energy very quickly after being excited. The usual flame emission source has an upward velocity of from 1 to 10 m/sec, so the excited atom will move only about 10^{-8}–10^{-9} m between the time of excitation and emission.

The monochromator will collect energy of emission from almost all excitations that occur in the optical path of the instrument.

In contrast to flame emission, atomic absorption relies on ground state atoms being excited by a photon of proper energy. Ground state atoms are stable and can remain in that condition almost indefinitely. If the energy emitted by the source is to decrease, it can do so only by interaction with a ground state atom to produce an excited atom. Since this process is inefficient, only a small percentage of the atoms in the optical path produce excited atoms, thus producing a beam of energy emerging from the atomic cloud having an intensity less than the entering beam.

Another important difference between flame emission and atomic absorption is the difference in their "signals." The analytical signal in flame emission is the sum of all energies emitted as excited atoms drop to the ground state. The signal comes entirely from the emitting atoms. In atomic absorption the "signal" is the *difference* between the intensity of the source in the absence of analyte atoms and the decreased intensity obtained when analyte atoms are present in the optical path. This basic difference in the two techniques also contributes to the fact that detection limits of many elements are quite similar in emission and absorption spectrometry.

In terms of detection limits atomic absorption and flame emission are quite similar; some elements have lower detection limits by flame emission and others by atomic absorption. For comparison purposes reference can be made to Appendix VIII. Table 10-1 lists elements in three categories: (1) those more sensitive by flame emission, (2) those of about equal sensitivity by emission and absorption, and (3) those more sensitive by atomic absorption.

TABLE 10-1
Comparison of Sensitive Elements by Flame Emission and Atomic Absorption[a]

More sensitive by flame emission			About equally sensitive		More sensitive by atomic absorption		
Al	Li	Sn	Cr	Pd	Ag	Hg	Sn
Ba	Lu	Tb	Cu	Rh	As	Ir	Te
Ca	Na	Tl	Dy	Sc	Au	Mg	An
Eu	Nd	Tm	Er	Ta	B	Ni	
Ga	Pr	W	Gd	Ti	Be	Pb	
Ho	Rb	Yb	Ge	V	Bi	Pt	
In	Rd		Mn	Y	Cd	Sb	
K	Ru		Mo	Zr	Co	Se	
La	Sm		Nb		Fe	Si	

[a] From E. E. Pickett and S. R. Koirtyohann, Emission Flame Photometry—A New Look at an Old Method, *Anal. Chem.*, **41**, 28A (1969). Used by permission of the American Chemical Society.

ABSORPTION SPECTROSCOPY

Although atomic absorption is not entirely analogous to absorption of radiant energy by, for example, a colored solution, the Beer–Lambert relation is valid for atomic absorption processes. The Beer–Lambert law relating absorption of radiant energy to path length and concentration is

$$P_\lambda = P_{0\lambda} e^{-k_\lambda l c} \tag{10-1}$$

where $P_{0\lambda}$ is the power of the source at wavelength λ, P_λ is the power of radiation at wavelength λ after passage through the sample, k_λ is the absorption coefficient, l is the path length through the sample, and c is the concentration of the atomic vapor.

Equation (10-1) may be rewritten as

$$\log \frac{P_{0\lambda}}{P_\lambda} = alc \tag{10-2}$$

where a replaces the constant k_λ, and includes the transformation from natural to common logarithms; it is called the absorptivity.

Since absorbance is defined as $A = \log(P_{0\lambda}/P_\lambda)$, then

$$A = alc \tag{10-3}$$

Equation (10-3) suggests that the best way to plot atomic absorption calibration curves is to use absorbance vs. concentration since l, the path length, and a, the absorptivity, are constant for a given system. Such a plot of data should, therefore, result in a straight line passing through the origin in the absence of interference effects.

A common procedure to determine absorbance is to obtain data in terms of percentage transmission. If the transmission of radiant energy through the blank is adjusted to 100% transmission, the signal, after passing through to sample, will produce a reading of less than 100%. If these conditions are met, the absorbance A is

$$A = \log \frac{100}{\%T}$$

or

$$A = 2 - \log \%T \tag{10-4}$$

where $\%T$ is the percentage transmission through the sample. Zero percent transmission should be established when no radiant energy enters the monochromator slit.

2. INSTRUMENTATION REQUIREMENTS

Figure 10-2 is a block diagram of the instrumentation necessary for atomic absorption measurements. There are several basic variations to this instrumental arrangement. The components used in atomic absorption

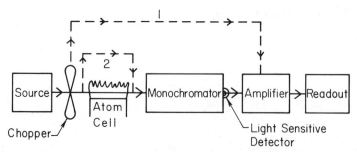

FIGURE 10-2. Block diagram of atomic absorption instrumentation.

include a source, which is usually a hollow cathode, but may be a microwave-excited discharge lamp or an Osram or Philips lamp; the sample cell, which is usually a flame, but also can be a furnace, carbon rod, or tantalum strip; a monochromator; a light-sensitive detector; an amplifier, which may be a dc or an ac amplifier depending on other components used; and an appropriate read-out device.

The simplest arrangement to obtain atomic absorption data dispenses with the chopper, uses a hollow cathode powered by a dc source, and a dc amplifier. Such a unit will receive a dc signal from the flame cell due to spectral emission lines. Accurate readings of the absorption signal thus depend on obtaining a steady emission state in the flame as well as a steady absorption state.

Use of a chopper and an amplifier sensitive only to the frequency of the chopped signal from the hollow cathode will eliminate dc signals and random noise produced in the sample cell. Two types of ac amplifiers are used, one is called a "broadband" amplifier, which responds to a chopped signal over a wide frequency range, and the other a "locked-in" amplifier, which responds only to a narrow frequency range and thus must be "locked in" to the frequency produced by the chopper.

Still another variation of instrumentation is the double-beam system, where light from the source is split with a beam splitter. Part of the energy passes through the sample and part bypasses the sample as shown at 2 of Figure 10-2. A ratio of the two signals is then obtained electronically and recorded. Such a system reduces the effects of source variation to a minimum.

3. RADIATION SOURCES

The source unit for atomic absorption should emit stable, intense radiation of the analyte element, usually of a resonance line of the element. Preferably the radiation should be a sharp line, with a width no greater than the width of the absorption line. There should be no general background or other extraneous lines emitting within the band pass of the monochromator.

3.1. Hollow Cathode Lamps

In the original paper by Walsh describing the atomic absorption process he advocated use of hollow cathode sources since this source seemed to meet most of the criteria for a good source as described above. Such lamps had been developed prior to the advent of atomic absorption spectroscopy and were known to produce sharp, intense line spectra.

Figure 10-3 shows the basic features of a hollow cathode lamp source. Here A is the anode (the plus electrode) and C is the cathode, terminated in the lamp as a hollow cup. The anode can be a wire, such as tungsten, and the cathode cup may be constructed from the element whose spectrum is desired or it may be an inert material into which the desired element or a salt of the desired element is placed. The lamp envelope is made of glass and W is a window of suitable properties. If an ultraviolet line spectrum is desired, the window may be quartz or a high silica glass. The hollow cathode has an inert gas present, usually neon or argon, at low pressure.

When a dc voltage of 300–500 V is applied between the anode and the cathode the inert fill-gas is ionized, and a current, usually 5–30 mA, flows through the tube. The current is carried by electrons and positive ions. The positive ions include fill-gas ions and metal ions from the hollow cathode. Collisions of high-velocity, positive fill-gas ions with metal atoms of the hollow cathode dislodge some of the metal atoms from the surface of the hollow cathode. The low fill-gas pressure and the low vapor pressure of the metal atoms results in a minimum of pressure broadening, and, with low lamp currents resulting in a low temperature, Doppler broadening also is minimized. The rate of vaporization (sputtering) of the cathode material depends on the mass of the carrier gas ions, the potential difference between the cathode and anode, the distance between the cathode and the anode, and the nature of the cathode material.

Further collisions of the vaporized metal atoms from the cathode with energetic fill-gas ions result in excited metal atoms. Some metal ions and excited metal ions also may be produced. The emission spectrum produced is therefore a sharp line spectrum of the cathode material and the fill-gas.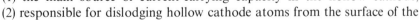

The fill-gas fulfills three functions in the hollow cathode lamp. It is (1) the main source of current-carrying capacity in the hollow cathode, (2) responsible for dislodging hollow cathode atoms from the surface of the

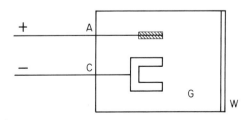

FIGURE 10-3. Basic elements of a hollow cathode lamp.

cathode, and (3) primarily responsible for excitation of the ground state metal atoms.

The fill-gas must possess certain properties to meet the above requirements. It should (1) be chemically inert, (2) have a high ionization potential, (3) have a large ionic radius, and (4) produce a simple spectrum, thus minimizing spectral interference possibilities with the hollow cathode metal. No one gas is optimum in all these requirements. The two most-used gases are neon and argon. Atomic lines are generally more intense with neon and it has a simpler spectrum than argon. Argon may be used in cases where neon lines would cause spectral interference with the hollow cathode metal.

Fill-gas pressures are important where the best range is from 1 to 5 Torr. At higher pressures the discharge tends to be unstable and at lower pressures vaporization of the hollow cathode metal increases, as does the operating temperature.

Each hollow cathode combination of physical size, fill-gas, fill-gas pressure, and hollow cathode metal has an optimum current. Figure 10-4 illustrates this. The absorbance reaches a maximum and then decreases as the current increases. The optimum hollow cathode current for one element is not optimum for another. Furthermore, although the absorbance increases and then decreases as the cathode current increases, the optimum current produces a relatively narrow peak for copper and a broad plateau for

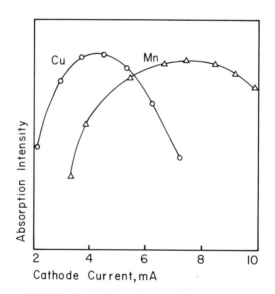

FIGURE 10-4. Effect of hollow cathode lamp current on the absorbance of a spectral line. [From a Kansas State University Ph.D. dissertation by Donald E. Smith.]

ABSORPTION SPECTROSCOPY

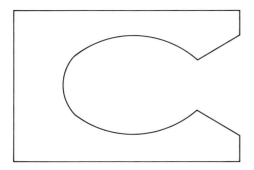

FIGURE 10-5. Design of a hollow cathode.

manganese. Increasing the hollow cathode current broadens the spectral line and ultimately produces a self-absorbed line, thus reducing its effectiveness as a sharp line source for atomic absorption.

The dimensions and shape of the hollow cathode can affect lamp performance. The most common shape is cylindrical; however, with a shape such as shown in Figure 10-5 it was found that line reversal and cathode erosion were minimized. The interior of the hollow cathode erodes with usage, and is one factor ultimately leading to decreased emission intensity and lamp failure.

3.1.1. High-Intensity Lamps

It is evident that high-intensity hollow cathode lamps are desirable, provided narrow spectral linewidths can be maintained. Sullivan and Walsh succeeded in designing such a lamp, as shown in Figure 10-6. The lamp is provided with additional electrodes as shown, where a glow discharge occurs. Separate power is supplied the additional electrodes, which are coated with an electron-producing substance. This produces a region of high electron density and results in a system that can operate at low voltage and high current. Very intense resonance lines are thus produced. Intensities as much as 100 times greater than those produced by conventional hollow cathodes may be obtained.

3.1.2. Multiple-Element Lamps

If a laboratory requires equipment to determine a large number of elements by atomic absorption, the expense involved in providing hollow cathode lamps can be substantial. As a result, attempts to construct multiple-element hollow cathode lamps have been made, with some success. One method is to construct the hollow cathode of a metal alloy; thus spectra of all the elements in the alloy would be produced. Sintered hollow cathodes

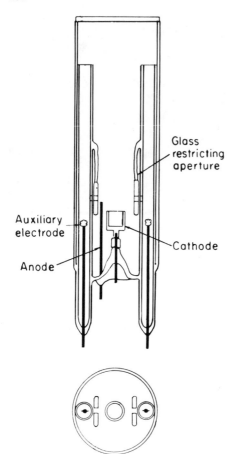

FIGURE 10-6. High-intensity hollow cathode lamp. [From J. V. Sullivan and A. Walsh, High Intensity Hollow Cathode Lamps, *Spectrochim. Acta*, **21**, 721 (1965). Used by permission of Pergamon Press.]

also have been partially successful. These cathodes are prepared by mixing the metals in powder form and heating until an aggregate is formed that can be machined into a hollow cathode.

Difficulties occur with multiple-element hollow cathodes and only certain combinations of elements have been successful. The principal source of difficulty is due to different volatilities of elements. The more easily volatilized element is preferentially vaporized and, on cooling, deposits over the entire hollow cathode surface. The intensities of emission of other metals in the hollow cathode thus are seriously reduced.

Another approach is to introduce two or more hollow cathodes, each of a single element, into one envelope. This technique also is subject to the problem of selective volatilization but at a lower rate than the single hollow cathode design.

ABSORPTION SPECTROSCOPY

The intensity of emission from multiple-element hollow cathodes is generally lower than single-element tubes, although several successful combinations are possible if proper choice of elements is made, taking into account volatilization rates and noninterfering emission spectra. For example, a multiple-element combination of magnesium, calcium, and aluminum is available, as is a combination of silver, lead, and zinc.

3.1.3. Demountable Lamps

Another approach to the problem of hollow cathode lamp sources is the use of a demountable unit. Such a lamp can be used for a variety of elements simply by changing the hollow cathode, reassembling, evacuating, and introducing the fill-gas at a predetermined pressure. Intensities and stabilities comparable with individually sealed lamps can be obtained. These lamps have the handicap of the inconvenience involved, the need for a good vacuum pump, fill-gas, and a method of introducing the fill-gas to the demountable hollow cathode at a prescribed pressure. They have the advantage of being especially useful for research purposes and are inexpensive if many different metal element determinations are required. A commercially produced demountable hollow cathode lamp is available from Barnes Engineering together with a large number of replaceable hollow cathodes.

3.2. Gaseous Discharge Lamps

Gaseous discharge lamps which contain internal electrodes also can serve as sources for atomic absorption. They are variously called arc lamps, spectral lamps, vapor lamps, and by the name of the manufacturer, such as Osram lamps and Philips lamps. Gaseous discharge lamps contain an inert gas at low pressure and a metal or metal salt. They are especially suited to metals of relatively high vapor pressure, such as the alkali metals and some other metals such as mercury, cadmium, and lead.

Excitation occurs by ionization of the inert gas followed by collisions with the metal atoms. The ionization energy is supplied by internal electrodes and the voltage may be from a low-voltage source, in the range of 150–200 V, or a high voltage of 1000 V or more, depending on the kind of electrodes and their spacing. Gaseous discharge lamps produce very intense spectral lines of the excited metal atoms. Relatively few lines of excited metal ions occur.

Spectral lines produced by gaseous discharge lamps are subject to line reversal and line broadening. Reduction of the operating voltage results in some improvement; however, lowering the voltage must be carefully controlled to avoid instability of the discharge. The radiation intensity also may change with temperature and the discharge is less stable than with hollow cathode lamps.

Gaseous discharge lamps have been useful as sources, particularly for alkali metals, because hollow cathode sources for these elements are troublesome. Recent improvements in alkali metal hollow cathode sources have led to their increasing use.

3.3. Electrodeless Discharge Lamps

Electrodeless discharge lamps (EDL), powered with energy in the radiofrequency range, were used as early as 1928 by Jackson, and in 1948 Meggers used them to determine hyperfine structure of atomic spectra. These lamps produce narrow-line, high-intensity spectra with little self-absorption. They would appear, therefore, to be promising sources for atomic absorption.

The EDL are usually constructed of quartz or glass, depending on the spectral region desired, and are 3–5 cm long and about 1 cm in diameter. A metal or metal salt is placed in the lamp, along with an inert gas at low pressure. When the lamp is placed in a radiofrequency field, excitation of the metal occurs.

The amount of the metal or metal salt added to the tube does not appear to be critical. Metal halides, particularly the iodide, have been shown to be very useful for metals of very low vapor pressure. Metals of high vapor pressure can be placed in the tube in metal form.

The intensity of the spectrum of an EDL is dependent, in part, on the pressure of the inert fill-gas. Neon and argon are the commonly used fill-gases. The optimum pressure for maximum spectral intensity depends on the metal element involved and is determined experimentally in each case. If fill-gas pressure is too low, spectral emission may become erratic and tube life is decreased.

Lamp dimensions also should be optimized. If the lamp is of too large a diameter, some spectral line reversal can occur; if the diameter is too small, the lifetime of the lamp is decreased. If the length of the lamp is too large, temperature gradients can occur within the lamp that affect the lamps adversely. In fact, temperature gradients may be sufficiently large to vaporize metal in one region of the lamp and recondense the metal on the lamp wall in a cooler region. Lamps of a diameter of 8–12 mm and a length of 2–5 cm have been found to perform best.

Lamp intensity also is a function of temperature, where the optimum temperature will be different for different substances. Variations of lamp temperature will cause fluctuations in the emission intensities of the lamps. Figure 10-7 illustrates (curve B) the stability of a zinc EDL.

Several methods have been devised to control EDL temperature. One technique is to surround the EDL with a glass or quartz cylinder open at both ends. This system will protect the lamp from temperature variations

ABSORPTION SPECTROSCOPY

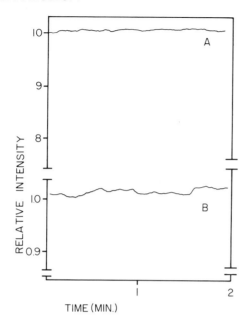

FIGURE 10-7. Stability and intensity of a zinc electrodeless discharge lamp. [From W. G. Schrenk, S. E. Valente, and H. R. Bohman, Microwave Discharge Tubes for Atomic Absorption Spectroscopy, *Trans. Kan. Acad. Sci.*, **74**, 249 (1971). Used by permission of the Kansas Academy of Science.]

caused by movement of air around the lamp. Another technique is to surround the EDL with a vacuum jacket as shown in Figure 10-8. The vacuum jacket effectively prevents lamp intensity fluctuations due to ambient temperature variations. It also permits the temperature within the EDL to increase. In the case of a zinc EDL, the rise in temperature increases the intensity of the emission spectrum of zinc. This is illustrated by curve A of Figure 10-7, which also shows a significant increase in stability as compared with an unjacketed lamp.

Winefordner has developed a thermostatically controlled device to regulate the temperature at which the EDL operates, as shown in Figure 10-9. Heated air is blown around the EDL, thereby controlling its operating temperature. Each element exhibits its own optimum operating temperature. For example, maximum spectral emission intensity for mercury occurred at 50°C, for cadmium at 250°C, and for iron at 380°C.

Coupling RF Energy to the Lamp. Radiofrequency energy, in the range of 10–3000 MHz, can be used to excite an electrodeless discharge lamp. The medical diathermy frequency of 2450 MHz is commonly used because

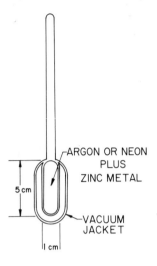

FIGURE 10-8. Vacuum-jacketed zinc electrodeless discharge lamp. [From W. G. Schrenk, S. E. Valente, and H. R. Bohman, Microwave Discharge Tubes for Atomic Absorption Spectroscopy, *Trans. Kan. Acad. Sci.*, **74**, 249 (1971). Used by permission of the Kansas Academy of Science.]

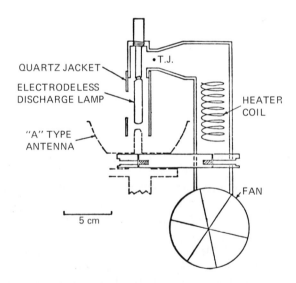

FIGURE 10-9. Diagram of a device to control the operating temperature of an electrodeless discharge lamp. T. J., thermocouple junction position. All-brass construction. [From R. F. Browner, B. M. Patel, T. H. Glenn, M. E. Rietta, and J. D. Winefordner, A Device for Precise Temperature Control of Electrodeless Discharge Lamps, *Spectroscopy Letters*, **5** (9), 311 (1972). Used by permission of *Spectroscopy Letters*.]

ABSORPTION SPECTROSCOPY

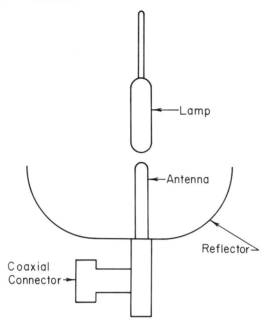

FIGURE 10-10. Type A antenna coupled to an electrodeless discharge lamp.

of the availability of these units. Some commercial units constructed expressly for gaseous excitation and using the 2450 MHz frequency also are available. Power output up to 100 W is sufficient for exciting EDL's. A magnetron usually is used in the frequency-generating circuit.

The output of the microwave generator is coupled to an antenna or microwave cavity via a coaxial cable. A common type of antenna, known as a Type A, is shown in Figure 10-10. The EDL is mounted axially immediately above the antenna. The microwave power output usually can be varied at the generating unit. The generator also should include a reflected power meter. Power reflected from the load (the EDL) back to the generator can damage the magnetron. When the antenna is tuned to resonance, minimum power will be reflected back to the generator.

Microwave cavities also are useful devices to couple RF energy to EDL's. One of the most popular is the Evenson one-fourth wave cavity pictured in Figure 10-11. The cavity is tunable to resonance and thus can provide a system that can be adjusted to produce minimum reflected power. As the EDL heats, however, retuning of the cavity may become necessary. Some three-fourths wave cavities have been used and are available. The three-fourths wave cavities seem to offer no substantial advantage over the one-fourth wave cavities.

Electrodeless discharge lamps have been successfully prepared for many

FIGURE 10-11. One-fourth wave Evenson Cavity for use at 2450 MHz. [Courtesy Opthos Instrument Company.]

elements. They produce intense, sharp line spectra. Their stability is not as good as that of a hollow cathode and they require the use of an RF generator as an energy source. In many cases they provide lower detection limits than hollow cathode sources.

3.4. Flame Emission Sources

Use has been made of flame emission spectra as sources for determination of lithium, copper, and some of the rare earth elements. The sensitivity is lower than for a hollow cathode source and the stability also is less.

3.5. Continuous Sources

Walsh investigated the possibility of using a continuum as a source for atomic absorption spectroscopy and came to the conclusion that it was not a desirable source. He noted that a resolution of over 500,000 would be required and that the intensity of a very narrow segment of a continuum would be extremely low. He therefore advocated the use of the hollow cathode source. Reference to Figure 10-1 will serve as a reminder concerning the problem of measuring the decrease in signal intensity caused by line absorption from a continuum.

In spite of the apparent problems associated with the use of a continuum as an atomic absorption source, some useful absorbance data have been obtained. Sensitivities have generally been considerably poorer than with hollow cathodes by a factor of about ten.

Several types of continuous source have been used, depending on the spectral region under consideration. Tungsten filaments can be used in the

ABSORPTION SPECTROSCOPY

infrared and visible spectral regions. Quartz halide lamps are good to about 3000 Å. Hydrogen and xenon lamps can serve for ultraviolet radiation.

The continuous source is quite useful for certain purposes if it is intense and a monochromator of high resolution is available. Photographic recording of absorption spectra can be made in the same manner as arc or spark emission spectra are recorded. In this manner atomic absorption spectra are readily available for the study of a number of spectral absorption lines, in contrast to the single-line absorption usually obtained with a hollow cathode source.

4. PRODUCTION OF THE ATOMIC VAPOR

The function of the sample cell is to convert the sample into ground state atoms in the optical path of the atomic absorption system. The ground state atoms must be obtained from the sample in a reproducible manner if quantitative data are to be obtained. The sample is usually in liquid form, although recent developments with carbon rods and tantalum boats have made possible the use of solid samples.

The steps involved in producing an atomic vapor from a liquid sample as outlined in Chapter 9 for flame emission are the same for atomic absorption.

4.1. Nebulization of the Sample

The formation of small droplets from the liquid sample is called nebulization and is the first step in the production of an atomic vapor. A common method of nebulization is by use of a gas moving at high velocity, called pneumatic nebulization. The Beckman total-consumption burner illustrated in Figure 9-4 and the Jarrell-Ash burner illustrated in Figure 9-5 have been commonly used for emission and atomic absorption measurements. In these burners, a back pressure as high as 250 Torr can occur at the tip of the burner due to the high velocity of the aspirating gas as it emerges from the orifice. The liquid sample is drawn up the capillary and broken into droplets by the high velocity gas stream. The formation of the liquid droplets requires energy to overcome the forces of surface tension and viscosity; thus it is important in this type of system to maintain gas velocities constant and to have samples and standards of the same viscosity and surface tension. In burners of this type (i.e., Beckman total-consumption and Jarrell-Ash Triflame) the nebulization occurs as the sample enters the flame of the burner.

Other burners separate the nebulization process and the flame by producing small liquid droplets in a nebulizing chamber before the sample enters the flame. Figure 10-12 shows this process in the Perkin-Elmer unit.

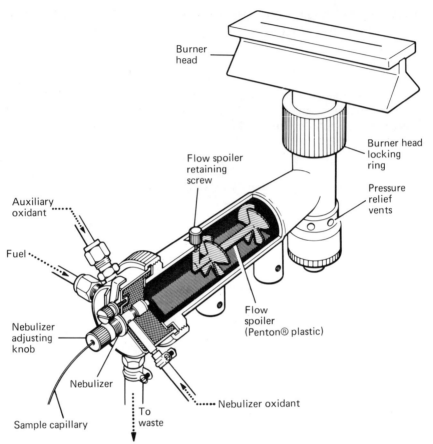

FIGURE 10-12. Perkin-Elmer atomic absorption sample cell showing nebulizing chamber and burner. [Courtesy the Perkin-Elmer Corp.]

The sample is nebulized pneumatically at the orifice. It is then mixed with fuel and oxidant. "Spoilers" render the fuel–air–sample droplet mixture uniform before the mixture enters the burner. The large droplets fall to the bottom of the mixing chamber and exit through the drain. This method produces a more uniform droplet size to enter the flame than does the total-consumption type of burner. Almost all commercial nebulizers for liquid samples use modifications of the two systems described, i.e., total consumption as illustrated with the Beckman burner (Figure 9-4), or the separate nebulizing-gas mixing chamber illustrated in Figure 10-12.

The Jarrell-Ash Tri-flame burner provides a system that can be used with either technique. When the Hetco total-consumption unit (Figure 9-5) is attached to the mixing chamber and burner head (see Figure 9-6), it serves

ABSORPTION SPECTROSCOPY

as a nebulizing unit. A glass "diverter" and spoiler reduce the droplets to a uniform mist and the larger droplets are removed via the vent.

4.1.1. Ultrasonic Nebulization

Ultrasonic wave generators have been proposed as nebulization units. In this type of system, shown in Figure 10-13, an ultrasonic transducer serves to generate high-frequency standing waves in the liquid sample. Small droplets are formed on the surface of the liquid and are carried into a mixing chamber prior to entering the flame. Droplet size is a function of the ultrasonic frequency and range from 2.0 μm at 3 MHz to about 20 μm at 100 kHz. Thus a frequency of about 500 kHz should be used if droplets are to be in

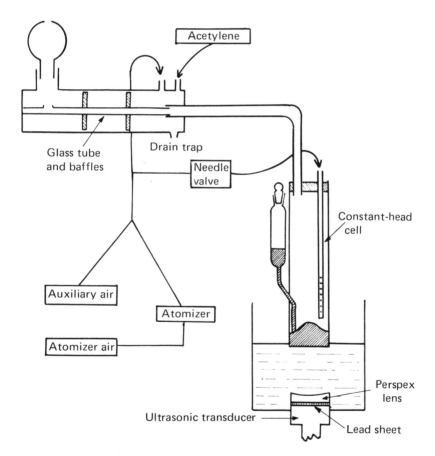

FIGURE 10-13. Ultrasonic nebulizing unit. [From H. C. Hoare, R. A. Mostyn, and B. T. N. Newland, An Ultrasonic Atomizer Applied to Atomic Absorption Spectrophotometry, *Anal. Chim. Acta,* **40**, 181 (1968). Used by permission of Elsevier Press.]

the same size range as those produced by pneumatic nebulization. Ultrasonic nebulization is inconvenient because of the need for the sonic generator, the complicated construction of the nebulizing unit, and the problem of contamination from sample to sample.

4.2. Flame Systems

Two types of flames are in general use to convert the liquid sample droplets into the atomic vapor required for atomic absorption measurements. In the *total-consumption burner* the fuel and oxidant gases are mixed in the flame. The result is a diffusion flame with considerable turbulence and associated noise. Such burners also are referred to as direct injection burners since the liquid droplets are formed at the entry point into the flame. This type of burner cannot cause "flashback" as can occur with the premixed burners. Figures 9-4 and 9-5 are typical of this type of unit.

The total-consumption burner can be adjusted to produce a fuel-rich or oxidant-rich environment very easily since danger of an explosion is very low. All the nebulized sample enters the flame; however, droplet size is quite variable and some droplets pass through the flame without complete evaporation and dissociation into atoms. The flame, because of its turbulence and high velocity, will entrain air surrounding the flame, which may react with sample elements and other constituents in the flame. Some use of sheathed turbulent flow burners has been made. The flame is surrounded by a sheath of inert gas to prevent entrainment of air into the flame. Such flames are said to provide greater flame stability and higher flame temperatures than unsheathed flames.

The *slot-type burner* is commonly used with a separate nebulizer and a mixing chamber. The mixture of nebulized sample, fuel, and oxidant enters a slot from which the gases exit to form the flame. These burners are often referred to as laminar flow burners. The burning proceeds quietly and in a laminar fashion. The optical path lies immediately above the slot and therefore advantage is taken of a long path length through the flame and the atomic vapor. The flame is more nearly homogeneous than with a turbulent flow system. Figure 10-14 is a photograph of the Perkin-Elmer slot burner.

The slot-type premix burner is subject to "flashback." The mixed gases must emerge through the slot at a velocity exceeding the "flashback" velocity of the fuel–oxidant mixture. If this does not occur, an explosion will be produced in the mixing chamber. If manufacturer's instructions on operating the burners are followed, the chances of an explosion occurring are minimized. The slot of the burner is 0.4–0.6 mm wide and 5–10 cm long. If the burner is operated with a concentrated sample, buildup of solid may occur along the edges of the slot. For good results this substance must be removed from the burner slot or flame instability will occur.

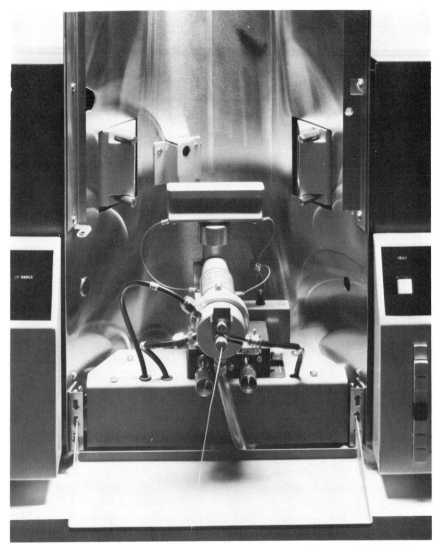

FIGURE 10-14. Perkin-Elmer premixed laminar flow burner. [Courtesy the Perkin-Elmer Corp.]

The Boling three-slot burner head, shown in Figure 10-15, offers several advantages over the conventional single-slot burner. It is designed to minimize the effect of entrained air since the two outside slots protect the center slot from the atmosphere. Analytical calibration curves are more nearly linear and flame noise is decreased. The Boling burner is especially useful for samples requiring a reducing (fuel-rich) flame.

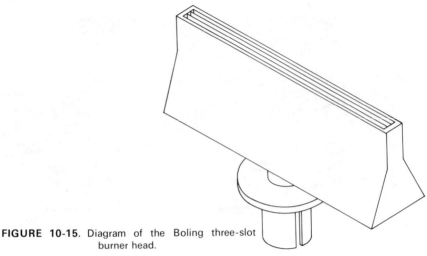

FIGURE 10-15. Diagram of the Boling three-slot burner head.

The Wang burner head can be obtained to fit most of the popular slot burners currently used in atomic absorption spectroscopy. This burner head, shown in Figure 10-16, has a slot with a row of small openings on each side of the slot. Metal fins are mounted along the burner to remove excess heat from the burner head. Flashback difficulty is reduced; however, the burner head ends are plug-in devices to prevent damage to the burner if flashback does occur. The unit can be used with both oxidizing and reducing flames.

The holes on each size of the slot form a shield for the slot flame, thus providing an operational condition similar to the three-slot burner. The optical path is through the central part of the flame formed by the slot. Increased signal stability is attained and some increase in sensitivity occurs. Figure 10-17 shows the calibration curves obtained for copper with the Wang burner head and with a single-slot burner on the same burner unit. Fuel and oxidant flow rates were maintained constant in both cases. The increased slope of the calibration curve is typical of the response of the Wang burner head when compared with a single-slot burner of the same dimensions operating under similar conditions. Flame noise also is decreased with the Wang burner head when compared with a standard single-slot burner.

5. FUELS AND OXIDANTS

5.1. Atomic Distribution in Flames

The neutral atom distribution is quite variable in the flames of both premixed and total-consumption burners. Therefore the maximum absorption signal from a flame cell will only be obtained if the optical path traverses

ABSORPTION SPECTROSCOPY

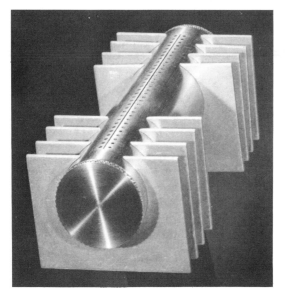

FIGURE 10-16. Wang laminar flow burner head. [Courtesy Mid-Continent Scientific, Inc.]

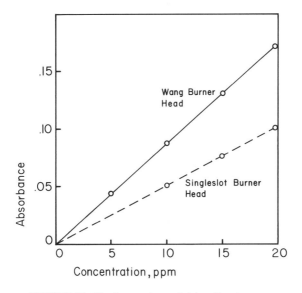

FIGURE 10-17. Comparison of the calibration curves of copper using the Wang burner head and a standard single-slot burner.

the flame through the region of maximum neutral atom population. The optimum path through the flame is different for different elements and also varies with the fuel–oxidant combination used as well as with the fuel-to-oxidant ratio. Figure 10-18 shows the atomic absorption intensity profiles of two elements, rhenium and zinc, when using a total-consumption turbulent-flow burner. The profiles are quite different and illustrate the need for careful optical alignment, especially in the case of rhenium. The optical alignment is not nearly so critical for zinc.

A large number of fuel–oxidant combinations have been used for atomic absorption spectroscopy, including air–propane, air–acetylene, air–hydrogen, oxygen–propane, oxygen–acetylene, oxygen–hydrogen, and nitrous oxide–acetylene. Many variations in the ratio of fuel to oxidant also have been used.

Flame temperatures and fuel-to-oxidant ratios are two of the most important parameters to consider when using a flame as an atomic absorption sample cell. Table 10-2 lists maximum temperatures attainable with various combinations of fuel and oxidant.

Early work in atomic absorption spectroscopy used the lower temperature flames; thus the method was restricted to those elements that could be converted to atoms at lower temperatures. Since some metals, such as molybdenum, rhenium, and tin, are only partly converted to gaseous atoms in low temperature flames, higher temperature flames were developed with success.

The high-temperature flames, however, did not solve all the problems of atomic absorption spectroscopy for many elements. If excess oxygen is present in a flame, a significant fraction of certain metal elements is converted to oxides. If the oxide is particularly stable, it is not redissociated into atoms. Thus the refractory oxides, such as MgO, CaO, AlO, MoO, and others, resist decomposition even in high temperature flames. To inhibit formation of these oxides, it is possible to use a fuel-rich flame to produce a "reducing"

TABLE 10-2
Maximum Temperatures of Various Fuel–Oxidant Combinations

Fuel	Oxidant	Maximum temperature, °C
Propane	Air	1725
Hydrogen	Air	2025
Acetylene	Air	2300
Hydrogen	Oxygen	2650
Propane	Oxygen	2900
Acetylene	Oxygen	3050
Acetylene	Nitrous oxide	2950

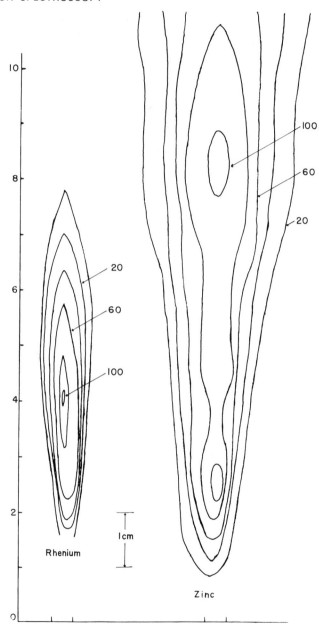

FIGURE 10-18. Atomic absorption flame profiles for rhenium and zinc. [From W. G. Schrenk, D. A. Lehman, and L. Neufield, Atomic Absorption Characteristics of Rhenium, *Appl. Spectrosc.*, **20**, 389 (1966). Used by permission of *Applied Spectroscopy*.]

environment in the flame. It is possible to satisfactorily determine many of the refractory oxide metals with such a flame.

Amos and Willis suggested another combination of fuel and oxidant, acetylene and nitrous oxide, as another approach to the analysis of refractory elements. The combination of acetylene and nitrous oxide produces a high-temperature flame (2950°C) with little free oxygen to react with the metal elements. This flame is now very successfully used in atomic absorption spectroscopy and permits satisfactory atomic absorption analysis for many refractory elements. Use of nitrous oxide and acetylene requires a burner head that will withstand the temperatures produced by the flame. A common burner head for this combination of fuel and oxidant is 5 cm long and 0.05 cm wide.

Flashback is more likely with an acetylene–oxygen mixture than with any of the other fuel–oxidant combinations in common use. Flame propagation in this system is at 1130 cm/sec. This imposes severe limitations on the use of acetylene and oxygen with premixed burners. The relatively low flame propagation rate of nitrous oxide and acetylene (180 cm/sec) is a definite advantage for the system.

6. NON-FLAME ABSORPTION CELLS

Non-flame atomization has been developed in recent years and has advantages over the flame under certain circumstances. Generally, non-flame methods have greater sensitivities and require smaller samples than flames. A number of non-flame techniques have been developed that can produce an atomic vapor from a liquid or solid sample.

6.1. Hollow Cathodes

Walsh suggested that a hollow cathode could serve as a sample cell as well as a source lamp, as shown in Figure 10-19. The sample is in the form of a hollow cylinder, which forms the cathode of the discharge system. The light beam is passed through the cylinder and the absorption is measured when an electrical discharge is passed through the tube. Walsh was able to determine successfully phosphorus and silver in copper, and silicon in aluminum and steel, using this method.

The atomic vapor is produced by an action similar to the action of a conventional hollow cathode source. Positive ions of the fill-gas strike the cathode, dislodging metal atoms into the optical path. The sensitivity is a function of the discharge current. Walsh used currents of up to 60 mA, but currents of 200–600 mA have been used by others. Higher currents cause some additional heating of the cathode and some additional atomic vapor is produced by the heating process.

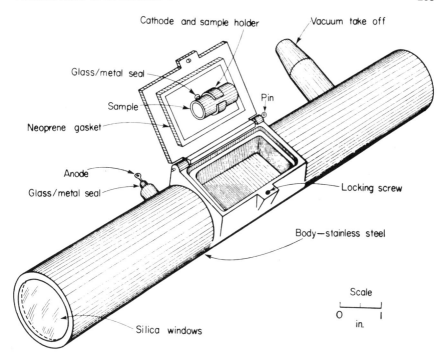

FIGURE 10-19. Hollow cathode sputtering chamber. [From B. M. Gatehouse and A. Walsh, Analysis of Metal Samples by Atomic Absorption Spectroscopy, *Spectrochim. Acta*, **16**, 602 (1960). Used by permission of Pergamon Press.]

6.2. L'vov Furnace

L'vov developed a long-path graphite furnace for the production of an atomic vapor. His device is shown in Figure 10-20. The graphite tube is heated with electrical energy. The sample is dried on the end of the vertical electrode. The electrode is raised to the graphite tube, and is heated with an electric arc between the electrode and the tube. The graphite tube is either made of or coated with pyrolytic graphite to minimize diffusion of the atomic vapor into the graphite tube. Time required for the evaporation of the sample is about 0.1 sec. Sensitivities are claimed to be about two orders of magnitude better than with conventional flame methods.

6.3. Woodriff Furnace

The Woodriff furnace (Figure 10-21) uses an electrically heated graphite absorption tube about 6 mm in I.D. and 30 cm long. It is enclosed in a second graphite tube, which is surrounded by graphite felt insulation. The

270   CHAPTER 10

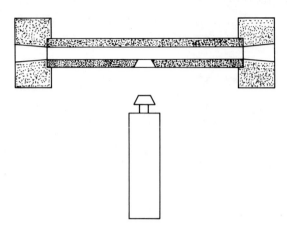

FIGURE 10-20. The L'vov graphite furnace cell.

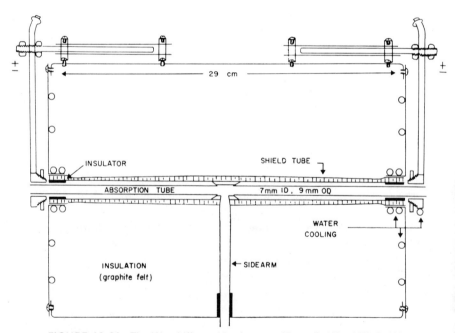

FIGURE 10-21. The Woodriff graphite furnace. [From R. Woodriff, R. W. Stone, and A. M. Held, Electrothermal Atomization for Atomic Absorption Analysis, *Appl. Spectrosc.*, **22**, 409 (1968). Used by permission of *Applied Spectroscopy*.]

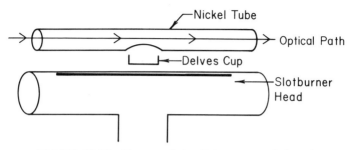

FIGURE 10-22. Diagram of the Delves cup technique for vaporizing solid samples.

current to the absorption tube can be adjusted to produce any desired temperature up to about 3000°C. The liquid sample is nebulized and carried into the absorption tube by an inert gas such as argon. A variation in the design divides the absorption tube into two parts supported at the center. In this unit a solid sample can be placed in a graphite cup on the end of a graphite rod and placed in contact with the center support. The sample vaporizes rapidly and enters the light beam passing through the absorption tube.

Water as a solvent causes rapid deterioration of the graphite absorption tube. Methanol is a more satisfactory sample solvent. Each element has its own optimum furnace temperature; for example, the best temperature for cadmium is 1000°C and for aluminum is 2140°C. The estimated detection limit for cadmium is 0.0018 μg/ml.

6.4. Delves Cup

Delves[1] reported on a sampling cup technique to determine lead in blood. The blood is placed in a nickel cup, dried, and oxidized. The sample then is volatilized, using an air–acetylene flame, into a nickel absorption tube mounted parallel to and just above the air–acetylene flame. The instrumental arrangement is shown in Figure 10-22. Delves has reported a sensitivity of 1×10^{-10} g lead at 1% absorption and a standard deviation of $\pm 4\%$ at the 3-ng level of lead. The sample size most commonly used was 50 μl. The nickel absorption tube was 100 mm long with a 12.5 mm inside diameter.

Application of the Delves cup technique to other metals in aqueous solution has been investigated by Kerber and Fernandez,[2] who found the method useful particularly with metal elements of relatively high volatility. Matrix effects were not studied sufficiently to draw any general conclusions about their effects. Table 10-3 presents absolute detection limits for a series of elements using the Delves cup technique to volatilize the samples.

[1] H. T. DELVES, *Analyst*, **95**, 431 (1970).
[2] J. D. KERBER and F. J. FERNANDEZ, *Atomic Absorption Newsletter*, **10**, 78 (1971).

TABLE 10-3
Absolute Detection Limits for a Series of Elements
Using the Delves Cup Technique[a]

Element	Detection limit, g	Element	Detection limit, g
Ag	0.1×10^{-9}	Pb	0.1×10^{-9}
As	20×10^{-9}	Sc	100×10^{-9}
Bi	2×10^{-9}	Te	30×10^{-9}
Cd	0.005×10^{-9}	Tl	1×10^{-9}
Hg	10×10^{-9}	Zn	0.005×10^{-9}

From J. D. Kerber and F. J. Fernandez, The Determination of Trace Metals in Aqueous Solution with a "Delves Sampling Cup" Technique, *Atomic Absorption Newsletter*, **10**, 78 (1971). Used by permission of the Perkin-Elmer Corporation.

6.5. Carbon Rod Analyzers

Several variations of electrically heated carbon rods have been used to convert a solid sample into an atomic vapor. One such system is shown in Figure 10-23. The carbon rod (Figure 10-23a) has a center section of reduced diameter, a 3-mm hole longitudinally through the rod, and a small hole for insertion of the liquid sample. The rod is reduced in diameter to concentrate the electrical heating around the sample. A small current is applied to the rod to evaporate the solvent prior to vaporization of the sample by a larger current.

The rod is mounted in the sample cell as shown in Figure 10-23b. It is necessary to exclude oxygen from the vicinity of the carbon rod to prevent its rapid deterioration. Windows for the optical path are silica, to allow use of spectral lines in the ultraviolet region. Air and sample solvent are swept out of the chamber by an inert gas. The carbon rod is heated by an electrical current, up to about 150 A, to produce a temperature of 2500°C. A timing circuit permits the current to flow for a controllable period of from 2 to 10 sec.

The carbon rod or carbon tube analyzers offer two distinct advantages over conventional flame methods: (1) The sample can be very small, from 1 to 2 μl, and (2) the absolute detection limits are of the order of 10^{-12} g for many elements. The carbon rod can be heated to as high as 3000°C, but at this temperature the life of the rod is limited to 20–40 determinations. Light scattering due to incandescent carbon particles also is serious at 3000°C, contributing to a high background.

One disadvantage of the carbon rod analyzer is the tendency of certain metals, such as barium, strontium, and tungsten, to form refractory carbides at high temperatures. If metal carbides are formed, the metal is removed from the sample vapor and a loss of sensitivity for the metal occurs.

ABSORPTION SPECTROSCOPY

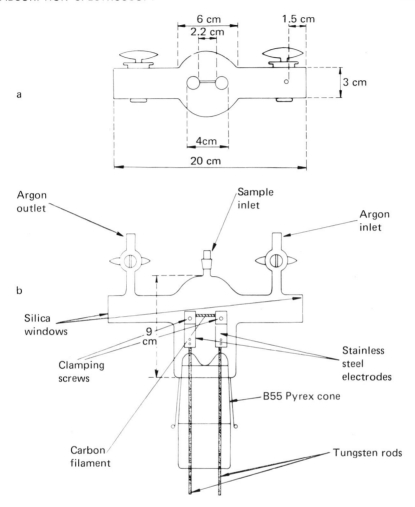

FIGURE 10-23. The carbon rod analyzer. [From T. S. West and X. K. Williams, Atomic Absorption and Fluorescence Spectroscopy with a Carbon Filament Atomic Reservoir, *Anal. Chim. Acta,* **45**, 26 (1969). Used by permission of Elsevier Press.]

6.6. Tantalum Boat Analyzer

Another technique to produce an atomic vapor from a solid sample is the use of a tantalum boat. A tantalum boat is electrically heated, in a manner similar to the carbon rod system, to produce an atomic vapor. The high melting point of tantalum (2996°C) and its ability to be shaped into a boat make it useful for this purpose. Figure 10-24 shows a tantalum boat and a sample cell that can be used for this purpose. The boat is shaped with a small

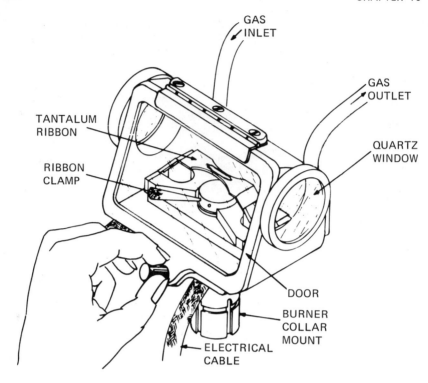

FIGURE 10-24. The tantalum boat analyzer. [From J. H. Hwang, P. A. Ullucci, and S. B. Smith, Jr., A Simple Flameless Atomizer, American Laboratory, August 1971. Used by permission of International Scientific Communications.]

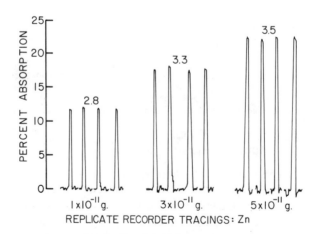

FIGURE 10-25. Replicate recorder tracings of zinc using the tantalum boat analyzer.

ABSORPTION SPECTROSCOPY

TABLE 10-4
Absolute Detection Limits Using a Tantalum Boat Atomizing System

Element	Wavelength, Å	Boat temperature, °C	Detection limit, g
Aluminum	3093	2000	3×10^{-10}
Calcium	4227	2200	1×10^{-12}
Copper	3248	2000	3×10^{-11}
Iron	2483	2400	1×10^{-10}
Magnesium	2852	1400	1×10^{-13}
Sodium	5890	1600	1×10^{-12}
Vanadium	3184	2300	4×10^{-10}
Zinc	2139	1200	1×10^{-12}

depression to serve as the sample holder. The width of the tantalum ribbon is decreased adjacent to the sample-holding depression to concentrate the heating effect in the region near the sample. An electrical system with a timer is needed and is used in a manner similar to its use with the carbon rod analyzer. Currents from 10 to 60 A are allowed to pass through the tantalum for a specified length of time, usually from $\frac{1}{2}$ to 5 sec. The time and current are optimized for each different element.

The analysis is performed in an inert atmosphere. A liquid sample of from 1 to 25 μl is placed in the boat and a small current (~ 5 A) is used to evaporate the solvent. The sample is then atomized by passage of a high current through the boat. The atomic vapor appears as a spike on a strip chart, developing and decaying quite rapidly. The optical path should be just above the surface of the tantalum ribbon for best results.

The tantalum boat atomization technique provides very low detection limits coupled with small sample size. The reproducibility of the sample size is apparently the limiting factor on precision of results. Figure 10-25 shows results obtained on replications of zinc samples. Relative standard deviations of between 2.8 and 3.4% were obtained for zinc concentrations between 1×10^{-11} and 5×10^{-11} g. Some typical absolute detection limits with the tantalum boat systems are given in Table 10-4. They range from 10^{-10} to 10^{-13} g.

6.7. Other Non-Flame Cells

The *Massman*[3] solid state sampling cell is a graphite cylinder about 50 mm long and 8 mm in diameter. Sample is introduced as a liquid through

---

[3] H. MASSMAN, *Spectrochim. Acta*, **23B**, 215 (1968). [Used by permission of Pergamon Press.]

a small hole in the center of the tube. The sample is dried and ashed (if necessary) electrically, and then is atomized by use of an electrical current of up to 400 A. The entire process is carried out in an atmosphere of argon. Sample size can vary from 1 to 100 μl and absolute detection limits are in the range of 10^{-12} g. Pye Unicam, Ltd., has a commercial version of the Massman cell as an accessory for their atomic absorption spectrometers. The Perkin-Elmer Corporation also has a commercial version of the Massman cell available. Their most recent unit is shown in Figure 10-26.

Varian-Techtron uses a so-called carbon rod atomizer, which consists of short carbon rod with a hole axially through it and a smaller, transverse hole through the center for introduction of the sample solution. The atomizer is clamped between two other carbon rods, which also serve as electrical connectors. Sample is introduced as a liquid through the small hole in the rod. The drying, ashing, and vaporizing processes are accomplished by electrical heating. The Varian unit is not enclosed but a flow of inert gas emerges from an opening below the carbon rod atomizer and effectively prevents the oxygen from the atmosphere from reacting with the hot carbon

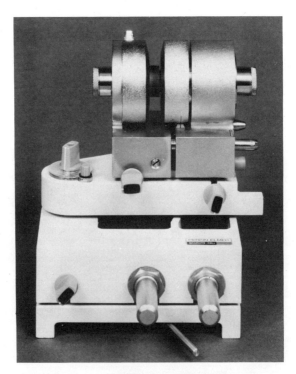

FIGURE 10-26. Graphite furnace atomizer. [Courtesy the Perkin-Elmer Corp.]

ABSORPTION SPECTROSCOPY

FIGURE 10-27. The carbon rod atomizer. [Courtesy Varian-Techtron, Ltd.]

atomizer. A photograph of the Varian Techtron unit is shown in Figure 10-27, and shows clearly the method of mounting the graphite rod cell between the two carbon supports. The optical path of the monochromator is directed through the axial hole in the carbon rod.

Instrumentation Laboratories has recently modified their tantalum boat unit to also accommodate a graphite cuvette. The cell is enclosed so atomization of the sample can occur in an inert atmosphere. A temperature-sensing device is included to permit accurate temperature measurements and temperature control.

The unit, with pyrolytic graphite cuvettes, can be programmed to provide an atomization temperature up to approximately 3500°C. The tantalum boats are limited to temperatures below the melting point of tantalum (2995°C). A photograph of the Instrumentation Laboratories flameless atomizer is shown in Figure 10-28.

Electrical energy is used almost exclusively for the drying, ashing, and vaporization steps in flameless atomization. In commercially available instrumentation these steps are programmed sequentially, with each step in the process being adjustable. Readout of the analytical signal can be either

FIGURE 10-28. Flameless atomizer for interchangeable graphite cell or tantalum boat vaporizer unit. [Courtesy Instrumentation Laboratories, Inc.]

peak height or peak area. Peak area is preferred in most cases because strip-chart recorders frequently do not respond to the rapid formation and decay of the atomic vapor produced during atomization of the sample. Woodriff[4] has recently reviewed the basic designs of atomization chambers for solid state sampling.

6.8. Special Systems

Several atomic absorption analytical problems are best handled by producing a gaseous sample external to the sampling cell. Elements frequently sampled in this manner include mercury, arsenic, selenium, and antimony.

Mercury has an appreciable vapor pressure at room temperature. It is possible, therefore, to place mercury vapor in a sample cell in the optical path of an atomic absorption instrument and obtain a usable absorption signal. The mercury must be separated from the sample material prior to introduction into the sample cell.

[4]R. WOODRIFF, *Appl. Spectrosc.* **28**, 413 (1974).

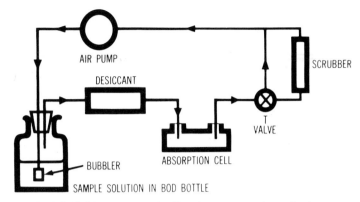

FIGURE 10-29. Apparatus for flameless mercury determination. [Courtesy the Perkin-Elmer Corp.]

To detect low levels of mercury, it is necessary to reduce all the mercury in a sample to the metallic state. Stannous chloride usually is used as the reducing agent. After the mercury is reduced, it is removed from the reducing chamber by bubbling air through the chamber. The air–mercury vapor mixture is dried by passage through a calcium chloride drying column and then enters an absorption cell in the light path of an atomic absorption spectrometer for absorbance measurements. The mercury resonance line at 2537 Å is used for analysis. No heating of the absorption cell is needed.

Figure 10-29 is an example of a commercially available system for the determination of mercury. The sample is placed in the sample bottle, diluted to about 100 ml, and treated with stannous chloride. The sample bottle is placed into the system as shown in the figure. The air pump is turned on and in about 30 sec an equilibrium concentration of mercury is produced in the absorption cell and mercury absorbance is then determined. After the absorbance is read, the T valve is adjusted so the air stream passes through the scrubber. The scrubber cleans the system to prepare it for another mercury determination.

The method is very sensitive, with a detection limit of about 1.0 ppb in solution. The method can be adapted to a variety of samples and has been used for biological tissues, geological samples, and water analysis. For details of the procedure reference should be made to Hatch and Ott[5], who describe the procedure in detail, and more recent papers dealing with applications of the technique.

Arsenic, antimony, and selenium can be determined by first converting them to AsH_3, SbH_3, and H_2Se, each of which is a gas at room temperature.

---

[5] W. R. HATCH, and W. L. OTT, *Anal. Chem.*, **40**, 2085 (1968).

One method used has been to collect the hydrogen and the hydride, generated by treatment of the sample with zinc and hydrochloric acid, in a balloon and then sweep the contents of the balloon into a slot burner with a flow of argon. Arsenic is determined using the spectral line at 1937 Å, for selenium the 1961 Å line is used, and for antimony the 2176 Å line is used. Flame absorption at the wavelengths listed above is much less with an argon–hydrogen–entrained air mixture than with the more conventional air–hydrogen or air–acetylene system, and use of the argon–hydrogen–entrained air increases significantly the sensitivity of detection. An absolute detection limit of 0.02 μg arsenic is possible. The method is described by Fernandez and Manning.[6]

Goudden and Brooksbank[7] describe an automated system for determination of arsenic, antimony, and selenium in water that utilizes the same technique for producing AsH_3, SbH_3, and H_2Se but introduces the gaseous compounds into a heated furnace mounted in the light path of an atomic absorption spectrometer. The furnace decomposes the hydrides and increases the sensitivity of analysis. They report detection limits of 0.1 μg/liter for arsenic and selenium and 0.5 μg/liter for antimony.

7. MONOCHROMATORS

Reference should be made to Chapters 3 and 4 regarding the theory and performance of spectral isolation devices and monochromator design. The discussion here will be limited to specific applications for atomic absorption spectroscopy.

Most monochromators for atomic absorption are grating devices, although a few prism monochromators are in use. The monochromator must isolate the desired spectral line from the source and no other spectral lines should occur within the spectral band pass of the monochromator. Commonly used atomic absorption sources emit very narrow lines, usually of one element only. The spectrum therefore usually is not highly complex; thus very high resolving power usually is not required. Some exceptions to these generalizations occur with some elements.

Commercially packaged atomic absorption instrumentation commonly includes a monochromator of about $\frac{1}{2}$ m focal length with a linear reciprocal dispersion in the range 16–35 Å/mm.

If a nonabsorbing spectral line occurs within the band pass of the monochromator along with the absorbing line, useful analytical data can frequently still be obtained. The sensitivity of the determination will be reduced and a nonlinear absorbance vs. concentration calibration curve will result.

[6]F. Fernandez, and D. C. Manning, *Atomic Absorption Newsletter*, **10**(4), 86 (1971).
[7]P. D. Goudden, and P. Brooksbank, *Anal. Chem.*, **46**, 1431 (1974).

ABSORPTION SPECTROSCOPY

The power supply to the hollow cathode source is modulated and an ac detection system is used. This arrangement prevents any radiation from the flame or resonance detector from producing an output signal. Random noise is less troublesome than in a conventional spectrophotometer. The resonance detector must, of course, produce a cloud of atomic vapor of the same element being aspirated into the flame. The hollow cathode source also must emit resonance lines of the same element. Analytical calibration curves closely parallel those obtained with conventional atomic absorption systems and sensitivities and detection limits are similar.

9. AMPLIFIERS

The output from the photomultiplier detector is further amplified before the signal is fed to a read-out device. There are many variations of design used for amplification of the photomultiplier signal. Our intention is to describe these devices in a general way rather than detail the electronics involved.

Alternating current amplification is preferred over dc since it is much less subject to drift. Direct current amplifiers are usually limited to one stage since any variation or instability in the first stage is amplified by following stages. An ac signal usually is produced by either modulating the hollow cathode source or mechanically chopping the signal from the hollow cathode. The ac signal can then be amplified and any dc components from the flame or sample cell will be rejected. Usually the output of the amplifier is rectified and filtered to produce a dc signal. The dc output of the amplifier allows the use of simple read-out devices and a dc output proportional to the ac input is obtained.

Amplifiers may be "broadbanded," that is, the amplifier may respond to a wide frequency range. If the amplifier is constructed to respond to a narrow band of frequencies, it is called a "tuned" ac amplifier. Successive stages of amplification, all tuned to the same frequency, will narrow the bandpass width. The modulation of the source or the mechanical chopper used in the optical path must maintain the frequency to which the amplifier responds; thus a narrow-bandpass ac amplifier imposes more stringent requirements on the stability of the modulation devices.

"Lock-in" amplifiers provide a very narrow frequency band pass and thus achieve an excellent signal-to-noise ratio. With suitable resistance–capacitance filters, an effective bandpass width of 1 Hz can be obtained. The lock-in amplifier uses a chopper to modulate the energy source. A reference source, usually a small flashlight bulb, is modulated at exactly the same frequency. The two signals are combined in a synchronous detector to produce "sum" and "difference" frequencies; the "sum" frequency will be twice the chopping frequency and the "difference" frequency will be zero.

The "difference" frequency therefore will be a dc signal. A low pass filter removes the "sum" frequency and a dc signal is amplified by a dc amplifier and then goes to the read-out. A block diagram of a lock-in amplifier is shown in Figure 10-32. The use of a lock-in amplifier is an excellent method to obtain a good signal-to-noise ratio at the output of the dc amplifier and is beginning to replace more conventional dc, ac, and tuned ac amplifiers in the better instrumentation for atomic absorption and flame emission spectroscopy.

10. READ-OUT DEVICES

The simplest read-out device presently in use for atomic absorption is a visually observed meter with a galvanometer-type movement, usually used with a series resistance to read voltage. By using variable potentiometers in the output circuit of the amplifier, it is simple to adjust the meter to full-scale reading on a sample blank and zero when no signal enters the entrance slit of the monochromator. The signal produced by the sample can then easily be obtained in terms of percentage transmittance if the meter scale is 0–100.

Chart recorders also frequently are used as read-out devices for atomic absorption. Most chart recorders are potentiometers using a servomotor to move the recording pen. The pen displacement usually is proportional to the input voltage, although logarithmic recorders are available. Some recorders have multiple input ranges, frequently from a 1 mV full-scale to 10 mV full-scale deflection. Recorders provide a permanent record of the measurement and allow a recheck of results if needed. A recorder also can show the magnitude of the random noise associated with the analytical determination and provides a convenient method to optimize the signal-to-noise ratio. Another very useful feature of a recorder is its ability to provide a permanent record by scanning a wavelength region. To perform this function, the monochromator also must have provision for spectral scanning.

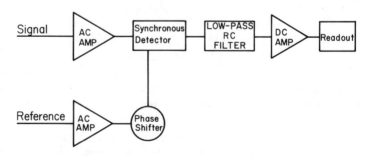

FIGURE 10-32. Block diagram of a lock-in amplifier.

Digital read-out devices also are used in atomic absorption spectroscopy. The digital voltmeter frequently uses a "nixie" tube display. These devices call for the operator to record a visually displayed number.

Read-out devices that produce printed data also are available. These devices produce a printed report of the analytical results which can be treated by any desired computer calculation to present final results of the analysis. The data are not subject to "human" error and a permanent record is obtained.

Signal integrators also are available and are useful to obtain a reading of the total absorption signal, as contrasted with the more common peak height technique of relating element concentration to a read-out system. Signal integrators have particular application to signals that are produced rapidly and decay rapidly, as is the case with some solid sampling methods, since the usual strip-chart recording of pen response is time limited by the inertia of the system.

11. INTERFERENCES IN ATOMIC ABSORPTION

When analytical atomic absorption was in its early stages of development much information was circulated concerning the freedom of the method from interference effects. This led early users of the technique to assume no interferences occurred in atomic absorption. In fact, atomic absorption has as many types of interferences as flame emission, although in some cases the magnitude of the interference is smaller. Any factor that affects the ground state population of the analyte element can be classed as an interference, since, in atomic absorption, the concentration of the analyte element in the sample is considered to be proportional to the ground state atom population in the flame. Any other factor that affects the ability of the atomic absorption instrument to read this parameter also can be classed as an interference and proper control of these effects is necessary to obtain correct analytical results. The common types of interferences that occur in atomic absorption are the topics of this section.

11.1 Spectral Interferences

Spectral interferences occur whenever any radiation overlaps that of the analyte element. The interfering radiation may be an emission line of another element, radical, or molecule, unresolved band spectra, or general background radiation from the flame, solvent, or analytical sample. If the spectral interference does not coincide or overlap the analyte element, spectral interference may still occur if the resolving power and spectral band pass of the monochromator permit the undesired radiation to reach the photoreceptor.

Several examples of possible spectral line interferences include sodium at 2852.8 Å with magnesium at 2852.1 Å, iron at 3247.3 Å with copper at 3247.3 Å, and iron at 3524.3 Å with nickel at 3524.5 Å. Spectral interferences also are possible from hollow cathode lamps. The fill-gas of a hollow cathode lamp is commonly argon or neon and the lamps emit the line spectrum of the fill-gas as well as that of the hollow cathode material. The fill-gas therefore must be one that does not produce an emission line at the desired wavelength of the hollow cathode element.

Koirtyohann and Pickett first reported on molecular spectral interferences caused by alkali halides. CaOH absorption occurs in the region of the barium line at 5535.6 Å; thus the presence of calcium in the analytical sample can interfere with the determination of barium.

General radiation background interference can be caused by some organic solvents and also from the presence of solid luminescent particles in the flame. Solid particles of carbon can be produced in a highly fuel-rich flame and care should therefore be exercised to prevent their occurrence in such flames.

11.2 Ionization Interferences

Ionization interferences are more troublesome in flame emission; however, they can also occur in atomic absorption. In the equilibria

$$MX \leftrightarrows M^0 + X^0$$
$$\hookrightarrow M^* \leftrightarrows M^+ + e^- \qquad (10\text{-}5)$$

the analyte substance MX is dissociated to atoms. Some ground state atoms will be excited to M^* and, if M^0 is easily ionized, some will form ions M^+. In atomic absorption, since we measure the concentration of M^0, any process that changes this concentration will affect analytical results. For easily ionizable elements, such as alkali metals, a high flame temperature can actually decrease the sensitivity since the population of M^0 decreases and M^+ increases.

If a second easily ionized element N is added to sample solution, it will ionize thus

$$N \leftrightarrows N^+ + e^- \qquad (10\text{-}6)$$

The extra electron from the reaction (10-6) therefore will tend to reverse the equilibrium in (10-5) and increase the concentration of M^0 in the system. This effect is observable in Figure 10-33. A calibration curve for sodium alone is shown and another for sodium in the presence of 100 ppm cesium. The ionization of sodium is reduced in the presence of cesium, resulting in an increase in the population of Na^0 and a calibration curve for sodium with an

ABSORPTION SPECTROSCOPY

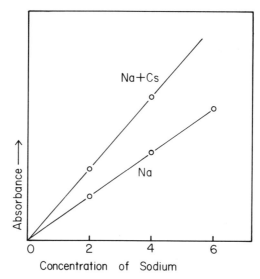

FIGURE 10-33. Effect of cesium ionization on the atomic absorption calibration curve of sodium.

increased slope. The addition of an easily ionizable "buffer," such as a cesium salt, frequently is used to stabilize the ionization effect illustrated in Figure 10-33 and also to take advantage of the increased sensitivity that results.

11.3. Chemical Interferences

Chemical interferences occur both in flame emission and atomic absorption. This occurs whenever some chemical reaction changes the concentration of the ground state analyte element in the flame. Reference to Chapter 9, Figure 9-10, shows processes that occur in the flame that affect both atomic absorption and flame emission spectroscopic measurements. If the analyte element can form stable oxides or hydroxides, some of the ground state atoms of the element are not available for absorption of energy. The result is an absorption signal of decreased intensity; thus oxide formation removes ground state atoms from the flame.

Chemical reactions between two or more species from the analytical sample also can occur. The calcium–phosphorus system has been one of those most extensively studied. Calcium, in the presence of phosphate, produces a stable calcium–phosphorus compound that removes large numbers of calcium atoms from the flame. The result is a severe depression of the calcium absorption signal. The same effect is observed with strontium in the presence of phosphate.

Cations also can mutually interfere in atomic absorption measurements, although the effect usually is less pronounced than in flame emission work. Aluminum will interfere with the alkaline earth elements, as will boron with calcium and magnesium with calcium.

11.4. Interferences with Flameless Sampling

The most recent development in sampling, that of rapidly heating a solid sample in a graphite furnace or metal boat to vaporize and dissociate the sample into the optical path, also is subject to interference problems, although flame gases, oxidants, fuels, and products of the combustion processes are not present. Effects of certain ions on rhenium are shown in Figure 10-34. The rhenium concentration was constant at 50 μg and the ions listed were added at concentrations of 50, 500, and 5000 μg.

The flameless atomization technique also gives slightly different results depending on the anion associated with the metal ion. Some of these results are shown in Figure 10-35. If zinc chloride is used as a source of zinc, the calibration curve rises more steeply than if either zinc nitrate or zinc sulfate is the source of zinc. Other elements that behave similarly are copper, manganese, and iron. As the temperature of the tantalum boat increases, the calibration curves move closer together; thus this effect is, in part, temperature dependent.

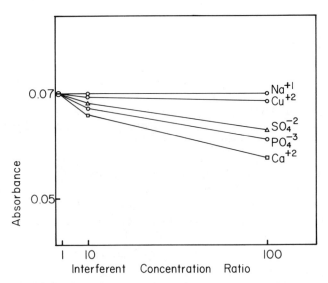

FIGURE 10-34. Interferences to rhenium with the tantalum boat analyzer.

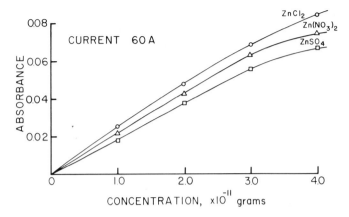

FIGURE 10-35. Calibration curves for zinc in the presence of chloride, nitrate, or sulfate ions using the tantalum boat analyzer.

12. CONTROL OF INTERFERENCES

For analytical purposes it is essential that interference effects in atomic absorption spectroscopy be eliminated or minimized. If they cannot be eliminated, it is necessary to compensate adequately for their presence through use of proper standards or other compensating techniques. This section deals with this problem. Reference also should be made to Chapter 9, as many techniques used for flame emission also apply to atomic absorption.

12.1. Spectral Interference Control

Spectral line interference is less critical in atomic absorption than it is in flame emission. This is due to the fact that absorption is usually concerned with one spectral line only for each element and that, by proper modulation of the source signal, extraneous spectral lines that do not actually overlap the desired line are not detected. It is wise, however, to use as narrow a slit width as possible to keep the spectral band pass of the monochromator to a minimum. Actual spectral line overlap cannot be corrected by this means. If spectral line overlap occurs, as might happen, e.g., with palladium at 3404.6 Å and cobalt at 3405.1 Å, the only solutions are (1) to use another spectral line of the element or (2) remove the offending element from the analytical sample.

Spectral bands or a general continuum can be troublesome; for example, the copper line at 3274.0 Å and the OH band with a band head at 3274.2 Å mutually interfere. The best system for control of the OH band interference is to use a narrow slit and read the magnitude of the OH band contribution immediately above and below the copper signal at 3274.0 Å. This technique also is described in Chapter 9 and can be applied to atomic absorption as well.

Spectral band interferences should, theoretically, be eliminated by proper modulation of the source; however, a strong band or continuum may overload the photomultiplier and require treatment as explained above.

12.2. Ionization Interference Control

Since ionization affects the population of ground state atoms, it is desirable to suppress the ionization process. The simplest procedure is to stabilize the electron concentration in the flame. The best method of control is to "flood" the flame with electrons from an easily ionized added element. The result is to shift the equilibrium

$$MX \leftrightarrows M^0 + X^0 \leftrightarrows M^+ + e^- \qquad (10\text{-}7)$$

to the left, by addition of electrons from, for example, a cesium salt added to the sample solution. An added advantage is an enhanced signal due to an increase in concentration of M^0. This technique is successfully used in the determination of the alkali metals, as well as in elements in other families where ionization occurs to a lesser degree than is found with alkali metals.

12.3. Chemical Interference Control

A variety of techniques can be used to control chemical interference effects. Since chemical interferences affect the population of the ground state atoms in the instrumental optical path, they cannot be controlled by modulation of the source signal. Some of the more common methods for control of chemical interferences follow.

12.3.1. Flame Temperature

Many interferences that occur in low-temperature flames are reduced or do not occur at higher temperatures. In atomic absorption this effect is not as noticeable as in flame emission. For atomic absorption the best flame condition would be to use the lowest temperature that completely dissociates the analyte substance(s) to ground state atoms.

If the analyte element, after dissociation, reacts with oxygen in the flame, it may form oxides, which are often very stable. A high flame temperature frequently can be used to dissociate the oxide to reform ground state atoms. Thus one method for at least partial control of stable oxide formation is to use a high-temperature flame.

12.3.2. Fuel-to-Oxidant Ratio

Another approach to the problem of stable oxide formation is to use a fuel-rich flame. The fuel-rich flame, with its limited oxygen supply, limits formation of oxides, thus keeping the population of ground state atoms at a higher level. Use also can be made of the acetylene–nitrous oxide flame. This mixture combines a high temperature with a limited supply of free oxygen. The result is a fuel–oxidant mixture capable of keeping a large percentage of even refractory oxide-forming elements in the uncombined state. The nitrous oxide–acetylene flame will eliminate most oxide interferences in the alkaline earth family and reduce the magnitude of the interferences in other cases, such as with aluminum and boron. Some loss of sensitivity due to ionization is encountered with the nitrous oxide–acetylene flame. This can be reduced by addition of an ionization buffer, such as cesium, to the analytical sample.

12.3.3. Flame Region

Interference effects differ at different locations in the flame. The region of maximum absorption usually is quite small, especially in stoichiometric flames. In fuel-rich flames the region of maximum absorption usually is more diffuse.

In some cases a region higher in the flame than normally used provides a region of less interference. In an oxygen–acetylene flame the interference of phosphate on calcium can be minimized by observing the calcium absorption signal 2 or 3 cm above the burner tip in a total-consumption burner. With lower temperature fuel–oxidant combinations there is no area in the flame where the calcium–phosphate interference disappears. Thus a combination of a high-temperature flame and proper choice of flame area can reduce this type of interference to zero or to a very low level.

12.3.4. Releasing and Chelating Agents

Releasing and chelating agents are substances which, when added in sufficient quantity, in the presence of an interferent, will restore the absorption signal to the value obtained in the absence of the interferent. Lanthanum is a classic example of a releasing agent as used for calcium in the presence of phosphate. In general, releasing agents are considered to preferentially combine with the interferent, in this case the phosphate ion, leaving the calcium free to produce its absorption signal. Lanthanum also is effective against the interference of the sulfate ion with calcium. It now is common practice to add lanthanum, usually in the form of the chloride, to the test solution for determination of calcium by either emission or atomic absorption measurements. Another example is the use of copper as a releasing agent in the determination of the noble metals in an air–propane flame, since Sb, Co, Sn, and Ni all affect the absorption signals of Pt, Au, and Ag.

Chelating agents also are effective in certain cases. Ethylenediaminetetraacetic acid (EDTA) forms strong chelates with alkaline earth elements and can be used to "protect" magnesium and calcium from reaction with phosphate, sulfate, selenium, boron, aluminum, and silicon. 8-Hydroxyquinoline has been used to protect magnesium and calcium from aluminum.

The mechanisms whereby releasing and chelating agents function are not clear. It has been suggested that the greater thermal stability of $LaPO_4$ than of $Ca_3(PO_4)_2$ or possibly $Ca_2P_2O_7$ accounts for the effectiveness of lanthanum as a releasing agent. Chelating agents (EDTA) are known to form very stable complexes with either calcium or magnesium, thus preferentially protecting these elements from further reactions with phosphate or sulfate ions.

12.3.5. Chemical Separations

Chemical treatment of a sample also may be used to control interferences. A number of techniques are available, including separation by precipitation, solvent–solvent extraction, and ion exchange techniques. For example, 8-hydroxyquinoline can be used to precipitate and separate a number of elements, such as manganese, copper, cobalt, nickel, and iron, from other major matrix elements. The test solution therefore has fewer extraneous elements present and fewer possibilities of interference exist.

Some elements, such as copper, manganese, and iron, can be separated on an anion exchange resin as complex chloride anions and then removed by eluting with water. A less complex solution is therefore available for absorption spectral measurements.

For other separation and concentration techniques the reader is referred to standard analytical separation texts and current publications in analytical chemical techniques.

12.3.6. Background Correction

If resonance lines of short wavelengths (below about 2800–3000 Å) are used for atomic absorption measurements, there is a possibility of nonspecific light absorption contributing to the total measured absorption of the

ABSORPTION SPECTROSCOPY

desired signal. Any absorption not caused by the element of interest can produce an erroneous signal. Light losses of this kind are called background absorption. Background absorption increases as the wavelength decreases and is essentially negligible at wavelengths above 3000 Å. Concentrated salt solutions produce absorption due to molecular absorption, or light scattering, due to solid solute particles in the flame. Several methods of background correction are possible and three of them will be discussed.

Blank solutions can be used for background correction under certain circumstances. For this purpose a blank solution is prepared containing the same matrix elements as the unknown and in the same concentrations. The blank solution should then produce the same background absorption as the unknown and its absorption can be used as a correction for the measured absorption of the unknowns. The method is accurate and simple to use if a large number of samples with the same matrix composition are to be analyzed.

Two-line compensation can be used under certain circumstances. This technique makes use of a nonabsorbing line nearby in wavelength to the analysis line. It may be a line from the same source lamp or from another lamp, and absorption measurements of both lines are necessary. The absorbing line (resonance line) measurement will include absorbance due to the analyte element and also the background absorbance. Measurement on the nearby nonresonance line will produce only background absorbance. The difference between the two measurements will be the absorbance due only to the analyte element.

The two-line method has been used successfully for nickel in oil, using the resonance spectral line at 2320 Å with the nonabsorbing line at 2316 Å. Lead also can conveniently be determined by this method. The method cannot be used if the two spectral lines are too far separated in wavelength, since background absorption varies with wavelength. The two-line method also requires two separate measurements taken at separate times or it requires two-channel instrumentation. The method also has limited application because of the difficulty of suitable fixed line pairs for the measurement.

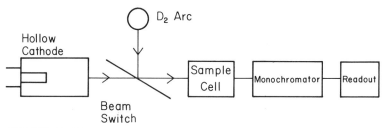

FIGURE 10-36. Optical diagram of a deuterium arc background correction system.

A deuterium lamp correction technique also can be used for background correction. A deuterium arc lamp produces a continuum in the 2000–3500 Å spectral region. The optical arrangement required to use a deuterium lamp for background correction is shown in Figure 10-36. A rotating chopper or sector allows energy from the hollow cathode lamp to pass through the sampling cell alternately with the deuterium arc. The undesired background absorption affects both signals in the same manner. The analyte element can absorb only a very narrow wavelength band. It therefore absorbs energy from the hollow cathode line source but practically no radiation from the deuterium arc continuum. The two beams are subtracted electronically and the effect of the background absorption is eliminated.

The deuterium lamp method for correction of background absorption also is very useful for solid vaporization of the sample using a graphite furnace. If the sample contains traces of organic material some solid carbon particles also may be formed as the sample is vaporized. This type of effect can be corrected using a deuterium lamp if the amount of carbon is not too great. Background correction also is useful if matrix material is not completely vaporized during the heating process since the solid particles thus produced contribute to a general background absorption.

A hydrogen-filled hollow cathode also can be used for background correction. Hydrogen also produces a continuum in the region from about 3500 to 2000 Å, although the continuum is not as intense as that produced by deuterium. An advantage of the hydrogen-filled hollow cathode is that it does not require a separate power supply as does the deuterium arc.

13. ANALYTICAL TREATMENT OF DATA

Reference should be made to Section 7 in Chapter 9, since many of the techniques described there also are applicable to atomic absorption spectroscopy. As in flame emission, a variety of read-out devices are used for atomic absorption. These include visual meters, chart recorders, digital read-out, devices that convert data to absorbance units, and even digital devices that can be calibrated to present analytical data directly in concentration units. The method used to process the read-out data depends on the nature of the read-out signal. In the case of direct read-out in concentration units the data require no further processing. Such data may be fed to the input of a computer for a direct printing of analytical results.

If data are obtained either from a visual meter or a chart recorder, they most often are available as percentage transmittance. The simpler and less costly atomic absorption spectrometers usually provide the data in this form. The following section describes the transformation of these data into a working curve or calibration curve.

13.1. The Working Curve

In atomic absorption the attenuation of a signal from the source is a measure of the concentration of the analyte element in the sample solution. The signal attenuation obeys Beer's law, given in equation (10-1). The form of Beer's law that is more useful for preparation of analytical working curves is that presented in equation (10-3). Since a and l of equation (10-3) are constants for any particular instrument and element, the equation reduces to

$$A = mc \qquad (10\text{-}8)$$

where m is the slope of the calibration curve relating absorbance to concentration. The read-out device should be adjusted to 100% transmittance with a blank and 0% transmittance when no radiant energy enters the monochromator slit. The same type of adjustment can be made if the read-out device is a chart recorder. The monochromator must previously have been centered on the proper wavelength for the desired signal. A series of standards of proper concentration is then successively aspirated into the burner and percentage transmittance data are obtained. These data are then converted

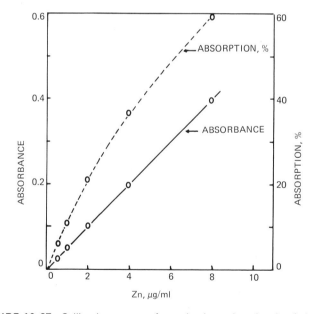

FIGURE 10-37. Calibration curves of atomic absorption signals of zinc using absorbance and percentage absorption versus zinc concentration in $\mu g/ml$. [From W. G. Schrenk, Evaluation of Data, in *Flame Emission and Atomic Absorption Spectroscopy*, Vol. 2, Edited by J. A. Dean and T. C. Rains, Marcel Dekker, New York (1971), Chap. 12. Used by permission of Marcel Dekker, Inc.]

TABLE 10-5
Atomic Absorption Calibration Data for Zinc (2138.6 Å)[a]

Zn, µg/ml	Percentage absorption	Percentage transmittance	Absorbance $(2 - \log \% T)$
0.50	6.1	93.9	0.027
1.00	11.1	88.9	0.051
2.00	21.6	78.4	0.106
4.00	36.9	63.1	0.201
8.00	59.3	40.7	0.390

[a]From W. G. Schrenk, Evaluation of Data, in *Flame Emission and Atomic Absorption Spectrometry*, Vol. 2, Edited by J. A. Dean and T. C. Rains, Marcel Dekker, New York (1971), Chapter 12. Used by permission of Marcel Dekker, Inc.

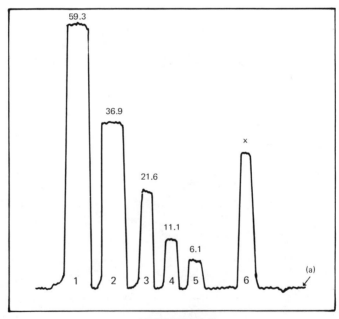

ABSORPTION, %

FIGURE 10-38. Atomic absorption signals of zinc standards (2138.6 Å) and unknown X. [From W. G. Schrenk, Evaluation of Data, in *Flame Emission and Atomic Absorption Spectroscopy*, Vol. 2, Edited by J. A. Dean and T. C. Rains, Marcel Dekker, New York (1971), Chapter 12. Used by permission of Marcel Dekker, Inc.]

to absorbances. Appendix V can be used to convert percentage transmittance to absorbance. Table 10-5 illustrates such data for zinc, using the zinc resonance line at 2138.6 Å. The absorbance values are plotted vs. concentration as shown in Figure 10-37. If a linear calibration results, the slope of the calibration curve can be obtained and use made of the equation $A = mc$ to calculate the concentrations of unknown solutions. For comparison purposes Figure 10-37 also includes a plot of percentage absorption vs. zinc concentration. Note that the plot is nonlinear, in contrast to the linear plot obtained when using absorbance.

A strip chart recording of the data given in Table 10-5 is shown in Figure 10-38. The baseline at (a) is 100% transmission. Zero percent transmission is vertically above the figure and is not shown. The numbers above each absorption signal are percentage absorption. The baseline is obtained with a blank sample solution (0.00 µg/ml Zn) and signals 1-5 correspond to 8.00-0.50 µg/ml zinc as given in Table 10-5. These results produce a calibration curve identical to that shown in Figure 10-37.

Use of a recorder with an amplifier allows for "scale expansion," a technique for amplifying weak signals for ease of reading, but which places zero absorbance off the chart. Such a technique may make the determination of precise absorbance units indefinite. However, data also may be plotted in terms of signal intensity or in some arbitrary units, such as millimeters on the chart vs. concentration. If these data are plotted on semilogarithmic paper, a nearly linear and very useful calibration results. When scale expansion is used, noise also is amplified, so measurements of signal intensities will be less accurate.

13.2. Analytical Procedures

Special methods for handling analytical problems frequently are useful. Most of these are described in Chapter 9 and are mentioned here as a reminder. Some of the more common special techniques that can be used, in addition to "scale expansion" described above, include sampling bracketing, standard addition, and dilution methods.

14. SIMULTANEOUS MULTIELEMENT ANALYSIS

Multielement atomic absorption spectroscopy is complicated by the need for a multielement emission source. Some multielement hollow cathode lamps are available and continuum sources are possibilities. Another method is to place several hollow cathode lamps along the focal plane of a spectrometer, at positions corresponding to the desired wavelengths, thus permitting all the radiation to emerge from a single slit. In this approach the optics is the reverse of the usual multielement read-out commonly used in emission spectroscopy.

Mitchell et al.[8] describe a multielement atomic absorption system using a multielement hollow cathode source and a vidicon detection system. The monochromator is an 0.25-m grating that permits scanning a spectral range of 1680 Å.

Using a spectral region from 2320 to 3281 Å, Mitchell et al. were able to detect eight elements (Zn, Cd, Ni, Co, Fe, Mn, Cu, and Ag) simultaneously. Such a system may well be developed to provide quantitative results in the future.

Selected Reading

Atomic Absorption Spectroscopy, Symposium sponsored by ASTM, June 1968, ASTM Publication 433.

Christian, Gary D., and Feldman, Frederic J., *Atomic Absorption Spectroscopy, Applications in Agriculture, Biology and Medicine*, Wiley–Interscience, New York (1970).

Dean, John A., and Rains, Theodore C., Editors, *Flame Emission and Atomic Absorption Spectrometry*, Vol. 1, *Theory* (1969), Vol. 2, *Components and Techniques* (1971), Marcel Dekker, New York.

Elwell, W. T., and Gidley, J. A. F., *Atomic Absorption Spectrophotometry*, Macmillan, New York (1962).

Hoda, K., and Hasegawa, T., *Atomic Absorption Spectroscopic Analysis*, Genshi Kyuko Bunseki, Kondanska, Tokyo (1972).

Hubbard, D. P., *Annual Reports on Analytical Atomic Spectroscopy*, Vol. 1 (1971), Vol. 2 (1972), Society of Analytical Chemistry, London.

Mavrodineanu, R., Editor, *Analytical Flame Spectroscopy*, Springer-Verlag, New York (1969).

Pinta, M., *Spectrometrie d'absorption atomique*. Tome I: *Problems generaux*. Tome II: *Application à l'analyse chimique*, Masson et Cie, Paris (1971).

Price, W. J., *Analytical Atomic Absorption Spectroscopy*, Heyden and Sons, London (1972).

Ramirez-Munos, Juan, *Atomic Absorption Spectroscopy*, Elsevier, New York (1968).

Robinson, James W., *Atomic Absorption Spectroscopy*, Marcel Dekker, New York (1966).

Rubeska, I., and Muldan, B., *Atomic Absorption Spectrophotometry*, English Translation edited by P. T. Woods, CRC Press, Cleveland, Ohio (1969).

Slavin, Walter, *Atomic Absorption Spectroscopy*, Interscience, New York (1968).

Welz, B., *Atomic Absorption Spectroscopy*, Verlag Chemie, GMBIT (1972).

---

[8] D. G. Mitchell, K. W. Jackson, and K. M. Aldous, *Anal. Chem.*, **45**, 1215A (1973).

Chapter 11
Atomic Fluorescence Spectroscopy

Although fluorescence has been known for a number of years (see Chapter 1), the application of atomic fluorescence to chemical analysis had its start in 1964, when Winefordner and Vickers[1] published a paper describing the principles involved in the method, and Winefordner and Staab[2] published another describing use of atomic fluorescence for the determination of small amounts of zinc, cadmium, and mercury.

The spectral mechanisms involved in atomic fluorescence have been described in Chapter 2 and reference to that chapter should be made to review the various types of atomic fluorescence. Resonance fluorescence is most frequently used for analytical purposes, although other fluorescence mechanisms also are occasionally used.

In analytical atomic fluorescence the sample is reduced to an atomic vapor and excited by radiant energy of a suitable wavelength. The excited atoms emit energy when they return to a lower excited state or the atomic ground state. The intensity of emitted fluorescence energy is measured and is a function of the concentration of the atoms in the sample. Fluorescence spectra of atoms are simple and uncomplicated, in contrast to fluorescence spectra of molecules, since atomic spectra are not affected by vibrational and rotational energy states that exist in molecules.

1. THEORETICAL BASIS OF ANALYTICAL ATOMIC FLUORESCENCE SPECTROSCOPY

Winefordner and Vickers (see footnote 1) have presented the theoretical basis for the use of atomic fluorescence as an analytical technique. This development closely follows their original work.

[1] J. D. WINEFORDNER, and T. J. VICKERS, *Anal. Chem.*, **36**, 161 (1964).
[2] J. D. WINEFORDNER, and R. A. STABB, *Anal. Chem.*, **36**, 165 (1964).

The intensity of the fluorescence emission is proportional to the quantity of absorbed radiant energy, so

$$P_F = \phi P_{abs} \tag{11-1}$$

where P_F is the intensity of fluorescence, P_{abs} is the quantity of radiant energy absorbed, and ϕ is the quantum efficiency for that process.

The relation that exists between the absorbed power and incident power is

$$P_{abs} = P^0(1 - e^{-k_0 L})\Delta v \quad \text{watts} \tag{11-2}$$

where P^0 is the intensity of the incident radiation, k_0 is the atomic absorption coefficient at the center of the absorption line, L is the length of the absorption path, and Δv is the half-width of the absorption line profile.

The half-width Δv can be evaluated if the absorption line profile is assumed to be triangular. Since a line profile is a Gaussian distribution, this becomes a good approximation. Under these conditions, the relation

$$\Delta v = \frac{\pi^{1/2}}{2(\ln 2)^{1/2}} \Delta v_G \tag{11-3}$$

holds, where Δv_G is the half-intensity spectral linewidth.

Combining (11-1) and (11-2), and including a term to correct for self-absorption in the sample, we obtain

$$P_F = \phi P^0 \Delta v (1 - e^{-k_0 L}) e^{-k_0 L/2} \cosh(k_0 L/2) \tag{11-4}$$

The self-absorption term was obtained by Kolb and Streed[3] and is the quantity, $e^{-k_0 L/2} \cosh(k_0 L/2)$.

Equation (11-4) can be written in terms of the fluorescence intensity I_F by dividing P_F by the area A_F of the fluorescence cell from which the fluorescence energy is emitted and by 4π steradians. Then we obtain

$$I_F = \frac{\phi P^0 \Delta v}{4\pi A_F}(1 - e^{-k_0 L}) e^{-k_0 L/2} \cosh(k_0 L/2) \tag{11-5}$$

Equation (11-5) is valid in the case of resonance radiation but must be expanded for other types of fluorescence, or if several absorption lines contribute to the intensity of the fluorescence radiation.

The intensity of the fluorescence I_F is proportional to the number of atoms absorbing energy N. As N increases, the term $(1 - e^{-k_0 L})$ approaches unity. Therefore I_F goes through a maximum as N increases; then quenching becomes appreciable, causing I_F to decrease. A theoretical curve of I_F versus N is shown in Figure 11-1. Mathematically, at low concentrations (small

---

[3] A. C. Kolb and E. R. Streed, *J. Chem. Phys.*, **20**, 1872 (1952).

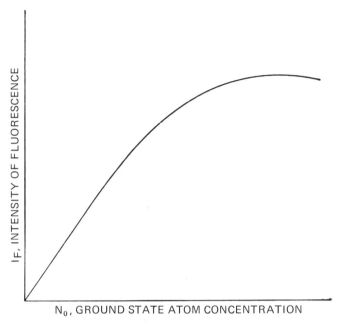

FIGURE 11-1. Theoretical working curve for fluorescence intensity versus atom concentration. [From J. D. Winefordner and T. J. Vickers, Atomic Fluorescence Spectrometry as a Means of Chemical Analysis, *Anal. Chem.*, **36**, 161 (1964). Used by permission of the American Chemical Society.]

values of N), equation (11-5) reduces to

$$I_F = \frac{\phi P^0 \, \Delta v \, k_0 L}{4\pi A_F} \tag{11-6}$$

Mitchell and Zemansky[4] derived an expression for k^0 as follows:

$$k_0 = \frac{(\ln 2)^{1/2} \lambda^2 g_1}{4\pi^{3/2} \Delta v_D g_2} N_0 A_t \delta \tag{11-7}$$

where Δv_D is the Doppler half-width of the absorption line, g_1 and g_2 are the statistical weights of atoms in states 1 and 2, A_t is the transition probability, δ is a term to correct for the relative half-widths of the exciting and emitting spectral lines, and λ is the wavelength of the absorbing line.

Substituting equation (11-7) into (11-6) gives

$$I_F = \left(\frac{(\ln 2)^{1/2} \phi \, \Delta v \, L \lambda^2 g_1 A_t \delta}{16\pi^{5/2} A_F \, \Delta v_D \, g_2} \right) P^0 N_0 \tag{11-8}$$

[4] A. C. G. MITCHELL and M. W. ZEMANSKY, *Resonance Radiation and Excited Atoms*, University Press, Cambridge (1961).

The quantity in parenthesis in equation (11-8) is a constant for any given experimental arrangement and any given spectral line; therefore equation (11-8) reduces to

$$I_F = CP^0 N_0 \tag{11-9}$$

This indicates that a linear relation exists between the intensity of fluorescence and the concentration of the fluorescing species and provides the theoretical basis for analytical atomic fluorescence spectroscopy.

2. ADVANTAGES AND LIMITATIONS OF ATOMIC FLUORESCENCE

Several advantages and disadvantages to the use of atomic fluorescence are apparent from the theoretical development given in the preceding section. Some of the advantages include the following.

1. Increasing the radiation intensity P^0 should linearly increase the fluorescence intensity I_F.
2. A long-path sample cell should increase the fluorescence intensity I_F if suitable means of collecting the fluorescence signal is available.
3. Fluorescence intensity, as a function of concentration of fluorescing atoms, is linear at low concentration levels, making the procedure especially useful for trace element determinations.
4. Fluorescence spectra are simple, so high resolution spectrometers are unnecessary.
5. In contrast to atomic absorption, a radiation source producing narrow spectral lines is not required.

Some of the limitations of atomic fluorescence include the following.

1. Self-absorption occurs at higher concentrations, producing a non-linear response with concentration.
2. Reactions in a flame sample cell are similar to those observed in atomic absorption and can cause problems in the preparation of standard analytical curves.
3. Quenching of fluorescence in certain sample systems can reduce the sensitivity of the method.
4. The quantum efficiency ϕ varies with flame temperature and flame composition; thus, as with any analytical method based on comparison with a standard, adequate control of this factor must be observed.

3. INSTRUMENTATION

The basic instrumentation requirements for analytical atomic fluorescence are quite modest. A simple instrumental arrangement is shown in Figure 11-2, and includes (1) an excitation source, (2) a sample cell, (3) a monochromator, (4) a detector, and (5) a read-out system. Mirrors and lenses can be used at appropriate positions along the optical path to collect the radiant energy and thereby increase its intensity.

The fluorescence signal usually is observed at a right angle to the exciting radiation to decrease the possibility of scattered radiation from the excitation source entering the monochromator slit. This arrangement is particularly important in the case of resonance fluorescence when the excitation radiation is of the same wavelength as the fluorescence radiation.

Use of a chopper and an amplifier that will respond to the frequency of the chopped signal is highly desirable. If the amplifier does not respond to a dc signal, thermal emission from the sample cell will not be detected by the amplifier. Thermal emission frequently is more intense than the fluorescence signal. Unmodulated fluorescence signals can be observed and measured but some means of subtracting thermal emission in the sample cell from the fluorescence signal is required.

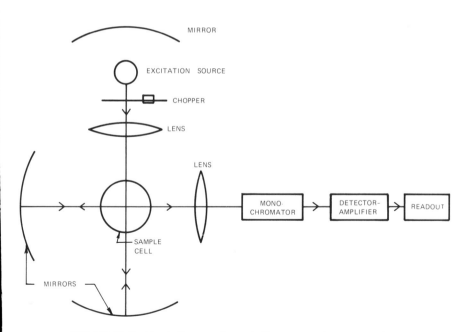

FIGURE 11-2. Block diagram of atomic fluorescence instrumentation.

3.1. Excitation Sources

A variety of excitation sources may be used for atomic fluorescence since the prime requirements are high intensity and stability. A narrow spectral line source, such as those used for atomic absorption, is not essential. Some of the sources that have been used successfully for atomic fluorescence include the following.

3.1.1. Hollow Cathode Lamps

Hollow cathode lamps of the type commonly used in atomic absorption do not have sufficient intensity to be useful for atomic fluorescence. Increasing the current to a hollow cathode lamp is not sufficient for atomic fluorescence since very high currents may actually result in decreased emission intensity due to a high degree of self-absorption. The lifetime of the lamp also is reduced when high currents are used.

Some of the more recently developed "high intensity" hollow cathode lamps are useful. Sullivan and Walsh[5] developed such lamps but they require two power supplies since two sets of independently controlled electrodes are required. One set of electrodes controls the sputtering action and the second set controls the excitation process. These lamps have been used to a limited extent in atomic fluorescence.

3.1.2. Metal Vapor Lamps

Philips and Osram spectral discharge lamps have been used as spectral sources for analytical atomic fluorescence. These lamps have internal electrodes and produce intense spectral lines. The spectral lines, however, are subject to line reversal and the lamps are available only for a limited number of elements. Use of Philips and/or Osram lamps require careful control of input energy to produce maximum intensity without line reversal. Under these conditions they have produced satisfactory atomic fluorescence signals for some elements, including cadmium, mercury, zinc, and thallium.

A mercury metal vapor lamp emits a very intense line spectrum. It is possible to use the line spectrum of mercury to excite the fluorescence spectra of elements other than mercury if line overlap exists. Omenetto and Rossi[6] have been able, by this technique, to produce fluorescence spectra of iron, manganese, nickel, chromium, thallium, copper, and magnesium. Table 11-1 illustrates some of these results and also gives detection limits obtained by this method.

---

[5] J. V. SULLIVAN, and A. WALSH, *Spectrochim. Acta,* **21**, 721 (1965).
[6] N. OMENETTO and G. ROSSI, *Anal. Chem.,* **40**, 195 (1968).

TABLE 11-1
Atomic Fluorescence of Several Elements Excited by
a Mercury Discharge Lamp[a]

Element	Fluorescence line, Å	Hg line, Å	Detection limit, µg/ml
Fe	2483.27	2482.72	1
Tl	2775.72	3776.26	0.3
Cr	3593.49	3593.48	5
Mg	2852.13	2852.42	0.5
Mn	2794.82	Continuum	0.5
Ni	2320.03	Continuum	3
Cu	3247.54	Continuum	0.1

[a]From N. Omenetto and G. Rossi, Atomic Fluorescence Flame Spectrometry Using a Mercury Line Source, *Anal. Chimica Acta*, **40**, 195 (1968). Used by permission of Elsevier Scientific Publishing Co.

3.1.3. Electrodeless Discharge Lamps

Electrodeless discharge lamps, excited with microwave energy, appear to offer considerable promise for atomic fluorescence analytical techniques. They combine high intensity with relatively little line reversal. Properly made, they have a long lifetime and can be constructed for many different elements. Electrodeless discharge lamps are sensitive to temperature variations. Each element has its own optimum temperature; thus temperature control of the lamp is desirable. Browner et al.[7] have developed a thermostatically controlled air-flow device to aid in temperature control of electrodeless discharge lamps. Lamps usually are excited with radiofrequency energy at 2450 MHz, the diathermy frequency, since RF generators for this frequency are readily available. Other radiofrequencies appear to be equally suitable, however, and may be used if available. Additional discussion of electrodeless discharge lamps is found in Chapter 10, since they also have application in atomic absorption spectroscopy.

3.1.4. Continuous Sources

Continuous spectral emission sources also are useful in atomic fluorescence if they have sufficient intensity. The most commonly used continuous source has been the xenon arc lamp, the 450-W xenon lamp being especially useful. These lamps emit a continuum in the visible and near ultraviolet. Their intensities, however, decrease rapidly below about 2500 Å,

---

[7]R. F. Browner, B. M. Patel, T. H. Glenn, M. E. Rutta, and J. D. Winefordner, *Spectrosc. Letters*, **5**, 311 (1972).

a region with many useful spectral lines for atomic fluorescence. The light intensity of the xenon lamp is quite stable and special means for temperature control usually is not required.

3.1.5. Laser Sources

Laser sources for analytical atomic fluorescence offer several advantages over more conventional sources, most of them related to the very high intensity emission and the pulsing characteristic of the laser signal. The development of the tunable dye laser removes the wavelength restrictions associated with earlier lasers and can be used at any wavelength in the region from 3600 to 6500 Å by suitable choice of the dye and grating angle. The bandwidths of the laser pulse in this region can be restricted to 1–10 Å. Since atomic fluorescence does not require as narrow a bandwidth as atomic absorption, the laser provides the high emission energy necessary to achieve low detection limits.

Tunable dye lasers provide, in addition to high peak power levels and narrow spectral bandwidth, a pulse half-width of 2–8 nsec and a repetition rate of 1–30 Hz, depending on the dye used. Since the energy pulse of the laser is of very short duration, the effects of random noises from the flame and also electronic noise are greatly reduced.

The fluorescence produced by the laser signal, because of its pulse width and repetition rate, requires a fast response photomultiplier tube and an integrator capable of responding to the laser pulses. The data then can be processed and recorded by use of a chart recorder.

Fraser and Winefordner,[8] using a system as described above and a modified Jarrell-Ash Tri-flame burner, have been able to lower the detection limits of aluminum, calcium, chromium, and indium by a factor of approximately 2–3 over the best previously reported values for these elements. Detection limits for gallium, iron, and manganese were not as low as previously reported values. Analytical curves were linear over a range of 3–4 orders of magnitude. Tunable dye lasers also permit scanning of a narrow wavelength region, thus making possible a chart-type correction for any background radiation.

The high radiant intensities provided by lasers make possible the practical use of nonresonance fluorescence processes for analytical purposes. Nonresonance fluorescence also reduces scattered radiation effects to practically zero, making line scanning generally unnecessary. Detection limits for nonresonance fluorescence include (see footnote 8) aluminum (3961 Å) at 3×10^{-2} μg/ml, cobalt (3575 Å) at 5×10^{1} μg/ml, indium (4105 Å) at 2×10^{-3} μg/ml, and nickel (3610 Å) at 2 μg/ml.

[8]L. M. FRASER and J. D. WINEFORDNER, *Anal. Chem.*, **44**, 1444 (1972).

4. THE SAMPLE CELL

The sample cell for atomic fluorescence should possess most of the same characteristics as a cell for atomic absorption. The purpose of the sample cell is to convert the analyte into an atomic ground state vapor with maximum efficiency. The temperature of the cell should be sufficient to dissociate the sample but not produce excitation of the dissociated atoms. The chemical composition of the flame should favor retaining the ground state atoms at a steady concentration level for as long as possible. The flame itself should produce low background emission and no particulate matter. The rate of introduction of the sample into the flame should be steady.

It should be obvious that no flame will meet all the above requirements; in practice, therefore, an attempt is made to approach this ideal situation as closely as possible. Present practice is to adapt flame cells in use for flame emission and atomic absorption for use in atomic fluorescence spectroscopy.

4.1. Total-Consumption Aspirator Burners

Turbulent-flow total-consumption burners were the standard for flame emission spectroscopy for a number of years. In spite of certain deficiencies, they also can serve as sample cells for atomic fluorescence. The most common types of total-consumption burners include a built-in sample aspirator system. Figures 9-4 and 9-5 show their general constructional features. Winefordner and Staab (see footnote 2) used this type of burner for their early studies on analytical atomic fluorescence.

The total-consumption burners have been used with a variety of fuel–oxidant combinations. The hydrogen–oxygen system produces a noisy flame with low sensitivity to atomic fluorescence. Use of a hydrogen–argon flame with entrained oxygen as the oxidant produces a much steadier flame with good atomic fluorescence sensitivity. Such a flame maintains reducing conditions in the center of the flame, thus minimizing the reoxidation of the ground state analyte atoms. The use of argon presumably increases the quantum efficiency of the fluorescence process. Air–propane mixtures produce relatively cool flames and are useful for easily atomized elements. An entrained air and hydrogen flame forms a good system since it also is cool and the entrained nitrogen (from the air) serves a purpose similar to argon in increasing the quantum efficiency of the atomic excitation process.

4.2. Laminar Flow Burners

The laminar flow premixed flame systems are characterized by their reduced noise and light scattering properties as compared with turbulent flow burners. A choice of fuel–oxidant combinations also is available for use with laminar flow burners. The choice is not as great as with turbulent flow burners due to the possibilities of explosions occurring in the mixing chamber if the rate of the gas flow exiting from the mixing chamber is too slow. Dagnall et al.[9] studied propane–air, acetylene–air, and hydrogen–air systems for suitability for atomic fluorescence. They found the propane–air system most satisfactory of those studied. Bratzel et al.[10] compared the turbulent flow and laminar flow burner systems and found only minor differences in them. The turbulent flow flames were preferred since they produced equal or better detection limits, are explosion-free, are simple to use, and are less costly. The turbulent flames produce more intense thermal emission. The use of a modulated source signal with accompanying ac amplification results in removal of the dc emission signal from the detector.

4.3. Non-Flame Sample Cells

Only a few attempts have been made to use a non-flame sample cell for atomic fluorescence spectroscopy. Massman[11] used a heated graphite tube furnace and an argon atmosphere. Detection limits were obtained for zinc, cadmium, antimony, iron, thallium, lead, magnesium, and copper and ranged from 2×10^{-9} g for thallium to 4×10^{-14} g for zinc. West and Williams[12] used a carbon filament and vaporized the sample into an argon atmosphere. They reported a detection limit of 3×10^{-11} g for silver and 1×10^{-16} g for magnesium. Bratzel et al.[13] have placed a liquid sample on a platinum loop. The sample is dried and then heated electrically in an argon atmosphere to vaporize it. Detection limits of 10^{-14} g for cadmium, 10^{-8} g for mercury, and 10^{-7} g for gallium were obtained. The analytical curves were linear over three to four orders of magnitude. Interferences were small. Although it has not yet been studied in atomic fluorescence, the tantalum boat technique for atomic absorption also should show promise as a non-flame sample cell for atomic fluorescence spectroscopy.

5. MONOCHROMATORS

Atomic fluorescence spectra are simple, with relatively few spectral lines; therefore monochromators of extremely high resolution are not required. Either grating or prism instruments may be used. A monochromator

---

[9] R. M. DAGNALL, K. D. THOMPSON, and T. S. WEST, *Anal. Chim. Acta*, **36**, 269 (1966).
[10] M. P. BRATZEL, R. M. DAGNALL, and J. D. WINEFORDNER, *Anal. Chem.*, **41**, 713 (1969).
[11] H. MASSMAN, in *XIII Colloquium Sectroscopicum Internationale*, Ottawa, Canada, June 1967.
[12] T. S. WEST, and X. K. WILLIAMS, *Anal. Chim. Acta*, **41**, 1476 (1969).
[13] M. P. BRATZEL, R. M. DAGNALL, and J. D. WINEFORDNER, *Anal. Chim. Acta*, **48**, 197 (1969); M. P. BRATZEL, R. M. DAGNALL, and J. D. WINEFORDNER, *Appl. Spectrosc.*, **24**, 518 (1970).

of large light-gathering ability is desirable to provide the greatest signal intensity possible. A short path length, such as those occurring in the smaller monochromators, also is desirable, especially if the spectral region of interest is in the 2000-Å range and lower. Absorption of radiation by air can be considerable in this spectral region.

Broadband ac amplifiers can be used to detect modulated source signals but will not respond to the dc emission of the flame. Modulation of the source can be accomplished by a mechanical chopper or by electronic modulation. Noise in the flame is of variable intensity and will not be entirely eliminated by a broadband ac amplifier. The bandwidth response of the amplifier is a factor under these conditions. The narrower bandwidth response amplifiers limit noise pickup only to those frequencies within the bandwidth.

Lock-in amplifiers are ac amplifiers of very narrow bandwidth response (about 1 Hz). Use of the lock-in amplifier requires a reference signal modulated at the same frequency as the source. These amplifiers can provide very high gain, excellent stability, and an excellent signal-to-noise ratio. They are highly desirable for general application in atomic fluorescence spectroscopy.

6. INTERFERENCES IN ATOMIC FLUORESCENCE

Since analytical atomic fluorescence depends on converting the desired analyte element into an atomic ground state vapor and maintaining it in that condition during the fluorescence measurement, it is apparent that any process that disturbs the ground state population of the analyte element will affect the intensity of the fluorescence signal. In addition to processes affecting the ground state population of atoms, the possibility of spectral interference also exists. Interferences to analytical atomic fluorescence therefore include (1) spectral interferences and (2) those related to the ground state atomic population of the flame. It is convenient to divide the second category into (a) physical effects and (b) chemical effects.

6.1. Spectral Interferences

Several different types of spectral interferences are possible in analytical atomic fluorescence. If a second, unwanted element emits a fluorescence radiation simultaneously with the analyte element and its wavelength is within the band pass of the monochromator slit width, interference occurs. Not many instances of this type of interference have been identified. Some known examples include cadmium at 2288.0 Å and arsenic at 2288.1 Å and mercury with iron, thallium, chromium, and magnesium. If such interferences occur, the result will be an erroneous increase in fluorescence signal strengths. Since this type of interference signal will have the same frequency as the desired signal, it cannot be eliminated by a tuned amplifier system.

A second type of spectral interference is due to thermal emission from the flame of the sample. For interference to occur in this case the thermally excited spectral lines also must lie within the spectral band pass of the monochromator slit width, and this also results in an erroneously enhanced signal. In this case, however, the interference can be eliminated by use of a modulated source and an amplifier tuned to the frequency of modulation since the undesired signal will be a steady (dc) signal.

Flame background or a continuum is a third type of spectral interference. It also can be controlled by use of an ac detection system. Random noise in the flame cell, since it contains a large number of ac components, cannot be entirely eliminated by use of an ac electronic system but the magnitude of this effect can be reduced substantially by use of an ac system, preferably the tuned (lock-in) type of system.

6.2. Chemical Interferences

Chemical interference effects in atomic fluorescence are similar to those observed in atomic absorption spectroscopy. In addition, any process that affects the quantum efficiency of the fluorescence or disrupts normal emission of the energy of the excited state also can be considered as a chemical interference.

Fluorescence signals for calcium and magnesium are decreased markedly by many ions.[14] Anions such as chloride, sulfate, nitrate, phosphate, and silicate cause severe decreases in the fluorescence signal. As with flame emission and atomic absorption, releasing agents such as lanthanum and EDTA reduce this effect. Addition of releasing agents, however, increases the scattered light. A correction for this effect usually can be obtained by proper use of a sample blank containing the releasing agent.

If refractory oxides are formed in the flame, the fluorescence signal will be decreased. This effect also is similar to that obtained in other analytical flame systems. The methods to minimize this effect that have been described in Chapters 9 and 10 also apply to atomic fluorescence. Reducing flames and very hot flames reduce this type of interference, but hot flames increase thermal excitation processes. Therefore it is important to determine optimum conditions for each analyte element being considered.

6.3. Physical Interferences

The most important physical interference in atomic fluorescence is that due to scattered radiation caused by solvent droplets and solid particles in

[14]D. R. DEMERS and D. W. ELLIS, *Anal. Chem.*, **40**, 860 (1968).

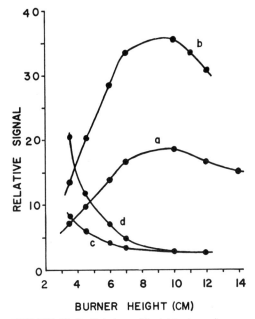

FIGURE 11-3. Relation of height above burner head to relative intensity of the fluorescence signal: Cadmium atomic fluorescence intensity profiles in premixed and unpremixed nitrous oxide–hydrogen flames. (*a*) Atomic fluorescence in optimized N_2O/H_2 premixed flame. (*b*) Atomic fluorescence in optimized N_2O/H_2 unpremixed flame. (*c*) Scatter in N_2O/H_2 premixed flame. (*d*) Scatter in N_2O/H_2 unpremixed flame. [From M. P. Bratzel, Jr., R. M. Dagnall, and J. D. Winefordner, Evaluation of Premixed Flames Produced Using a Total Consumption Nebulizer Burner in Atomic Fluorescence Spectrometry, *Anal. Chem.*, **41**, 1527 (1969). Used by permission of the American Chemical Society.]

the flame. Solid particles may result from incomplete vaporization of the solute in the flame. Another source of solid particles is the formation of refractory oxides in the flame. A highly reducing flame also may produce small carbon particles.

Scattered radiation interference is greater with a total-consumption burner than with premixed systems. Larger and more nonuniform droplet sizes are observed with total-consumption burner systems. The area of observation in a flame also is a factor in scattered radiation interference. Figure 11-3 illustrates this effect.[15] Scattering is greater in both turbulent

[15] M. P. Bratzel, R. M. Dagnall, and J. D. Winefordner, *Anal. Chem.*, **41**, 1527 (1969).

and premixed flames in the lower portion of the flame and is greater for the turbulent flow system than for the premixed flame.

Other physical interferences are similar to those observed in flame emission and atomic absorption spectroscopy. They include effects due to viscosity and temperature of the sample solution. Any factor that can alter the rate of uptake of the sample solution requires control. The best method to use to control these effects is to prepare a blank with physical properties similar to those of the test sample.

7. ANALYTICAL PROCEDURES

Before analytical atomic fluorescence data can be obtained several preliminary adjustments are required. The source lamp should be adjusted to produce maximum signal with minimum line reversal and with good stability. The electronic equipment (amplifier and photomultiplier power sources) also should be allowed to warm up. A solution containing the desired element should be aspirated into the sample cell and the monochromator adjusted to the wavelength of the fluorescence signal. Optical alignment of the system should then be made to produce maximum signal intensity. The area of observation in the sample cell should be adjusted also, since the maximum fluorescence signal will be obtained only from the proper portion of the flame.

The entrance and exit slits of the monochromator also should be adjusted if variable slit widths are available. Proper slit width adjustments can increase the signal-to-noise ratio as well as sensitivity. Relatively narrow slits usually provide best signal-to-noise conditions. The fuel-to-oxidant ratio also should be optimized for each element.

The procedure to obtain the fluorescence intensity for the desired element after the preliminary adjustments have been made depends in part on the type of equipment in use. If an ac system with a chopper or electrically modulated source is in use and is properly adjusted, any dc signal component of the flame cell can be ignored since the amplifier will not respond to the dc signal. This includes any continuum and any thermally excited spectral lines within the band pass of the monochromator.

If a dc amplifier is used, it is necessary to obtain a blank reading on the actual standards as well as the sample solution to account for any signal enhancement due to thermal emission. This can be obtained by aspirating the solution into the sample cell and interrupting the exciting signal with a shutter or other suitable device. Any thermal emission signal must be subtracted from the fluorescence signal due to the standard or the sample.

Any general background signal also should be subtracted from the fluorescence signal if one is present. Background can be obtained by use of

ATOMIC FLUORESCENCE SPECTROSCOPY

a blank solution using a blank that approximates the sample solution composition except for the desired element.

7.1. The Analytical Working Curve

When all instrumental parameters have been optimized an analytical working curve can be obtained by aspirating a series of standards into the sample cell. Precautions mentioned in the preceding section concerning the need for background correction and/or consideration of thermal emission effects must be observed and suitable corrections applied if necessary. The corrected fluorescence intensities can be plotted against concentration. A typical working curve is shown in Figure 11-4 for zinc (see footnote 2). Comparison of Figure 11-4 with Figure 11-1 indicates that the actual experimental working curve closely parallels the theoretical curve; thus the theory of atomic fluorescence appears to be confirmed. Analytical working curves for other elements are similar in shape to those in Figure 11-4.

The linear portion of the analytical working curve usually extends over a concentration range considerably greater than for atomic absorption or flame emission methods. Many working curves are linear over a concentration range of three to four orders of magnitude. The increased linearity is obtained primarily at low concentration levels, since detection limits are

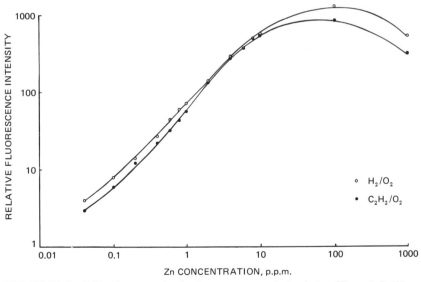

FIGURE 11-4. Calibration curve for the fluorescence analysis of zinc. [From J. D. Winefordner and R. A. Staab, Determination of Zinc, Cadmium, and Mercury by Atomic Fluorescence Spectrometry, *Anal. Chem.*, **36**, 165 (1964). Used by permission of the American Chemical Society.]

TABLE 11-2
Effect of the Nature of the Flame upon the Extent of the Linear Working Range of Magnesium[a]

Flame	Linear range, μg/ml
Air–H_2 not premixed	0–1
Air–H_2 premixed	0–5
O_2–H_2 not premixed	0–20
O_2–H_2 premixed	0–5
N_2O–H_2 not premixed	0–2
N_2O–H_2 premixed	0–2

[a] From A. Syty, Atomic Fluorescence Spectrometry, in *Flame Emission and Atomic Absorption Spectrometry,* Vol. 2, Edited by J. A. Dean and T. C. Rains, Marcel Dekker, New York (1971), Chapter 8. Used by permission of Marcel Dekker, Inc.

generally lower than for atomic absorption. The upper limit of linearity is determined primarily by self-absorption and this concentration is about the same for all flame methods.

The range of the linear portion of the working curve is affected by flame composition. This effect is shown in Table 11-2 for the element magnesium for a turbulent flame (see footnote 15).

7.2. Organic Solvents

Organic solvents have been used frequently in flame emission and atomic absorption to enhance the analytical signal. It has been shown that the intensity of a fluorescence signal also is enhanced by using organic solvents with premixed laminar flow burners, although the effect has not yet been extensively studied. Several examples of enhancement by use of organic solvents include an improvement in the detection limit for silver of about 40 (see footnote 11) and an improvement by a factor of five in the detection limit for zinc (see footnote 8).

With turbulent flow burners the enhancement effects are not so pronounced. Figure 11-5 illustrates the effect of isopropanol as a solvent on the fluorescence signal of thallium in a premixed and a turbulent-flow argon–entrained air–hydrogen flame.[16] The fluorescence signals of iron, cobalt, and nickel were studied using a series of organic solvents, including methanol, ethanol, propanol, acetone, diozane, and glycerol. The premixed air–hydrogen flame caused enhancement of the signal in each case, but the turbulent flame actually decreased the signal in some cases.[17]

[16] R. M. DAGNALL, M. R. G. TAYLOR and T. S. WEST, *Spectrosc. Letters,* **1**, 397 (1968).

7.3. Detection Limits

Table 11-3 lists atomic fluorescence detection limits for 39 elements. All data are for water solutions; however, the source units varied, as did flame conditions. The detection limits therefore may not be entirely consistent, nor can they be exactly applied to different instrumental or analytical situations. They represent the type of result that can be expected if care is taken to optimize operating conditions.

7.4. Sample Preparation

Except for a few attempts to vaporize directly a solid sample by a rapid heating technique, the sample for atomic fluorescence is a liquid solution that can be aspirated into a flame. The same type of sample preparation therefore is required as is necessary for flame emission and atomic absorption spectroscopy. The sample must be in solution, free of solid particles, and of low viscosity. The comparison standards should have the same physical characteristics as the sample solution. If organic solvents are used for the sample solution, they also must be used for the comparison standards.

Since some elements interfere with the fluorescence signals of other elements, it is necessary to compensate for this effect. Frequently this can be done by making comparison standards that closely approximate the sample composition.

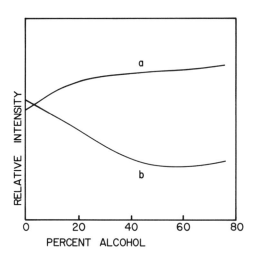

FIGURE 11-5. Effect of isopropanol on the fluorescence intensity of thallium. [From M. P. Bratzel, Jr., R. M. Dagnall, and J. D. Winefordner, A Comparative Study of Premixed and Turbulent Air–Hydrogen Flames in Atomic Fluorescence Spectrometry, *Anal. Chem.*, **41**, 713 (1969). Used by permission of the American Chemical Society.]

[17] J. MATOUREK. and V. SYRDRA, *Anal. Chem.*, **41**, 518 (1969).

TABLE 11-3
Some Experimental Detection Limits and Wavelengths of Detection in Atomic Fluorescence Spectroscopy[a]

Element	Wavelength	Detection limit, μg/ml
Ag	3281	0.0001
Al	3162	0.1
As	1937	0.1
Au	2676	0.005
Be	2349	0.01
Bi	2231	0.005
Ca	4227	0.02
Cd	2288	0.000001
Co	2407	0.005
Cr	3579	0.05
Cu	3247	0.001
Fe	2483	0.008
Ga	4172	0.01
Ge	2652	0.1
Hg	2537	0.0002
In	4511	0.1
Mg	2852	0.001
Mn	2795	0.006
Mo	3133	0.5
Ni	2320	0.003
Pb	4058	0.01
Pd	3405	0.04
Sb	2311	0.05
Si	2040	0.6
Se	1960	0.04
Sn	3034	0.05
Sr	4607	0.03
Te	2143	0.005
Tl	3776	0.008
V	3184	0.07
Zn	2138	0.00002

[a]Adapted from Table VI, J. D. Winefordner and R. C. Elser, Atomic Fluorescence Spectrometry, *Anal. Chem.*, **43**(4), 25A (1971). Used by permission of the American Chemical Society.

Concentration of the analytical sample can be done if needed. The same techniques of concentration that are used for flame emission and atomic absorption may be used. Most commonly these include precipitation and liquid–liquid extraction techniques.

Most of the processes in the flame that interfere with fluorescence intensity are similar to those that occur in flame emission and atomic absorption spectroscopy. The reader is referred to sections dealing with this topic in Chapters 9 and 10.

8. APPLICATIONS AND FUTURE DEVELOPMENTS

Atomic fluorescence has found application in a variety of situations and with quite different substances. Lubricating oils have been analyzed for silver, copper, iron, and magnesium.[18] Trace concentrations of zinc in copper have been determined.[19] Some of the other reported applications are the determination of bismuth in aluminum[20] and calcium, copper, magnesium, manganese, potassium, and zinc in soils.[21] Analytical applications of atomic fluorescence will increase as more attention is given to development of the methods applicable to this technique.

Some of the characteristics of atomic fluorescence that will influence its analytical use include its sensitivity and its large linear calibration range. For many elements present detection limits are an order of magnitude better than atomic absorption if high-intensity sources are used. The large linear calibration range provides a convenient system to use without the need for solution of samples.

Atomic fluorescence, since it makes use of ground state atoms for the analytical process, is much less temperature dependent than is flame emission. It is similar to atomic absorption in this respect. Atomic fluorescence is more subject to quenching effects than is flame emission. The effects of light scattering in atomic fluorescence can be a problem, although means to minimize this effect are available.

Instrumentation for atomic fluorescence is not expensive. A simple monochromator and detector are adequate. Alternating current amplification is highly desirable to aid in eliminating effects due to emission from the flame and to reduce the effects of scattered radiation. High-intensity sources are necessary to achieve low detection limits. The electrodeless discharge

---

[18] R. Smith, C. M. Stafford, and J. D. Winefordner, *Can. Spectry.*, **14**, 2 (1969).
[19] P. D. Warr, *Talanta*, **17**, 543 (1970).
[20] R. S. Hobbs, G. F. Kirkbright, and T. S. West, *Talanta*, **18**, 859 (1971).
[21] R. M. Dagnall, G. F. Kirkbright, T. S. West, and T. S. Wood, *Anal. Chem.*, **43**, 1765 (1971).

lamps appear to be best suited for this purpose at this time. Lasers may be the next source in an effort to provide increased intensities to be developed.

Selected Reading

Bratzel, M. P., Dagnall, R. M., and Winefordner, J. D., *Anal. Chim. Acta*, **48**, 197 (1969).
Bratzel, M. P., Dagnall, R. M., and Winefordner, J. D., *Anal. Chem.*, **41**, 713 (1969).
Bratzel, M. P., Dagnall, R. M., and Winefordner, J. D., *Anal. Chem.*, **41**, 1527 (1969).
Bratzel, M. P., Dagnall, R. M., and Winefordner, J. D., *Appl. Spectrosc.*, **24**, 518 (1970).
Browner, R. F., Patel, B. M., Glenn, T. H., Rutta, M. E., and Winefordner, J. D., *Spectrosc. Letters*, **5**, 311 (1972).
Dagnall, R. M., Thompson, K. C., and West, T. S., *Anal. Chim. Acta*, **36**, 269 (1966).
Dagnall, R. M., Taylor, R. G., and West, T. S., *Spectrosc. Letters*, **1**, 397 (1968).
Dagnall, R. M., Kirkbright, G. F., West, T. S., and Wood, T. S., *Anal. Chem.*, **43**, 1765 (1971).
Demers, D. R., and Ellis, D. W., *Anal. Chem.*, **40**, 860 (1968).
Fraser, L. M., and Winefordner, J. D., *Anal. Chem.*, **43**, 1693 (1971).
Fraser, L. M., and Winefordner, J. D., *Anal. Chem.*, **44**, 1444 (1972).
Hobbs, R. S., Kirkbright, G. F., and West, T. S., *Talanta*, **18**, 859 (1971).
Kolb, A. C., and Streed, E. R., *J. Chem. Phys.*, **20**, 1872 (1952).
Massmann, H., in *XIII Colloquium Spectroscopicum Internationalle*, Ottawa, Canada, June 1967.
Matourek, J., and Syrdra, V., *Anal. Chem.*, **41**, 518 (1969).
Mitchell, A. C. G., and Zemansky, M. W., *Resonance Radiation and Excited Atoms*, University Press, Cambridge (1961).
Omenetto, N., and Rossi, G., *Anal. Chem.*, **40**, 195 (1968).
Smith, R., Stafford, C. M., and Winefordner, J. D., *Can. Spectry.*, **14**, 2 (1969).
Sullivan, J. V., and Walsh, A., *Spectrochim. Acta*, **21**, 721 (1965).
Warr, P. D., *Talanta*, **17**, 543 (1970).
West, T. S., and Williams, X. K., *Anal. Chem.*, **40**, 335 (1968).
West, T. S., and Williams, X. K., *Anal. Chem. Acta*, **41**, 1476 (1969).
Winefordner, J. D., and Vickers, T. J., *Anal. Chem.*, **36**, 161 (1964).
Winefordner, J. D., and Staab, R. A., *Anal. Chem.*, **36**, 165 (1964).

Appendix I

Some Basic Definitions, Physical Constants, Units, and Conversion Factors

A. Some Basic Definitions

Absorbance A	Logarithm to the base 10 of the reciprocal of the transmittance, $A = \log_{10}(1/T)$
Absorptivity a	Absorbance divided by the product of sample path length l and the concentration of the absorbing material, $a = A/lc$
Frequency v	Cycles per second $\equiv$ hertz (Hz)
Spectrograph	Spectrometer equipped to photographically record a spectral region
Spectrometer	An instrument including an entrance slit and a spectral dispersing device, and which observes spectra at selected wavelengths or by scanning, utilizing some device other than photographic to determine radiant power
Spectrometry	The branch of science that deals with measurements of spectra
Transmittance T	Ratio of radiant power transmitted by a sample to the radiant power incident on the sample
Wavelength λ	Distance between two points on adjacent waves that are in phase
Wave number $\bar{v}$	Number of waves per unit length $= 1/\lambda$ (usually given in cm^{-1})

B. Selected Physical Constants

Avogadro's number	N	6.0223×10^{23} mole^{-1}
Boltzmann's constant	k	1.38×10^{-16} erg °K^{-1}
Planck's constant	h	6.6256×10^{-27} erg sec
Electron charge	e	1.6021×10^{-19} C
Velocity of light	c	2.9979×10^{10} cm sec^{-1}
Rydberg constant for hydrogen	R_H	109,737.31 cm^{-1}
Electron rest mass	m_e	0.9109×10^{-27} g
Proton rest mass	m_p	1.67239×10^{-24} g
Neutron rest mass	m_n	1.67470×10^{-24} g

C. Selected Units and Conversion Factors

Angstrom, $1 \text{ Å} = 10^{-10}$ m $= 10^{-8}$ cm
Nanometer, $1 \text{ nm} = 10^{-9}$ m $= 10 \text{ Å}$
Electron volt, $1 \text{ eV} = 1.6 \times 10^{-12}$ erg
Wavelength equivalent of one electron volt $= 1.2398 \times 10^{-4}$ cm
Standard of length, krypton 84 red line $= 6057.802106$ Å
(adopted October 1960 by the 11th General Conference on Weights and Measures, Paris)

Appendix II
Spectral Lines, Arranged by Wavelength, with *gf* and Intensity Values[1]

Appendices II and III include wavelengths, *gf* values, and intensity values of approximately ten of the most intense spectral lines for each of 70 elements. Appendices II and III should be useful for qualitative identification of elements. They also can serve as a source from which to select spectral lines for quantitative analysis. If more complete spectral line lists are needed the National Bureau of Standards Monographs cited in footnote 1 should be consulted.

---

[1] From *Tables of Spectral-Line Intensities*, NBS Monograph 32, W. F. Meggers, C. H. Corliss, and B. F. Scribner (1961) and *Experimental Transition Probabilities for Spectral Lines of Seventy Elements*, NBS Monograph 53, C. H. Corliss and W. R. Bozman (1962).

Wavelength	Element	gf	Intensity	Spectrum
1960.26	Se	0.59	34	I
1971.97	As	0.29	28	I
1994.2	Te	0.19	8	I
2002.0	Te	0.38	16	I
2003.34	As	0.83	28	I
2012.00	Au	0.098	8	I
2039.85	Se	0.78	40	I
2061.70	Bi	0.38	55	I
2062.79	Se	0.30	15	I
2074.79	Se	0.036	3.0	I
2081.03	Te	0.37	10	I
2088.93	B	0.099	7	I
2089.59	B	0.16	11	I
2135.47	P	0.035	1	I
2136.20	P	0.35	10	I
2138.56	Zn	1.3	1000	I
2142.75	Te	0.40	55	I
2144.38	Cd	7.3	60	II
2147.19	Te	0.40	11	I
2149.11	P	0.32	9	I
2152.95	P	0.051	1	I
2154.08	P	0.10	2	I
2175.81	Sb	0.18	38	I
2265.02	Cd	7.7	110	II
2288.02	Cd	0.92	1500	I
2288.12	As	1.2	44	I
2311.47	Sb	0.12	45	I
2348.61	Be	0.24	300	I
2349.84	As	2.1	85	I
2367.06	Al	0.15	18	I
2373.13	Al	0.31	36	I
2383.25	Te	0.55	55	I
2385.76	Te	0.70	70	I
2413.18	Ag	1.4	10	II
2417.37	Ge	1.9	85	I
2421.70	Sn	5.8	260	I
2427.95	Au	0.16	200	I
2429.49	Sn	2.5	420	I
2435.16	Si	0.67	26	I
2450.07	Ga	0.37	75	I
2456.53	As	0.84	36	I
2488.55	Os	8.4	380	I
2492.15	Cu	0.029	36	I
2492.91	As	0.93	44	I
2494.56	Be	1.5	70	I
2494.73	Be	2.1	100	I
2496.78	B	0.66	240	I
2497.73	B	1.3	480	I

APPENDIX II

Wavelength	Element	gf	Intensity	Spectrum
2497.96	Ge	0.22	90	I
2500.17	Ga	0.68	130	I
2500.70	Ga	0.063	12	I
2502.98	Ir	1.1	200	I
2506.90	Si	0.60	170	I
2514.32	Si	0.54	160	I
2516.11	Si	1.3	360	I
2519.21	Si	0.41	120	I
2524.11	Si	0.80	240	I
2528.51	Si	0.68	200	I
2528.52	Sb	5.4	320	I
2530.70	Te	0.071	12	I
2534.01	P	1.3	22	I
2535.65	P	3.6	60	I
2536.52	Hg	0.34	1500	I
2543.97	Ir	3.9	380	I
2546.55	Sn	0.37	240	I
2551.35	W	0.80	280	I
2553.28	P	2.3	38	I
2554.93	P	0.89	15	I
2560.15	In	0.46	110	I
2567.99	Al	0.10	24	I
2575.10	Al	0.21	48	I
2576.10	Mn	7.9	1200	II
2580.14	Tl	0.13	70	II
2592.54	Ge	1.1	500	I
2593.73	Mn	5.0	800	II
2598.05	Sb	6.4	600	I
2605.69	Mn	3.3	550	II
2614.18	Pb	2.7	700	I
2615.42	Lu	0.60	1200	II
2618.37	Cu	0.44	40	I
2627.91	Bi	1.2	70	I
2628.03	Pt	0.46	110	I
2631.28	Si	2.3	24	I
2637.13	Os	1.2	360	I
2641.49	Au	0.14	26	I
2647.47	Ta	0.87	280	I
2650.47	Be	3.1	140	I
2651.18	Ge	2.7	1200	I
2651.58	Ge	0.84	550	I
2652.49	Al	0.050	15	I
2653.27	Ta	1.6	300	I
2653.74	Yb	3.6	140	II
2656.61	Ta	0.66	220	I
2659.45	Pt	0.87	280	I
2659.87	Ga	0.087	34	I
2660.39	Al	0.068	20	I

Wavelength	Element	gf	Intensity	Spectrum
2661.34	Ta	2.6	180	I
2663.16	Pb	2.0	300	I
2664.79	Ir	0.65	200	I
2670.64	Sb	0.33	38	I
2675.95	Au	0.12	340	I
2677.16	Te		11	I
2681.41	W	1.2	260	I
2685.17	Ta	13.0	180	II
2691.34	Ge	0.80	500	I
2694.23	Ir	1.5	220	I
2700.89	Au	0.037	8	I
2702.40	Pt	0.68	200	I
2705.89	Pt	0.55	160	I
2706.51	Sn	1.1	700	I
2709.63	Ge	1.3	850	I
2710.26	In	0.81	160	I
2714.67	Ta	0.77	300	I
2718.90	W	1.0	260	I
2719.04	Pt	0.43	130	I
2719.65	Ga	0.22	80	I
2724.35	W	1.3	320	I
2733.96	Pt	0.56	180	I
2741.20	Li	0.046	5	I
2745.00	As	2.1	44	I
2748.26	Au	0.43	110	I
2750.48	Yb	4.7	180	II
2753.88	In	0.17	70	I
2754.59	Ge	1.1	650	I
2767.87	Tl	0.47	440	I
2769.95	Sb	0.86	90	I
2779.83	Mg	2.3	90	I
2780.22	As	6.9	140	I
2794.82	Mn	0.97	800	I
2795.53	Mg	1.1	1000	II
2798.27	Mn	0.78	650	I
2801.06	Mn	0.57	480	I
2801.99	Pb	5.1	1000	I
2802.70	Mg	0.62	600	II
2823.20	Pb	2.0	410	I
2824.37	Cu	0.37	50	I
2830.30	Pt	0.28	140	I
2833.06	Pb	0.22	950	I
2837.30	Th		110	II
2838.63	Os	3.9	480	I
2839.99	Sn	2.5	1400	I
2849.72	Ir	0.56	280	I
2850.98	Ta	1.9	220	I
2852.13	Mg	1.1	6000	I

APPENDIX II

Wavelength	Element	gf	Intensity	Spectrum
2860.44	As	4.1	90	I
2863.33	Sn	0.65	1000	I
2866.37	Hf	0.71	240	I
2873.32	Pb	1.2	280	I
2874.24	Ga	0.73	500	I
2877.92	Sb	0.70	140	I
2881.60	Si	1.9	260	I
2887.68	Re	4.8	260	I
2891.38	Yb	0.094	500	II
2894.84	Lu	5.0	420	II
2897.98	Bi	4.0	400	I
2898.26	Hf	1.1	200	I
2909.06	Os	1.4	900	I
2911.39	Lu	7.0	600	II
2916.48	Hf	2.1	220	I
2918.32	Tl	1.9	280	I
2921.52	Tl	0.29	44	I
2924.79	Ir	0.53	320	I
2929.79	Pt	0.26	170	I
2932.63	In	0.32	110	I
2933.55	Ta	0.30	200	I
2938.30	Bi	7.8	320	I
2940.77	Hf	0.55	220	I
2943.15	Ir	1.9	200	I
2943.64	Ga	1.5	950	I
2944.18	Ga	0.24	150	I
2944.40	W	0.69	300	I
2946.98	W	0.69	300	I
2961.16	Cu	0.13	24	I
2963.32	Ta	0.27	180	I
2964.88	Hf	0.73	160	I
2967.28	Hg	2.9	120	I
2970.56	Yb	0.044	280	II
2989.03	Bi	2.3	280	I
2993.34	Bi	0.57	70	I
2997.97	Pt	0.30	180	I
2999.60	Re	7.0	500	I
3002.49	Ni	0.90	320	I
3009.14	Sn	0.53	700	I
3012.54	Ta	11.0	240	II
3017.57	Cr	6.1	360	I
3018.04	Os	0.58	460	I
3021.56	Cr	6.4	360	I
3024.64	Bi	5.3	240	I
3029.20	Au	0.065	32	I
3029.83	Sb	1.4	50	I
3034.12	Sn	0.61	850	I
3039.06	Ge		750	I

APPENDIX II

Wavelength	Element	gf	Intensity	Spectrum
3039.36	In		800	I
3058.66	Os	1.0	900	I
3064.71	Pt	0.37	320	I
3067.72	Bi	0.99	3600	I
3072.88	Hf	0.45	240	I
3075.90	Zn	0.0018	26	I
3077.60	Lu	2.7	500	II
3082.16	Al	0.38	320	I
3092.71	Al	0.79	650	I
3093.11	V	6.1	500	II
3102.30	V	4.6	400	II
3122.78	Au	0.27	160	I
3125.66	Hg	1.1	40	I
3130.42	Be	10.0	480	II
3131.07	Be	6.7	320	II
3131.26	Tm	0.68	700	II
3131.55	Hg	0.92	32	I
3131.83	Hg	0.92	32	I
3132.59	Mo	1.4	1800	I
3133.32	Ir	2.2	340	I
3158.16	Mo	0.57	750	I
3170.35	Mo	0.82	1100	I
3175.05	Sn	0.49	550	I
3175.11	Te	0.94	10	I
3180.20	Th	0.19	75	II
3183.41	V	3.0	420	I
3183.98	V	5.3	700	I
3185.40	V	4.0	500	I
3193.97	Mo	0.68	950	I
3220.78	Ir	1.0	500	I
3229.75	Tl	0.43	120	I
3232.52	Sb	3.5	100	I
3232.61	Li	0.053	17	I
3234.52	Ti	2.6	550	II
3242.70	Pd	1.2	1200	I
3247.54	Cu	0.64	5000	I
3256.09	In	2.0	1300	I
3258.56	In	0.46	300	I
3262.34	Sn	1.8	550	I
3267.51	Sb	1.8	85	I
3269.49	Ge	0.31	110	I
3273.96	Cu	0.31	2500	I
3280.68	Ag	0.53	5500	I
3281.74	Lu	0.52	440	I
3282.33	Zn	0.64	20	I
3289.37	Yb	0.22	2600	II
3301.56	Os	0.59	800	I
3302.32	Na	0.11	30	I

APPENDIX II

Wavelength	Element	gf	Intensity	Spectrum
3302.59	Zn	2.9	90	I
3302.94	Zn	0.91	28	I
3302.99	Na	0.053	15	I
3312.11	Lu	0.23	360	I
3321.01	Be	1.3	60	I
3321.34	Be	2.1	100	I
3324.40	Tb		400	II
3345.02	Zn	4.7	140	I
3345.57	Zn	1.0	30	I
3349.41	Ti	3.9	1000	II
3350.47	Gd	2.4	550	II
3358.62	Gd	1.5	440	II
3359.56	Lu	0.46	440	I
3361.21	Ti	2.2	600	II
3362.23	Gd	2.0	550	II
3372.76	Er	2.2	750	II
3376.50	Lu	0.21	360	I
3380.71	Sr	2.4	65	II
3382.89	Ag	0.23	2800	I
3391.98	Zr	3.6	900	II
3392.03	Th	0.16	90	II
3397.21	Bi	0.66	55	I
3398.98	Ho		900	II
3399.30	Re	2.7	400	I
3399.80	Hf	0.18	260	II
3404.58	Pd	2.0	2600	I
3405.12	Co	2.7	700	I
3412.34	Co	1.9	420	I
3414.76	Ni	1.0	750	I
3421.24	Pd	1.5	1400	I
3422.47	Gd	3.3	700	II
3424.62	Re	5.2	800	I
3425.08	Tm	0.37	600	II
3434.89	Rh	0.73	700	I
3438.23	Zr	2.4	750	II
3440.61	Fe	0.50	400	I
3443.64	Co	2.4	550	I
3446.26	Ni	0.68	440	I
3447.12	Mo	6.1	400	I
3451.88	Re	0.36	1600	I
3453.50	Co	4.6	1300	I
3456.00	Ho		1800	II
3458.47	Ni	0.88	460	I
3460.46	Re	1.2	5500	I
3460.77	Pd	0.60	850	I
3461.65	Ni	0.57	460	I
3462.04	Rh	1.0	500	I
3462.20	Tm	0.44	800	II

Wavelength	Element	gf	Intensity	Spectrum
3464.36	Yb	0.13	340	I
3464.46	Sr	3.8	95	II
3464.73	Re	0.88	4000	I
3466.20	Cd	7.4	250	I
3467.66	Cd	2.4	80	I
3469.92	Tm		95	II
3474.02	Co	0.64	500	I
3481.15	Pd	2.0	1100	I
3484.84	Ho		700	II
3492.96	Ni	0.72	500	I
3496.21	Zr	1.6	650	II
3498.94	Ru	1.1	850	I
3499.11	Er	1.8	650	II
3502.28	Co	2.0	600	I
3502.52	Rh	0.47	500	I
3506.32	Co	1.7	440	I
3507.39	Lu	0.037	480	II
3509.17	Tb		600	II
3513.64	Ir	0.18	320	I
3515.05	Ni	0.83	600	I
3516.94	Pd	1.2	1300	I
3519.24	Tl	4.5	2000	I
3519.60	Zr	0.93	320	I
3524.54	Ni	0.85	750	I
3528.02	Rh	1.0	750	I
3529.43	Tl	1.1	500	I
3529.81	Co	1.8	460	I
3531.70	Dy	3.5	2000	I
3545.80	Gd	1.4	440	II
3547.68	Zr	0.92	280	I
3551.95	Zr	0.74	280	II
3553.08	Pd	3.4	1300	I
3556.60	Zr	2.1	340	II
3561.66	Hf	0.083	150	II
3561.74	Tb		340	II
3566.37	Ni	1.2	460	I
3568.27	Sm	1.6	350	II
3568.51	Tb		440	II
3569.38	Co	5.0	550	I
3570.10	Fe	3.4	400	I
3572.47	Zr	0.71	340	II
3572.53	Sc	1.1	1200	II
3578.69	Cr	1.6	2400	I
3580.27	Nb	1.2	600	I
3581.20	Fe	4.4	600	I
3584.96	Gd	1.7	550	II
3587.19	Co	5.0	420	I
3589.22	Ru	1.8	650	I

APPENDIX II

Wavelength	Element	gf	Intensity	Spectrum
3592.60	Sm	1.2	350	II
3593.02	Ru	1.7	700	I
3593.49	Cr	1.4	2100	I
3596.18	Ru	1.3	650	I
3597.15	Rh	1.0	500	I
3600.73	Y	1.0	1300	II
3601.19	Zr	2.0	550	I
3605.33	Cr	1.0	1600	I
3609.49	Sm	0.76	280	II
3609.55	Pd	1.8	2200	I
3610.51	Cd	12.0	360	I
3611.05	Y	0.70	1000	II
3612.88	Cd	2.3	70	I
3613.84	Sc	2.2	2500	II
3619.39	Ni	1.5	600	I
3630.75	Sc	1.5	1800	II
3633.12	Y	0.51	1000	II
3624.29	Sm	0.59	280	II
3634.70	Pd	1.2	2200	I
3639.58	Pb	0.26	550	I
3642.68	Ti	1.8	550	I
3642.79	Sc	0.98	1200	II
3645.41	Dy	1.9	1000	II
3646.19	Gd	2.1	600	II
3650.15	Hg	13.0	280	I
3653.50	Ti	2.0	600	I
3655.85	Ce	1.5	160	II
3657.99	Rh	0.82	700	I
3661.36	Sm	0.27	180	II
3670.07	U	0.38	160	II
3670.84	Sm	0.30	180	II
3676.35	Tb		380	II
3682.24	Hf	0.16	220	I
3683.48	Pb	0.64	1400	I
3692.36	Rh	0.58	800	I
3692.64	Er	1.5	700	II
3694.19	Yb	0.15	3200	II
3700.91	Rh	0.72	650	I
3702.85	Tb		460	II
3703.58	V	2.5	400	I
3710.30	Y	1.0	1500	II
3717.92	Tm	1.7	650	I
3719.94	Fe	0.52	600	I
3724.94	Eu	0.59	1700	II
3725.76	Re	38.0	400	I
3726.93	Ru	1.1	800	I
3728.03	Ru	0.96	1000	I
3734.87	Fe	4.2	700	I

APPENDIX II

Wavelength	Element	gf	Intensity	Spectrum
3737.13	Fe	0.32	340	I
3739.12	Sm		220	II
3739.95	Fe		280	I
3741.19	Th	0.10	90	II
3749.49	Fe	2.7	400	I
3752.52	Os	0.30	360	I
3768.39	Gd	1.8	850	II
3774.33	Y	0.69	1200	II
3775.72	Tl	0.22	1200	I
3782.84	U	0.25	140	II
3788.70	Y	0.45	850	II
3790.83	La	0.36	440	II
3794.78	La	0.50	460	II
3795.76	Tm	0.23	600	II
3796.37	Gd	0.90	500	II
3796.75	Ho		1000	I
3798.25	Mo	0.94	3200	I
3798.90	Ru	0.87	700	I
3799.35	Ru	0.62	700	I
3800.12	Ir	0.13	320	I
3801.53	Ce	5.9	200	II
3805.36	Nd		150	II
3810.73	Ho		1000	I
3812.00	U	1.7	140	I
3819.67	Eu	1.0	3400	II
3820.43	Fe	2.7	500	I
3829.35	Mg	2.4	140	I
3830.53	Er	0.52	320	II
3831.46	U		150	II
3832.31	Mg	5.1	300	I
3835.96	Zr	0.54	280	I
3838.26	Mg	8.6	500	I
3848.02	Tm	0.25	750	II
3848.76	Tb		340	II
3850.97	Gd	0.79	500	II
3851.66	Nd		140	II
3854.21	Sm		200	II
3854.66	U		180	II
3856.52	Rh	1.5	500	I
3859.58	U	0.58	360	II
3859.91	Fe	0.31	420	I
3862.82	Er		600	I
3863.33	Nd	0.35	220	II
3864.11	Mo	0.77	2800	I
3865.92	U	0.39	140	II
3871.64	La	0.26	340	II
3872.13	Dy	0.69	600	II
3874.19	Tb		320	II

APPENDIX II

Wavelength	Element	gf	Intensity	Spectrum
3885.29	Sm	0.87	280	II
3890.36	U	0.25	160	II
3891.02	Ho		1500	I
3891.78	Ba	3.3	140	II
3892.69	Er		340	I
3896.25	Er	0.71	420	II
3902.96	Mo	0.47	1800	I
3906.34	Er		850	II
3907.10	Eu	1.1	2400	II
3907.49	Sc	2.7	1800	I
3908.43	Pr	0.43	320	II
3911.81	Sc	3.3	2100	I
3930.48	Eu	1.2	2800	II
3933.67	Ca	0.21	4200	II
3942.75	Ce	4.4	190	II
3944.03	Al	0.15	450	I
3944.70	Dy	0.91	850	II
3949.10	La	1.2	900	II
3952.54	Ce	1.5	220	II
3961.53	Al	0.31	900	I
3968.42	Dy	1.1	1100	II
3968.47	Ca	0.11	2200	II
3971.96	Eu	0.84	2000	II
3987.98	Yb	0.38	1900	I
3988.52	La	0.55	440	II
3995.75	La	0.26	360	II
3998.64	Ti	1.5	650	I
3999.24	Ce	1.2	200	II
4000.48	Dy	0.83	650	II
4007.97	Er		1100	I
4008.75	W	0.44	950	I
4012.25	Nd	1.3	220	II
4012.39	Ce	2.1	190	II
4013.84	Ho		1000	I
4019.13	Th	0.16	300	II
4020.40	Sc	2.4	1800	I
4023.69	Sc	2.5	1800	I
4030.76	Mn	0.33	2000	I
4032.98	Ga	0.24	1000	I
4033.36	Mn		1400	I
4034.49	Mn	0.13	800	I
4040.80	Nd	0.36	180	II
4041.36	Mn	8.4	420	I
4042.91	La	1.2	300	II
4044.14	K	0.23	32	I
4045.99	Dy		1000	I
4046.56	Hg	8.8	180	I
4047.20	K	0.12	16	I

Wavelength	Element	gf	Intensity	Spectrum
4053.93	Ho	5.9	900	I
4054.85	Pr	0.43	200	II
4056.54	Pr	1.1	200	II
4057.83	Pb	2.3	3400	I
4058.94	Nb	1.9	1700	I
4061.09	Nd	1.1	280	II
4062.82	Pr	1.0	300	II
4074.36	W	0.24	550	I
4077.38	Y	1.1	950	I
4077.71	Sr	0.17	4600	II
4077.98	Dy	0.71	600	II
4079.73	Nb	1.2	1200	I
4086.72	La	0.25	550	II
4090.14	U	0.31	160	II
4094.19	Tm	1.3	750	I
4100.75	Pr	1.1	260	II
4100.92	Nb	0.64	700	I
4101.76	In	0.47	1700	I
4102.38	Y	1.3	1000	I
4105.84	Tm	1.2	700	I
4109.46	Nd	0.39	150	II
4111.78	V	2.8	700	I
4116.71	Th	0.21	75	II
4123.23	La	0.40	440	II
3123.81	Nb	0.46	550	I
4128.31	Y	1.1	900	I
4129.70	Eu	0.49	2200	II
4130.66	Ba	4.6	150	II
4133.80	Ce	3.7	190	II
4143.14	Pr	0.68	240	II
4151.10	Er		550	I
4152.58	Nb	0.43	460	I
4156.08	Nd	0.32	180	II
4163.03	Ho		900	I
4163.66	Nb	0.37	460	I
4164.66	Nb	0.36	420	I
4168.13	Nb	0.28	360	I
4172.06	Ga	0.53	2000	I
4177.32	Nd	0.19	140	II
4179.42	Pr	0.86	460	II
4186.60	Ce	4.6	250	II
4186.78	Dy		950	I
4187.62	Tm	1.0	650	I
4189.52	Pr	0.60	220	II
4199.90	Ru	2.6	700	I
4201.85	Rb	0.25	32	I
4205.05	Eu	0.84	4000	II
4206.74	Pr	0.88	220	II

APPENDIX II

Wavelength	Element	gf	Intensity	Spectrum
4210.94	Ag	0.68	9	I
4211.72	Dy		1300	I
4215.52	Sr	0.10	3200	II
4215.56	Rb	0.13	16	I
4222.98	Pr	0.43	340	II
4225.33	Pr	0.38	340	II
4226.73	Ca	0.28	1100	I
4227.46	Re	7.0	360	I
4246.83	Sc	1.2	1400	II
4254.35	Cr	0.54	1700	I
4260.85	Os	0.10	440	I
4274.80	Cr	0.41	1300	I
4289.72	Cr	0.26	850	I
4294.61	W	0.16	450	I
4302.53	Ca	2.0	110	I
4303.58	Nd	0.34	320	II
4305.92	Ti	5.2	500	I
4326.47	Tb	2.0	280	I
4333.74	La	0.25	460	II
4358.35	Hg	24.0	400	I
4374.94	Y	0.72	1200	II
4379.24	V	3.0	950	I
4381.86	Th	0.23	90	II
4384.72	V	1.7	550	I
4389.97	V	1.1	380	I
4391.11	Th	0.11	80	II
4420.47	Os	0.090	440	I
4424.34	Sm	0.38	200	II
4435.56	Eu	0.25	900	II
4454.78	Ca	2.2	140	I
4460.21	Ce	1.0	170	II
4511.31	In	0.66	1800	I
4513.31	Re	6.1	260	I
4518.57	Lu	0.057	340	I
4533.24	Ti	4.4	500	I
4554.03	Ba	0.28	6500	II
4555.36	Cs	0.42	40	I
4572.67	Be	1.7	12	I
4593.18	Cs	0.21	20	I
4594.03	Eu	2.1	750	I
4602.86	Li	0.59	13	I
4607.33	Sr	0.27	650	I
4607.34	Au	1.9	9	I
4627.22	Eu	1.8	650	I
4678.16	Cd	2.6	80	I
4680.14	Zn	1.9	40	I
4708.60	Be		7	BeO
4722.16	Zn	4.9	100	I

APPENDIX II

Wavelength	Element	gf	Intensity	Spectrum
4722.19	Bi	0.060	60	I
4799.92	Cd	4.9	140	I
4810.53	Zn	7.2	140	I
4934.09	Ba	0.068	2000	II
4962.26	Sr	1.8	80	I
4971.99	Li	0.29	8	I
4981.73	Ti	3.7	550	I
5085.82	Cd	12.0	280	I
5105.54	Cu	0.020	40	I
5167.34	Mg	0.48	75	I
5172.70	Mg	1.4	220	I
5183.62	Mg	2.6	400	I
5204.52	Cr	0.65	440	I
5206.04	Cr	1.0	700	I
5208.44	Cr	1.3	900	I
5209.07	Ag	6.1	100	I
5218.20	Cu	2.4	100	I
5350.46	Tl	0.92	1800	I
5460.74	Hg	38.0	320	I
5465.49	Ag	7.1	100	I
5471.55	Ag	0.71	10	I
5480.84	Sr	3.2	70	I
5506.49	Mo	0.90	480	I
5535.48	Ba	0.90	650	I
5556.48	Yb	0.010	140	I
5588.75	Ca	2.5	70	I
5682.66	Na	0.42	7	I
5688.22	Na	0.85	14	I
5782.13	Cu	0.027	40	I
5782.60	K	0.10	1	I
5801.96	K	0.14	1.4	I
5853.68	Ba	0.026	280	II
5889.95	Na	0.95	2000	I
5895.92	Na	0.47	1000	I
6010.33	Cs	1.0	8	I
6103.64	Li	7.5	320	I
6110.78	Ba	2.9	170	I
6141.72	Ba	0.21	2000	II
6160.76	Na	0.32	6	I
6162.17	Ca	1.0	140	I
6212.87	Cs	1.7	12	I
6298.33	Rb	0.42	4	I
6408.47	Sr	3.2	90	I
6439.07	Ca	2.0	70	I
6462.57	Ca	2.0	70	I
6496.90	Ba	0.094	1200	II
6498.76	Ba	2.5	160	I

APPENDIX II

Wavelength	Element	gf	Intensity	Spectrum
6707.84	Li	0.80	3600	I
6723.28	Cs	2.2	20	I
6911.30	K	0.20	2.5	I
6938.98	K	0.40	5	I
6973.29	Cs	2.4	20	I
7070.10	Sr	0.65	55	I
7280.00	Rb	0.29	3.5	I
7408.17	Rb	0.44	5	I
7618.93	Rb	0.57	7	I
7664.91	K	1.4	1800	I
7687.78	Ag	1.2	32	I
7698.98	K	0.70	900	I
7757.65	Rb	0.94	11	I
7800.23	Rb	2.7	3000	I
7947.60	Rb	1.2	1500	I
8015.71	Cs	1.4	6	I
8092.63	Cu	2.6	40	I
8126.52	Li	0.84	48	I
8183.27	Na	4.4	110	I
8194.81	Na	8.8	220	I
8273.52	Ag	2.3	50	I
8521.10	Cs	1.4	1500	I
8542.09	Ca	0.051	100	II
8761.38	Cs	5.0	55	I
8918.80	Se	20.0	5	I
8943.50	Cs	0.57	800	I

Appendix III
Spectral Lines, Arranged by Elements, with *gf* and Intensity Values[1]

Wavelength	Intensity	Spectrum	Wavelength	Intensity	Spectrum
Aluminum					
2367.06	18	I	2492.91	44	I
2373.13	36	I	2745.00	44	I
2567.99	24	I	2780.22	140	I
2575.10	48	I	2860.44	90	I
2652.49	15	I			
2660.39	20	I	Barium		
3082.16	320	I	3891.78	140	II
3092.71	650	I	4130.66	150	II
3944.03	450	I	4554.03	6500	II
3961.53	900	I	4934.09	2000	II
			5535.48	650	I
Antimony			5853.68	280	II
2175.81	38	I	6110.78	170	I
2311.47	45	I	6141.72	2000	II
2528.52	320	I	6496.90	1200	II
2598.05	600	I	6498.76	160	I
2670.64	38	I			
2769.95	90	I	Beryllium		
2877.92	140	I	2348.61	300	I
3029.83	50	I	2494.56	70	I
3232.52	100	I	2494.73	100	I
3267.51	85	I	2650.47	140	I
			3130.42	480	II
Arsenic			3131.07	320	II
1971.97	28	I	3321.01	60	I
2003.34	28	I	3321.34	100	I
2288.12	44	I	4572.67	12	I
2349.84	85	I	4708.60	7	BeO
2456.53	36	I			

[1] See introduction and footnote 1 to Appendix II.

APPENDIX III

Wavelength	Intensity	Spectrum	Wavelength	Intensity	Spectrum
Bismuth			4133.80	190	II
2061.70	55	I	4186.60	250	II
2627.91	70	I	4460.21	170	II
2897.98	400	I	Cesium		
2938.30	320	I	4555.36	40	I
2989.03	280	I	4593.18	20	I
2993.34	70	I	6010.33	8	I
3024.64	240	I	6212.87	12	I
3067.72	3600	I	6723.28	20	I
3397.21	55	I	6973.29	20	I
4722.19	60	I	8015.71	6	I
			8521.10	1500	I
Boron			8761.38	55	I
2088.93	7	I	8943.50	800	I
2089.59	11	I	Chromium		
2496.78	240	I	3017.57	360	I
2497.73	480	I	3021.56	360	I
Cadmium			3578.69	2400	I
2144.38	60	II	3593.49	2100	I
2265.02	110	II	3605.33	1600	I
2288.02	1500	I	4254.35	1700	I
3466.20	250	I	4274.80	1300	I
3467.66	80	I	4289.72	850	I
3610.51	360	I	5204.52	440	I
3612.88	70	I	5206.04	700	I
4678.16	80	I	5208.44	900	I
4799.92	140	I	Cobalt		
5085.82	280	I	3405.12	700	I
Calcium			3412.34	420	I
3933.67	4200	II	3443.64	550	I
3968.47	2200	II	3453.50	1300	I
4226.73	1100	I	3474.02	500	I
4302.53	110	I	3502.28	600	I
4454.78	140	I	3506.32	440	I
5588.75	70	I	3529.81	460	I
6162.17	140	I	3569.38	550	I
6439.07	70	I	3587.19	420	I
6462.57	70	I	Copper		
8542.09	100	II	2492.15	36	I
Cerium			2618.37	40	I
3655.85	160	II	2824.37	50	I
3801.53	200	II	2961.16	24	I
3942.75	190	II	3247.54	5000	I
3952.54	220	II	3273.96	2500	I
3999.24	200	II	5105.54	40	I
4012.39	190	II			

APPENDIX III

Wavelength	Intensity	Spectrum	Wavelength	Intensity	Spectrum
5218.20	100	I	3796.37	500	II
5782.13	40	I	3850.97	500	II
8092.63	40	I	Gallium		
Dysprosium			2450.07	75	I
3531.70	2000	I	2500.17	130	I
3645.41	1000	I	2500.70	12	I
3872.13	600	II	2659.87	34	I
3944.70	850	II	2719.65	80	I
3968.42	1100	II	2874.24	500	I
4000.48	650	II	2943.64	950	I
4045.99	1000	I	2944.18	150	I
4077.98	600	II	4032.98	1000	I
4186.78	950	I	4172.06	2000	I
4211.72	1300	I			
Erbium			Germanium		
3372.76	750	II	2417.37	85	I
3499.11	650	II	2497.96	90	I
3692.64	700	II	2592.54	500	I
3830.53	320	II	2651.18	1200	I
3862.82	600	I	2651.58	550	I
3892.69	340	I	2691.34	500	I
3896.25	420	II	2709.63	850	I
3906.34	850	II	2754.59	650	I
4007.87	1100	I	3039.06	750	I
4151.10	550	I	3269.49	110	I
Europium			Gold		
3724.94	1700	II	2012.00	8	I
3819.67	3400	II	2427.95	200	I
3907.10	2400	II	2641.49	26	I
3930.48	2800	II	2675.95	340	I
3971.96	2000	II	2700.89	8	I
4129.70	2200	II	2748.26	110	I
4205.05	4000	II	3029.20	32	I
4435.56	900	II	3122.78	160	I
4594.03	750	I	4607.34	9	I
4627.22	650	I	Hafnium		
Gadolinium			2866.37	240	I
3350.47	550	II	2898.26	200	I
3358.62	440	II	2916.48	220	I
3362.23	550	II	2940.77	220	I
3422.47	700	II	2964.88	160	I
3545.80	440	II	3072.88	240	I
3584.96	550	II	3399.80	260	II
3646.19	600	II	3561.66	150	II
3768.39	850	II	3682.24	220	I

Wavelength	Intensity	Spectrum	Wavelength	Intensity	Spectrum
Holmium			3949.10	900	II
3398.98	900	II	3988.52	440	II
3456.00	1800	II	3995.75	360	II
3484.84	700	II	4042.91	300	II
3796.75	1000	I	4086.72	550	II
3810.73	1000	I	4123.23	440	II
3891.02	1500	II	4333.74	460	II
4053.93	900	I	Lead		
4013.84	1000	I	2614.18	700	I
4163.03	900	I	2663.16	300	I
Indium			2801.99	1000	I
2560.15	110	I	2823.20	410	I
2710.26	160	I	2833.06	950	I
2753.88	70	I	2873.32	280	I
2932.63	110	I	3639.58	550	I
3039.36	800	I	3683.48	1400	I
3256.09	1300	I	3739.95	280	I
3258.56	300	I	4057.83	3400	I
4101.76	1700	I	Lithium		
4511.31	1800	I	2741.20	5	I
Iridium			3232.61	17	I
2502.98	200	I	4602.86	13	I
2543.97	380	I	4971.99	8	I
2664.79	200	I	6103.64	320	I
2694.23	220	I	6707.84	3600	I
2849.72	280	I	8126.52	48	I
2924.79	320	I	Lutetium		
2943.15	200	I	2615.42	1200	II
3133.32	340	I	2894.84	420	II
3220.78	500	I	2911.39	600	II
3513.64	320	I	3077.60	500	II
3800.12	320	I	3281.74	440	I
Iron			3312.11	360	I
3440.61	400	I	3359.56	440	I
3570.10	400	I	3376.50	360	I
3581.20	600	I	3507.39	480	II
3719.94	600	I	4518.57	340	I
3734.87	700	I	Magnesium		
3737.13	340	I	2779.83	90	I
3749.49	400	I	2795.53	1000	II
3820.43	500	I	2802.70	600	II
3859.91	420	I	2852.13	6000	I
Lanthanum			3829.35	140	I
3790.83	440	II	3832.31	300	I
3794.78	460	II	3838.26	500	I
3871.64	340	II			

APPENDIX III

Wavelength	Intensity	Spectrum	Wavelength	Intensity	Spectrum
			Nickel		
5167.34	75	I	3002.49	320	I
5172.70	220	I	3414.76	750	I
5183.62	400	I	3446.26	440	I
			3458.47	460	I
Manganese			3461.65	460	I
2576.10	1200	II	3492.96	500	I
2593.73	800	II	3515.05	600	I
2605.69	550	II	3524.54	750	I
2794.82	800	I	3566.37	460	I
2798.27	650	I	3619.39	600	I
2801.06	480	I			
4030.76	2000	I	Niobium		
4033.36	1400	I	3580.27	600	I
4034.49	800	I	4058.94	1700	I
4041.36	420	I	4079.73	1200	I
			4100.92	700	I
Mercury			4123.81	550	I
2536.52	1500	I	4152.58	460	I
2967.28	120	I	4163.66	460	I
3125.66	40	I	4164.66	420	I
3131.55	32	I	4168.13	360	I
3131.83	32	I			
3650.15	280	I	Osmium		
4046.56	180	I	2488.55	380	I
4358.35	400	I	2637.13	360	I
5460.74	320	I	2838.63	480	I
			2909.06	900	I
Molybdenum			3018.04	480	I
3132.59	1800	I	3058.66	900	I
3158.16	750	I	3301.56	800	I
3170.35	1100	I	3752.52	360	I
3193.97	950	I	4260.85	440	I
3447.12	400	I	4420.47	440	I
3798.25	3200	I			
3864.11	2800	I	Palladium		
3902.96	1800	I	3242.70	1200	I
5506.49	480	I	3404.58	2600	I
			3421.24	1400	I
Neodymium			3460.77	850	I
3805.36	150	II	3481.15	1100	I
3851.66	140	II	3516.94	1300	I
3863.33	220	II	3553.08	1300	I
4012.25	220	II	3609.55	2200	I
4040.80	180	II	3634.70	2200	I
4061.09	280	II			
4109.46	150	II	Phosphorus		
4156.08	180	II	2135.47	1	I
4177.32	140	II	2136.20	10	I
4303.58	320	II	2149.11	9	I

Wavelength	Intensity	Spectrum	Wavelength	Intensity	Spectrum
2152.95	1	I	3464.73	4000	I
2154.08	2	I	3725.76	400	I
2534.01	22	I	4227.46	360	I
2535.65	60	I	4513.31	260	I
2553.28	38	I	Rhodium		
2554.93	15	I	3434.89	700	I
Platinum			3462.04	500	I
2628.03	110	I	3502.52	500	I
2659.45	280	I	3528.02	750	I
2702.40	200	I	3597.15	500	I
2705.89	160	I	3657.99	700	I
2719.04	130	I	3692.36	800	I
2733.96	180	I	3700.91	650	I
2830.30	140	I	3856.52	500	I
2929.79	170	I	Rubidium		
2997.97	180	I	4201.85	32	I
3064.71	320	I	4215.56	16	I
Potassium			6298.33	4	I
4044.14	21	I	7280.00	3.5	I
4047.20	16	I	7408.17	5	I
5782.60	1	I	7618.93	7	I
5801.96	1.4	I	7757.65	11	I
6911.30	2.5	I	7800.23	3000	I
6938.98	5	I	7947.60	1500	I
7664.91	1800	I	Ruthenium		
7698.98	900	I	3498.94	850	I
Praseodymium			3589.22	650	I
3908.43	320	II	3593.02	700	I
4054.85	200	II	3596.18	650	I
4056.54	200	II	3726.93	800	I
4062.23	300	II	3728.03	1000	I
4100.75	260	II	3798.90	700	I
4143.14	240	I	3799.35	700	I
4179.42	460	II	4199.90	700	I
4189.52	220	II	Samarium		
4206.74	220	II	3568.27	350	II
4222.98	340	II	3592.60	350	II
4225.33	340	II	3609.49	280	II
Rhenium			3634.29	280	II
2887.68	260	I	3661.36	180	II
2999.60	500	I	3670.84	180	II
3399.30	400	I	3739.12	220	II
3424.62	800	I	3854.21	200	II
3451.88	1600	I	3885.29	280	II
3460.46	5500	I	4424.34	200	II

APPENDIX III

Wavelength	Intensity	Spectrum	Wavelength	Intensity	Spectrum
Scandium			Strontium		
3572.53	1200	II	3380.71	65	II
3613.84	2500	II	3464.46	95	II
3630.75	1800	II	4077.71	4600	II
3642.79	1200	II	4215.52	3200	II
3907.49	1800	I	4607.33	650	I
3911.81	2100	I	4962.26	80	I
4020.40	1800	I	5480.84	70	I
4023.69	1800	I	6408.47	90	I
4246.83	1400	II	7070.10	55	I
Selenium			Tantalum		
1960.26	34	I	2647.47	280	I
2039.85	40	I	2653.27	300	I
2062.79	15	I	2656.61	220	I
2074.79	3.0	I	2661.34	180	I
8918.80	5	I	2685.17	180	II
			2714.67	300	I
Silicon			2850.98	220	I
2435.16	26	I	2933.55	200	I
2506.90	170	I	2963.32	180	I
2514.32	160	I	3012.54	240	II
2516.11	360	I	Tellurium		
2519.21	120	I	1994.2	8	I
2524.11	240	I	2002.0	16	I
2528.51	200	I	2081.03	10	I
2631.28	24	I	2142.75	55	I
2881.60	260	I	2147.19	11	I
			2383.25	55	I
Silver			2385.76	70	I
2413.18	10	II	2530.70	12	I
3280.68	5500	I	2677.16	11	I
3382.89	2800	I	3175.11	10	I
4210.94	9	I			
5209.07	100	I	Terbium		
5465.49	100	I	3324.40	400	II
5471.55	10	I	3509.17	600	II
7687.78	32	I	3561.74	340	II
8273.52	50	I	3568.51	440	II
			3676.35	380	II
Sodium			3702.85	460	II
3302.32	30	I	3848.76	340	II
3302.99	15	I	3874.19	320	II
5682.66	7	I	4326.47	280	I
5688.22	14	I	Thallium		
5889.95	2000	I	2580.14	70	I
5895.92	1000	I	2767.87	440	I
6160.76	6	I	2918.32	280	I
8183.27	110	I			
8194.81	220	I			

Wavelength	Intensity	Spectrum	Wavelength	Intensity	Spectrum
2921.52	44	I	4533.24	500	I
3229.75	120	I	4981.73	550	I
3519.24	2000	I	**Tungsten**		
3529.43	500	I	2551.35	280	I
3775.72	1200	I	2681.41	260	I
5350.46	1800	I	2718.90	260	I
Thorium			2724.35	320	I
2837.30	110	II	2944.40	300	I
3180.20	75	II	2946.98	300	I
3392.03	90	II	4008.75	950	I
3469.92	95	II	4074.36	550	I
3741.19	90	II	4294.61	450	I
4019.13	300	II			
4116.71	75	II	**Uranium**		
4381.86	90	II	3670.07	160	II
4391.11	80	II	3782.84	140	II
Thulium			3812.00	140	I
3131.26	700	II	3831.46	150	II
3425.08	600	II	3854.66	180	II
3462.20	800	II	3859.58	360	II
3717.92	650	I	3865.92	140	II
3795.76	600	II	3890.36	160	II
3848.02	750	II	4090.14	160	II
4094.19	750	I			
4105.84	700	I	**Vanadium**		
4187.62	650	I	3093.11	500	II
Tin			3102.30	400	II
2421.70	260	I	3183.41	420	I
2429.49	420	I	3183.98	700	I
2546.55	240	I	3185.40	500	I
2706.51	700	I	3703.58	400	I
2839.99	1400	I	4111.78	700	I
2863.33	1000	I	4379.24	950	I
3009.14	700	I	4384.72	550	I
3034.12	850	I	4389.97	380	I
3175.05	550	I			
3262.34	550	I	**Ytterbium**		
Titanium			2653.74	140	II
3234.52	550	II	2750.48	180	II
3349.41	1000	II	2891.38	500	II
3361.21	600	II	2970.56	280	II
3642.68	550	I	3289.37	2600	II
3653.50	600	I	3464.36	340	I
3998.64	650	I	3694.19	3200	II
4305.92	500	I	3987.98	1900	I
			5556.48	140	I

Wavelength	Intensity	Spectrum	Wavelength	Intensity	Spectrum
Yttrium					
3600.73	1300	II	3345.02	140	I
3611.05	1000	II	3345.57	30	I
3633.12	1000	II	4680.14	40	I
3710.30	1500	II	4722.16	100	I
3774.33	1200	II	4810.53	140	I
3788.70	850	II	Zirconium		
4077.38	950	I	3391.98	900	II
4102.38	1000	I	3438.23	750	II
4128.31	900	I	3496.21	650	II
4374.94	1200	II	3519.60	320	I
Zinc			3547.68	280	I
2138.56	1000	I	3551.95	280	II
3075.90	26	I	3556.60	340	II
3282.33	20	I	3572.47	340	II
3302.59	90	I	3601.19	550	I
3302.94	28	I	3835.96	280	I

Appendix IV
Spectral Charts

Following are photographic charts of the iron spectrum from 2400 to 4400 Å with the positions of some of the more intense spectral lines of 50 elements identified. The charts can be used for qualitative identification of the 50 elements by comparing the spectrum of the unknown with the charts. It is most helpful if an iron spectrum is placed adjacent to the unknown spectrum to aid in locating the spectral lines of the unknown.

The iron spectrum was obtained using a 3.4-m focal length grating spectrograph.

APPENDIX IV

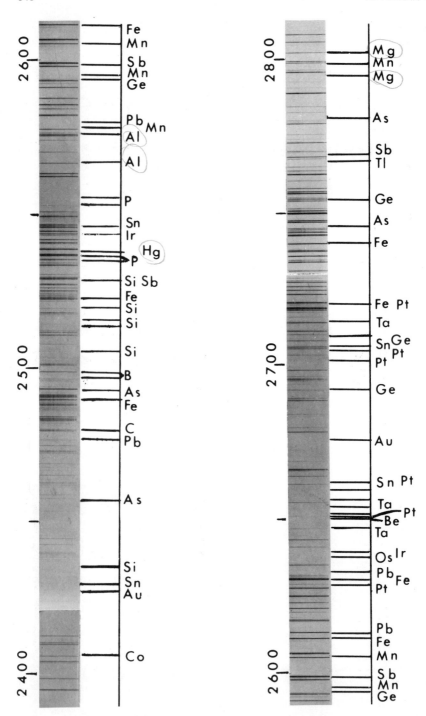

APPENDIX IV

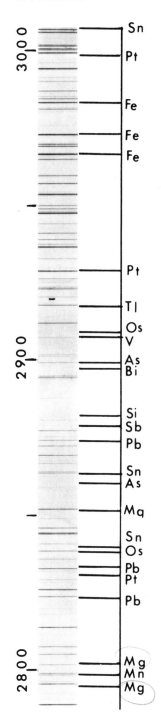
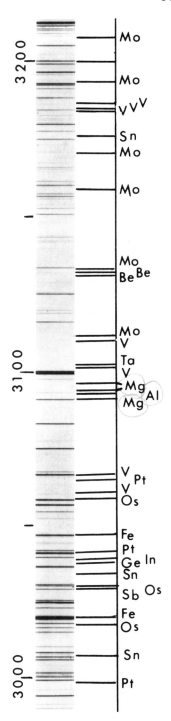

APPENDIX IV

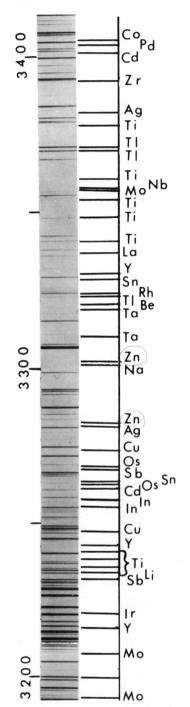
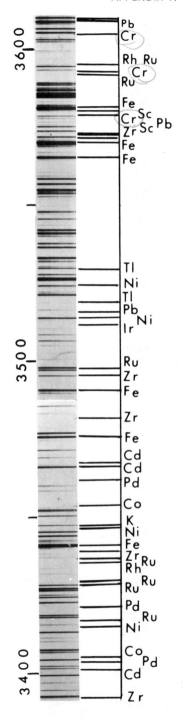

APPENDIX IV 351

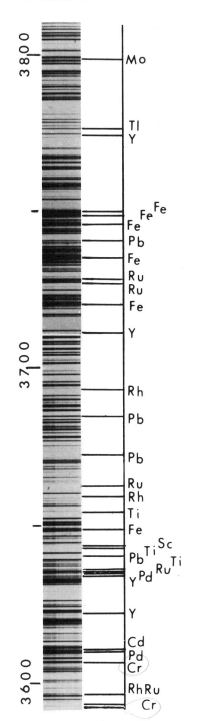
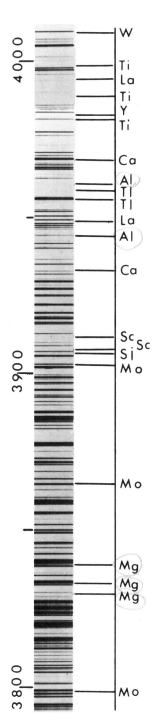

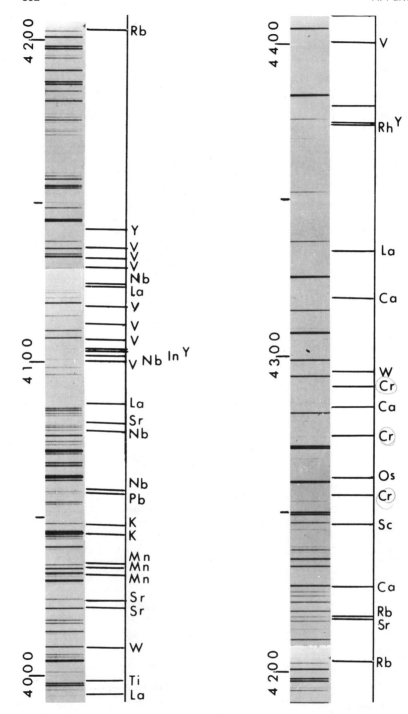

Appendix V
Absorbance Values Calculated from Percentage Transmittances

%T		1/4	1/2	3/4	%T		1/4	1/2	3/4
1	2.000	1.903	1.824	1.727	34	0.469	0.465	0.462	0.459
2	1.699	1.648	1.602	1.561	35	0.456	0.453	0.450	0.447
3	1.523	1.488	1.456	1.426	36	0.444	0.441	0.438	0.435
4	1.398	1.378	1.347	1.323	37	0.432	0.429	0.426	0.423
5	1.391	1.280	1.260	1.240	38	0.420	0.417	0.414	0.412
6	1.222	1.204	1.187	1.171	39	0.409	0.406	0.403	0.401
7	1.155	1.140	1.126	1.112	40	0.398	0.395	0.392	0.390
8	1.097	1.083	1.071	1.059	41	0.387	0.385	0.382	0.380
9	1.046	1.034	1.022	1.011	42	0.377	0.374	0.372	0.369
10	1.000	0.989	0.979	0.969	43	0.367	0.364	0.362	0.359
11	0.959	0.949	0.939	0.930	44	0.357	0.354	0.352	0.349
12	0.921	0.912	0.903	0.894	45	0.347	0.344	0.342	0.340
13	0.886	0.878	0.870	0.862	46	0.337	0.335	0.332	0.330
14	0.854	0.846	0.838	0.831	47	0.328	0.325	0.323	0.321
15	0.824	0.817	0.810	0.803	48	0.319	0.317	0.314	0.312
16	0.796	0.789	0.782	0.776	49	0.310	0.308	0.305	0.303
17	0.770	0.763	0.757	0.751	50	0.301	0.299	0.297	0.295
18	0.745	0.739	0.733	0.727	51	0.2924	0.2903	0.2882	0.2861
19	0.721	0.716	0.710	0.704	52	0.2840	0.2819	0.2798	0.2777
20	0.699	0.694	0.688	0.683	53	0.2756	0.2736	0.2716	0.2696
21	0.678	0.673	0.668	0.663	54	0.2676	0.2656	0.2636	0.2616
22	0.658	0.653	0.648	0.643	55	0.2596	0.2577	0.2557	0.2537
23	0.638	0.634	0.629	0.624	56	0.2518	0.2499	0.2480	0.2460
24	0.620	0.615	0.611	0.606	57	0.2441	0.2422	0.2403	0.2384
25	0.602	0.598	0.594	0.589	58	0.2366	0.2347	0.2328	0.2310
26	0.585	0.581	0.577	0.573	59	0.2291	0.2273	0.2255	0.2236
27	0.569	0.565	0.561	0.557	60	0.2218	0.2200	0.2182	0.2164
28	0.553	0.549	0.545	0.542	61	0.2147	0.2129	0.2111	0.2093
29	0.538	0.534	0.530	0.527	62	0.2076	0.2059	0.2041	0.2024
30	0.523	0.520	0.516	0.512	63	0.2007	0.1990	0.1973	0.1956
31	0.509	0.505	0.502	0.498	64	0.1939	0.1922	0.1905	0.1888
32	0.495	0.491	0.488	0.485	65	0.1871	0.1855	0.1838	0.1821
33	0.482	0.478	0.475	0.472	66	0.1805	0.1788	0.1772	0.1756

%T		1/4	1/2	3/4	%T		1/4	1/2	3/4
67	0.1739	0.1723	0.1707	0.1691	84	0.0757	0.0744	0.0731	0.0718
68	0.1675	0.1659	0.1643	0.1627	85	0.0706	0.0693	0.0680	0.0667
69	0.1612	0.1596	0.1580	0.1565	86	0.0655	0.0642	0.0630	0.0617
70	0.1549	0.1534	0.1518	0.1503	87	0.0605	0.0593	0.0580	0.0568
71	0.1487	0.1472	0.1457	0.1442	88	0.0555	0.0543	0.0531	0.0518
72	0.1427	0.1412	0.1397	0.1382	89	0.0505	0.0494	0.0482	0.0470
73	0.1367	0.1352	0.1337	0.1322	90	0.0458	0.0446	0.0434	0.0422
74	0.1308	0.1293	0.1278	0.1264	91	0.0410	0.0398	0.0386	0.0374
75	0.1249	0.1235	0.1221	0.1206	92	0.0362	0.0351	0.0339	0.0327
76	0.1192	0.1177	0.1163	0.1149	93	0.0315	0.0304	0.0292	0.0281
77	0.1135	0.1121	0.1107	0.1093	94	0.0269	0.0257	0.0246	0.0235
78	0.1079	0.1065	0.1051	0.1037	95	0.0223	0.0212	0.0200	0.0188
79	0.1024	0.1010	0.0096	0.0982	96	0.0177	0.0166	0.0155	0.0144
80	0.0969	0.0955	0.0942	0.0928	97	0.0132	0.0121	0.0110	0.0099
81	0.0915	0.0901	0.0888	0.0875	98	0.0088	0.0077	0.0066	0.0055
82	0.0862	0.0848	0.0835	0.0822	99	0.0044	0.0033	0.0022	0.0011
83	0.0809	0.0796	0.0783	0.0770	100	0.0000	0.0000	0.0000	0.0000

Appendix VI

Numerical Values of the Seidel Function

%T	$(d_0/d) - 1$	%T	$(d_0/d) - 1$	%T	$(d_0/d) - 1$
97	0.031	65	0.538	33	2.030
96	0.042	64	0.563	32	
95	0.053	63	0.587	31	
94	0.064	62	0.613	30	2.333
93	0.072	61	0.639	29	2.449
92	0.087	60	0.667	28	2.571
91	0.099	59	0.695	27	2.704
90	0.111	58	0.724	26	2.846
89	0.124	57	0.753	25	3.000
88	0.136	56	0.786	24	3.167
87	0.149	55	0.818	23	3.348
86	0.163	54	0.852	22	3.545
85	0.176	53	0.887	21	3.762
84	0.190	52	0.923	20	4.000
83	0.205	51	0.961	19	4.263
82	0.220	50	1.000	18	4.556
81	0.235	49	1.041	17	4.882
80	0.250	48	1.083	16	5.250
79	0.266	47	1.128	15	5.667
78	0.282	46	1.174	14	6.143
77	0.299	45	1.222	13	6.692
76	0.316	44	1.273	12	7.333
75	0.333	43	1.326	11	8.091
74	0.351	42	1.381	10	9.000
73	0.370	41	1.439	9	10.111
72	0.389	40	1.500	8	11.500
71	0.408	39	1.564	7	13.286
70	0.429	38	1.632	6	15.667
69	0.449	37	1.703	5	19.000
68	0.471	36	1.778	4	24.000
67	0.493	35	1.857	3	32.333
66	0.515	34	1.941	2	49.000

Appendix VII

Four-Place Logarithm Table

	0	1	2	3	4	5	6	7	8	9
10	0000	0043	0086	0128	0170	0212	0253	0294	0334	0374
11	0414	0453	0492	0531	0569	0607	0645	0682	0719	0755
12	0792	0828	0864	0899	0934	0969	1004	1038	1072	1106
13	1139	1173	1206	1239	1271	1303	1335	1367	1399	1430
14	1461	1492	1523	1533	1584	1614	1644	1673	1703	1732
15	1761	1790	1818	1847	1875	1903	1931	1959	1987	2014
16	2041	2068	2095	2122	2148	2175	2201	2227	2253	2279
17	2304	2330	2355	2380	2405	2430	2455	2480	2504	2529
18	2553	2577	2601	2625	2648	2672	2695	2718	2742	2765
19	2788	2810	2833	2856	2878	2900	2923	2945	2967	2989
20	3010	3032	3054	3075	3096	3118	3139	3160	3181	3201
21	3222	3243	3363	3284	3304	3324	3345	3365	3385	3404
22	3424	3444	3464	3483	3502	3522	3541	3560	3579	3598
23	3617	3636	3655	3674	3692	3711	3729	3747	3766	3784
24	3802	3820	3838	3856	3874	3892	3909	3927	3945	3962
25	3979	3997	4014	4031	4048	4065	4082	4099	4116	4133
26	4150	4166	4183	3200	4216	4232	4249	4265	4281	4298
27	4314	4330	4346	4362	4378	4393	4409	4425	4440	4456
28	4472	4487	4502	4518	4533	4548	4564	4579	4594	4609
29	4624	4639	4654	4669	4683	4698	4713	4728	4742	4757
30	4771	4786	4800	4814	4829	4843	4857	4871	4886	4900
31	4914	4928	4942	4955	4969	4983	4997	5011	5024	5038
32	5051	5065	5079	5092	5105	5119	5132	5145	5159	5173
33	5185	5198	5211	5224	5237	5250	5263	5276	5289	5302
34	5315	5328	5340	5353	5366	5378	5391	5403	5416	5428
35	5441	5453	5465	5478	5490	5502	5514	5527	5539	5551
36	5563	5575	5587	5599	5611	5623	5635	5647	5658	5670
37	5682	5694	5705	5717	5729	5740	5752	5763	5775	5786
38	5798	5809	5821	5832	5843	5855	5896	5877	5888	5899
39	5911	5922	5933	5944	5955	5966	5977	5988	5999	6010

APPENDIX VII

	0	1	2	3	4	5	6	7	8	9
40	6021	6031	6042	6053	6064	6075	6085	6096	6107	6117
41	6128	6138	6149	6160	6170	6180	6191	6201	6212	6222
42	6232	6243	6253	6263	6274	6284	6294	6304	6314	6325
43	6335	6345	6355	6365	6375	6385	6395	6405	6415	6425
44	6435	6444	6454	6464	6474	6484	6493	6503	6513	6522
45	6532	6542	6551	6561	6571	6580	6590	6599	6609	6618
46	6628	6637	6646	6656	6665	6675	6684	6693	6702	6712
47	6721	6730	6739	6749	6758	6767	6776	6785	6794	6803
48	6812	6821	6830	6839	6848	6857	6866	6875	6886	6983
49	6902	6911	6920	6928	6937	6946	6955	6964	6972	6981
50	6990	6998	7007	7016	7024	7033	7042	7050	7059	7067
51	7076	7084	7093	7101	7110	7118	7126	7135	7143	7152
52	7160	7168	7177	7185	7193	7202	7201	7218	7226	7235
53	7243	7251	7259	7267	7275	7284	7292	7300	7308	7316
54	7324	7332	7340	7348	7356	7364	7372	7380	7388	7396
55	7404	7413	7419	7427	7435	7443	7451	7459	7466	7474
56	7482	7490	7497	7503	7313	7520	7528	7536	7543	7551
57	7559	7566	7574	7582	7589	7597	7604	7612	7619	7627
58	7634	7642	7649	7657	7664	7672	7679	7686	7694	7701
59	7709	7716	7723	7731	7738	7745	7752	7760	7767	7774
60	7782	7789	7796	7803	7810	7818	7825	7832	7839	7846
61	7853	7860	7868	7875	7882	7889	7896	7903	7910	7917
62	7924	7931	7938	7945	7952	7959	7966	7973	7980	7987
63	7993	8000	8007	8014	8021	8028	8035	8041	8048	8055
64	8062	8069	8075	8082	8089	8096	8102	8109	8116	8122
65	8129	8136	8142	8149	8156	8162	8169	8176	8182	8189
66	8195	8202	8209	8215	8222	8228	8235	8241	8248	8254
67	8261	8267	8274	8280	8287	8293	8290	8306	8312	8319
68	8325	8331	8338	8344	8451	8457	8363	8370	8376	8382
69	8388	8395	8401	8407	8414	8420	8426	8432	8439	8445
70	8451	8457	8463	8470	8476	8482	8488	8494	8500	8506
71	8513	8519	8525	8531	8537	8543	8549	8555	8561	8567
72	8573	8579	8585	8591	8597	8603	8609	8615	8621	8627
73	8633	8639	8645	8651	8657	8663	8669	8675	8681	8686
74	8692	8698	8704	8710	8716	8722	8727	8733	8739	8745
75	8751	8756	8762	8768	8774	8779	8785	8791	8797	8802
76	8808	8814	8820	8825	8831	8837	8842	8848	8854	8859
77	8865	8871	8876	8882	8887	8893	8899	8904	8910	8915
78	8921	8926	8932	8938	8943	8949	8954	8960	8965	8971
79	8976	8982	8987	8993	8998	9004	9009	9015	9020	9025
80	9031	9036	9042	9047	9053	9058	9063	9069	9074	9079
81	9085	9090	9096	9101	9106	9112	9117	9122	9128	9133
82	9138	9143	9149	9154	9159	9165	9170	9175	9180	9186
83	9191	9196	9201	9206	9212	9217	9222	9227	9232	9238
84	9243	9248	9253	9258	9263	9239	9274	8279	9284	9289

APPENDIX VII

	0	1	2	3	4	5	6	7	8	9
85	9294	9299	9304	9309	9315	9320	9325	9330	9335	9340
86	9345	9350	9355	9360	9365	9370	9375	9380	9385	9390
87	9395	9400	9405	9410	9415	9420	9425	9430	9435	9440
88	9445	9450	9455	9460	9465	9469	9475	9479	9484	9489
89	9494	9499	9504	9509	9513	9518	9523	9528	9533	9538
90	9542	9547	9552	9557	9562	9566	9571	9576	9581	9586
91	9590	9595	9600	9605	9609	9614	9619	9624	9628	9633
92	9638	9643	9647	9652	9657	9661	9666	9671	9675	9680
93	9685	9689	9694	9699	9703	9708	9713	9717	9722	9727
94	9731	9736	9741	9745	9750	9754	9759	9763	9768	9773
95	9777	9782	9786	9791	9795	9800	9805	9809	9814	9818
96	9823	9827	9832	9836	9841	9845	9850	9854	9859	9863
97	9868	9872	9877	9881	9886	9890	9894	9899	9903	9908
98	9912	9917	9921	9926	9930	9934	9939	9943	9948	9952
99	9956	9961	9965	9969	9974	9978	9983	9987	9991	9996

Appendix VIII

Detection Limits by Flame Emission and Atomic Absorption[1,2]

Element	Wavelength, Å	Flame emission		Atomic absorption	
		$N_2O-C_2H_2$	$Air-C_2H_2$	$N_2O-C_2H_2$	$Air-C_2H_2$
Ag	3280.7	0.02	—	—	0.005
Al	3961.5	0.005	—	—	—
	3092.8	—	—	0.1	—
As	1937.0	—	—	—	0.2
Au	2767.0	0.5	—	—	—
	2428.0	—	—	—	0.02
B	2496.8	—	—	6	—
Ba	5535.5	0.001	—	0.05	—
Be	4708.6 (BeO)	0.02	—	—	—
	2348.6	—	—	0.002	—
Bi	2230.6	—	—	—	0.05
Ca	4226.7	0.0001	0.005	—	0.002
Cd	3261.1	2	—	—	—
	2288.0	—	—	—	0.005
Co	3453.5	0.05	—	—	—
	2407.2	—	—	—	0.005
Cr	4254.4	0.005	—	—	—
	3578.7	—	—	—	0.005
Cu	3274.0	0.01	—	—	—
	3247.5	—	—	—	0.005

[1] From E. E. Pickett and S. R. Koirtyohann, Emission Flame Photometry—A New Look at an Old Method, *Anal. Chem.,* **41**, 28A (1969). Used by permission of the American Chemical Society.
[2] In µg/ml in aqueous solution, measured in nitrous oxide–acetylene or air–acetylene premixed flames formed by 5–10-cm slot burners.

| | | Flame emission | | Atomic absorption | |
Element	Wavelength, Å	N_2O–C_2H_2	Air–C_2H_2	N_2O–C_2H_2	Air–C_2H_2
Dy	4046.0	0.07	—	—	—
	4211.7	—	—	0.2	—
Er	4008.0	0.04	—	0.1	—
Eu	4594.0	0.006	—	0.008	—
Fe	3719.9	0.005	—	—	—
	2483.3	—	—	—	0.005
Ga	4033.0	0.01	—	—	—
Ga	2874.2	—	—	—	0.07
Gd	4401.9	2	—	—	—
	3684.1	—	—	4	—
Ge	2651.2	0.5	—	1	—
Hg	2536.5	—	—	—	0.5
Ho	4053.9	0.02	—	—	—
	4103.8	—	—	0.1	—
In	4511.3	0.002	—	—	—
	3039.4	—	—	—	0.05
Ir	3800.1	30	—	—	—
	2639.7	—	—	—	2
K	7664.9	—	0.0005	—	0.005
La	5501.3	8	—	—	—
	4418.2 (LaO)	0.1	—	—	—
	3927.6	—	—	8	—
Li	6707.8	0.00003	—	—	0.005
Lu	4518.6	1	—	—	—
	3312.1	—	—	50	—
Mg	2852.1	0.005	—	—	0.0003
Mn	4030.8	0.005	—	—	—
	2794.8	—	—	—	0.002
Mo	3903.0	0.1	—	—	—
	3132.6	—	—	—	0.03
Na	5890.0	—	0.0005	—	0.002
Nb	4058.9	1	—	3	—
Nd	2924.5	0.2	—	—	—
	4634.2	—	—	2	—
Ni	3414.8	0.3	—	—	—
	2320.0	—	—	—	0.005
Pb	4057.8	0.2	—	—	—
	2833.1	—	—	—	0.03
Pd	3634.7	0.05	—	—	—
	2476.4	—	—	—	0.02
Pr	4951.4	1	—	10	—
Pt	2659.5	2	—	—	0.1
Rb	7800.2	—	0.001	—	0.005
Re	3460.5	0.2	—	1.5	—
Rh	3692.4	0.02	—	—	—
	3434.9	—	—	—	0.03

APPENDIX VIII

Element	Wavelength, Å	Flame emission		Atomic absorption	
		$N_2O-C_2H_2$	$Air-C_2H_2$	$N_2O-C_2H_2$	$Air-C_2H_2$
Ru	3728.0	0.02	—	—	—
	3498.9	—	—	—	0.3
Sb	2175.8	—	—	—	0.1
Sc	4020.4	0.03	—	—	—
	3911.8	—	—	0.1	—
Se	1960.3	—	—	—	0.5
Si	2516.1	—	—	0.1	—
Sm	4760.3	0.2	—	—	—
	4296.7	—	2	—	—
Sn	2840.0	0.5	—	—	—
	2246.0	—	—	—	0.06
Sr	4607.3	0.0001	—	—	0.01
Ta	4812.8	5	—	—	—
	2714.7	—	—	5	—
Tb	4318.9	0.4	—	—	—
	4326.5	—	—	2	—
Te	2142.8	—	—	—	0.3
Ti	3998.6	0.2	—	—	—
	3642.7	—	—	0.1	—
Tl	5350.5	0.2	—	—	—
	2767.9	—	—	0.2	—
Tm	3717.9	0.02	—	—	—
	4105.8	—	—	1	—
V	3184.0	—	—	0.02	—
W	4008.8	0.5	—	3	—
Y	3620.9	0.4	—	—	—
	4077.4	—	—	0.3	—
Yb	3988.0	0.002	—	0.04	—
Zn	2138.6	—	—	—	0.002
Zr	3601.2	3	—	5	—

Appendix IX

Periodic Table of the Elements

IA																	0
1 H 1.0	IIA											IIIA	IVA	VA	VIA	VIIA	2 He 4.0
3 Li 6.9	4 Be 9.0											5 B 10.8	6 C 12.0	7 N 14.0	8 O 16.0	9 F 19.0	10 Ne 20.2
11 Na 23.0	12 Mg 24.3	IIIB	IVB	VB	VIB	VIIB	←—	VIIIB	—→	IB	IIB	13 Al 27.0	14 Si 28.1	15 P 31.0	16 S 32.1	17 Cl 35.5	18 Ar 40.0
19 K 39.1	20 Ca 40.1	21 Sc 45.0	22 Ti 47.9	23 V 50.9	24 Cr 52.0	25 Mn 54.9	26 Fe 55.8	27 Co 58.9	28 Ni 58.7	29 Cu 63.5	30 Zn 65.3	31 Ga 69.7	32 Ge 72.6	33 As 74.9	34 Se 79.0	35 Br 79.9	36 Kr 83.8
37 Rb 85.5	38 Sr 87.6	39 Y 88.9	40 Zr 91.2	41 Nb 92.9	42 Mo 95.9	43 Tc (99)	44 Ru 101.1	45 Rh 102.9	46 Pd 106.4	47 Ag 107.9	48 Cd 112.4	49 In 114.8	50 Sn 118.7	51 Sb 121.8	52 Te 127.6	53 I 126.9	54 Xe 131.3
55 Cs 132.9	56 Ba 137.3	71 *Lu 175.0	72 Hf 178.5	73 Ta 180.9	74 W 183.9	75 Re 186.2	76 Os 190.2	77 Ir 192.2	78 Pt 195.1	79 Au 197.0	80 Hg 200.6	81 Tl 204.4	82 Pb 207.2	83 Bi 209.0	84 Po (210)	85 At (210)	86 Rn (222)
87 Fr (223)	88 Ra (226)	103 †Lr (257)	104 (260)														

	57 *La 138.9	58 Ce 140.1	59 Pr 140.9	60 Nd 144.2	61 Pm (145)	62 Sm 150.4	63 Eu 152.0	64 Gd 157.3	65 Tb 158.9	66 Dy 162.5	67 Ho 164.5	68 Er 167.3	69 Tm 168.9	70 Yb 173.0
	89 †Ac (227)	90 Th 232.0	91 Pa (231)	92 U 238.0	93 Np (237)	94 Pu (242)	95 Am (243)	96 Cm (247)	97 Bk (247)	98 Cf (251)	99 Es (254)	100 Fm (257)	101 Md (256)	102 No (256)

Appendix X
Relative Atomic Weights[1]

Element	Symbol	Atomic number	Atomic weight	Element	Symbol	Atomic number	Atomic weight
Actinium	Ac	89	[227]	Europium	Eu	63	151.96
Aluminum	Al	13	26.98154	Fermium	Fm	100	[257]
Americum	Am	95	[243]	Fluorine	F	9	18.99840
Antimony	Sb	51	121.7_5	Francium	Fr	87	[223]
Argon	Ar	18	39.94_8*	Gadolinium	Gd	64	157.2_5
Arsenic	As	33	74.9216	Gallium	Ga	31	69.72
Astatine	At	85	[210]	Germanium	Ge	32	72.5_9
Barium	Ba	56	137.3_4	Gold	Au	79	196.9665
Berkelium	Bk	97	[247]	Hafnium	Hf	72	178.4_9
Beryllium	Be	4	9.01218	Helium	He	2	4.00260
Bismuth	Bi	83	208.9804	Holmium	Ho	67	164.9340
Boron	B	5	10.81*	Hydrogen	H	1	1.0079*
Bromine	Br	35	79.904	Indium	In	49	114.82
Cadmium	Cd	48	112.40	Iodine	I	53	126.9045
Calcium	Ca	20	40.08	Iridium	Ir	77	192.2_2
Californium	Cf	98	[251]	Iron	Fe	26	55.84_7
Carbon	C	6	12.011*	Krypton	Kr	36	83.80
Cerium	Ce	58	140.12	Lanthanum	La	57	138.905_5
Cesium	Cs	55	132.9055	Lawrencium	Lr	103	[256]
Chlorine	Cl	17	35.453	Lead	Pb	82	207.2*
Chromium	Cr	24	51.996	Lithium	Li	3	6.94_1*
Cobalt	Co	27	58.9332	Lutetium	Lu	71	174.97
Copper	Cu	29	63.54_6*	Magnesium	Mg	12	24.305
Curium	Cm	96	[247]	Manganese	Mn	25	54.9380
Dysprosium	Dy	66	162.5_0	Mendelevium	Md	101	[258]
Einsteinium	Es	99	[254]	Mercury	Hg	80	200.5_9
Erbium	Er	68	167.2_6	Molybdenum	Mo	42	95.9_4

[1] Carbon-12 = 12. Atomic weights listed are 1971 IUPAC values. Value in brackets denotes the mass number of the isotope of longest known half-life (or a better known one for Pu, Po, Pm, and Tc). An asterisk denotes that the atomic weight varies because of natural variation in isotopic composition.

Element	Symbol	Atomic number	Atomic weight	Element	Symbol	Atomic number	Atomic weight
Neodymium	Nd	60	144.2_4	Scandium	Sc	21	44.9559
Neon	Ne	10	20.17_9	Selenium	Se	34	78.9_6
Neptunium	Np	93	237.0482	Silicon	Si	14	28.08_6*
Nickel	Ni	28	58.7_1	Silver	Ag	47	107.868
Niobium	Nb	41	92.9064	Sodium	Na	11	22.98977
Nitrogen	N	7	14.0067	Strontium	Sr	38	87.62
Nobelium	No	102	[255]	Sulfur	S	16	32.06*
Osmium	Os	76	190.2	Tantalum	Ta	73	180.947_9
Oxygen	O	8	15.999_4*	Technetium	Tc	43	[99]
Palladium	Pd	46	106.4	Tellurium	Te	52	127.60
Phosphorus	P	15	30.97376	Terbium	Tb	65	158.9254
Platinum	Pt	78	195.0_9	Thallium	Ti	81	204.3_7
Plutonium	Pu	94	[242]	Thorium	Th	90	232.0381
Polonium	Po	84	[210]	Thulium	Tm	69	168.9342
Potassium	K	19	39.09_8	Tin	Sn	50	118.6_9
Praseodymium	Pr	59	140.9077	Titanium	Ti	22	47.9_0
Promethium	Pm	61	[147]	Tungsten	W	74	183.8_5
Protactinium	Pa	91	231.0359	Uranium	U	92	238.029
Radium	Ra	88	226.0254	Vanadium	V	23	50.941_4
Radon	Rn	86	[222]	Xenon	Xe	54	131.30
Rhenium	Re	75	186.2	Ytterbium	Yb	70	173.0_4
Rhodium	Rh	45	102.9055	Yttrium	Y	39	88.9059
Rubidium	Rb	37	85.467_8	Zinc	Zn	30	65.38
Ruthenium	Ru	44	101.0_7	Zirconium	Zr	40	91.22
Samarium	Sm	62	150.4				

Author Index

Ahrens, L. H., 156, 157, 181, 207
Aldous, K. M., 235, 298
Alkemade, C. T. J., 8, 9
Amos, M. D., 268
Ångstrom, A. F., 3

Balmer, J. J., 5, 14, 15
Bardocz, A., 204
Barnes, R. B., 7
Berry, J. W., 7
Bohman, H. R., 255, 256
Bohr, N., 5, 13, 17, 18, 19, 21, 25
Boling, E. A., 263
Boltzmann, L., 34, 220
Born, M., 25
Bozman, W. R., 36
Brackett, C. B., 16
Bratzel, M. P., 308, 311, 315
Brode, W., 27, 32, 155
Brooksbank, P., 280
Browner, R. F., 256, 305
Buell, B. E., 215
Bunsen, R. W., 4, 7, 13, 14
Busch, K. W., 146, 235, 236

Cave, W. T., 201
Champion, P., 6
Churchill, J. R., 182, 183
Clements, H. E., 173, 175
Corliss, C. H., 36, 156
Crosswhite, H. M., 181, 182
Czerny, M., 81, 92

Dagnall, R. M., 308, 311, 314, 315, 317
Davis, B. K., 203
Dean, J. A., 226, 236, 237, 238, 239, 240, 243, 295, 296
De Broglie, L., 19, 25
De Gramont, A., 169
Delves, H. T., 271
Demers, D. R., 310
Dickinson, G. W., 176
Dieke, G. H. 181
Dixon, D. E., 112
Doppler, C. J., 37, 38
Doty, M. E., 227

Eberhard, G., 130, 131, 137
Ebert, H., 78
Ellis, D. W., 310
Ells, V. R., 7
Elser, R. C., 314
Evens, F. M., 203
Evenson, K. M., 257, 258

Fassel, V. A., 111, 112, 176, 203
Fastie, W. G., 79
Fernandez, F. J., 271, 280
Feussner, O., 107
Fraser, L. M., 306
Fraunhofer, J., 3, 7, 60
Fred, M., 199

Gatehouse, B. M., 269

Gerlach, W., 170
Gibson, J. H., 111
Glenn, T. H., 139, 256, 305
Goto, H., 205
Goudden, P. D., 280
Goudschmidt, S., 29
Grenier, M., 6
Griggs, M. A., 7
Grotrian, W., 21, 24
Grove, E. L., 165, 203

Harrison, G. R., 66, 67, 156
Hart, L. P., 139
Hartley, W. N., 5, 169
Hartmann, L. M., 53, 54, 100, 101, 150, 159
Harvey, C. E., 157, 160, 161, 162, 163, 164, 173
Hatch, W. R., 279
Heisenberg, W., 25, 37
Held, A. M., 8, 270
Herschel, J. F., 3, 6
Herzberg, G., 22, 23, 30
Hoare, H. C., 261
Hobbs, R. S., 317
Horowitz, W., 208
Hood, R. L., 7
Howell, V. G., 235, 236
Howes, H. L., 9
Hwang, J. H., 274

Jackson, K. W., 235, 254, 298

Kayser, H., 5

AUTHOR INDEX

Kerber, J. D., 271, 272
Kirchhoff, G., 4, 7, 13, 14
Kirkbright, G. F., 317
Kniseley, R. V., 111, 112
Koirtyohann, S. R., 212, 246, 286
Kolb, A. C., 300

Langstroth, G. O., 174
Lehman, D. A., 267
Lichte, F. E., 111
Lockyer, R., 5
Loofbourow, J. R., 156
Lord, R. C., 156
Loseke, W. A., 203
Lundegardh, H., 6, 211
L'vov, B. V., 8, 269, 270
Lyman, T., 15, 65

Manning, D. C., 280
Mansfield, J. M., 9
Marshall, C. E., 7
Mascart, E., 5
Massman, H., 275, 308
Matourek, J., 315
Matz, G. J., 68, 69
McRae, D. R., 174
Meggers, W. F., 149, 156, 254
Michelson, A. A., 5, 59, 67
Milatz, J. M. W., 8
Mitchell, A. C. G., 301
Mitchell, D. G., 235, 298
Morrison, G. H., 146, 235, 236
Mostyn, R. A., 261

Namioka, T., 90, 92
Neufeld, L., 33, 267
Newland, B. T. N., 261
Newton, Isaac, 2
Nichols, E. L., 9

Omenetto, N., 304, 305
Ott, W. L., 279

Parrish, D. B., 224
Parsons, M. L., 9
Paschen, F., 5, 16, 85, 86
Patel, B. M., 256, 305
Pellet, H., 6

Pfund, A. H., 16
Pickett, E. E., 212, 246, 286
Planck, M., 13, 16, 17, 25, 35, 42
Plankey, F. W., 139
Pyke, R. E., 224

Rains, T. C., 226, 236, 237, 238, 239, 240, 241, 295, 296
Rayleigh, Lord, 55, 56, 57, 63
Raziunas, V., 203
Richardson, D., 7
Rietta, M. E., 256, 305
Ritz, W., 5, 16, 18
Rodden, C. J., 199
Rossi, G., 304, 305
Rowland, H. A., 5, 65, 66, 84, 85, 86, 87, 88
Rudolph, J. J., 165
Runge, C., 5, 85, 86
Russell, B. J., 8
Rutherford, E. M., 17
Rydberg, J. R., 5, 20, 28

Saha, M. N., 225
Schrenk, W. G., 173, 175, 224, 226, 236, 237, 238, 239, 240, 243, 255, 256, 267, 295, 296
Schrödinger, E., 6, 25, 26
Schumann, V., 5
Scott, R. H., 112
Scribner, B., 156, 199
Seeley, J. L., 203
Seidel, C., 187, 188
Seya, M., 90
Shankoff, A., 67
Shellenberger, T. E., 224
Shelton, J. P., 8
Skogerboe, R. K., 111, 203
Smith, D. E., 250
Smith, S. B., Jr., 274
Snell, W., 2, 53
Somerfeld, A., 6, 21, 23, 25, 26
Staab, R. A., 299, 307, 313
Stallwood, B. J., 116, 200

Stanley, R. W., 149
Stone, R. W., 8, 270
Streed, E. R., 300
Strock, L. W., 164
Sullivan, J. V., 8, 251, 252, 282, 304
Syrdra, V., 315

Talbot, W. H. F., 3, 6
Taylor, H. E., 111
Taylor, M. R. G., 314
Taylor, S. R., 156, 157, 181, 207
Thompson, K. D., 308
Turner, A. F., 81, 92

Uhlenbeck, G. E., 29
Ullucci, P. A., 274

Valente, S. E., 255, 256
Veillon, C., 9
Vickers, T. J., 9, 299, 301

Wadsworth, F. L. O., 88, 89
Walsh, A., 8, 144, 243, 249, 251, 252, 268, 269, 282, 304
Walters, J. P., 204, 205
Wang, M. S., 165, 201, 264, 265
Warr, P. D., 317
West, T. S., 9, 273, 308, 314, 317
Winefordner, J. D., 9, 139, 256, 299, 301, 305, 306, 307, 308, 310, 313, 315, 316, 317
Williams, X. K., 273, 308
Willis, J. B., 8, 267
Wollaston, W. H., 3, 7
Wood, R. W., 9, 67
Wood, T. S., 317
Woodriff, R., 8, 270, 278
Woodson, T. T., 8

Young, T., 3

Zeeman, P., 26, 31, 33
Zemansky, M. W., 301

Subject Index

Absorbance values, table of, 353-354
Absorption spectroscopy, discovery of, 8
Amplifiers
  broadband alternating current, 283
  direct current, 283
  lock-in, 283, 284
Arc and spark stand, 114-116
Atomic absorption
  detectors for, 281
    photomultipliers, 281-282
    resonance, 282
    vidicon, 298
  flame cells for, 259-264
  fuels and oxidants for, 264-267
  instrumentation requirements, 247
  monochromators for, 280
  multielement analysis, 297
  nebulization of sample, 259, 261
    ultrasonic nebulizers, 261
  nonflame cells for, 268-276
    carbon rod, 272
    Delves cup, 271
    hollow cathode, 269
    L'vov furnace, 269
    Massman cell, 275
    tantalum boat, 273
    Woodriff furnace, 270
  radiation sources for
    continuous sources, 258
    electrodeless discharge lamps, 254
      coupling RF energy to, 255
      microwave cavity, 257
      type A antenna, 257
    gaseous discharge lamps, 253
      Osram, 253

Atomic absorption (*cont'd*)
  radiation sources for (*cont'd*)
    gaseous discharge lamps (*cont'd*)
      Philips, 253
    hollow cathode, 248-253
      demountable, 253
      high-intensity, 251
      multiple element, 251
  read-out devices, 284-286
    chart recorders, 284
    digital devices, 285
    signal integraters, 285
  sensitivity of, 246
  special analytical techniques for, 278-280
    antimony, 279
    arsenic, 279
    mercury, 278
    selenium, 279
  treatment of analytical data, 294-297
  working curve, 294
Atomic fluorescence, 9, 299-317
  analytical procedures, 312-317
    applications, 317
    detection limits, 315
    sample preparation, 315
    working curve, 313
  excitation sources, 304-307
    continuous sources, 305
    electrodeless discharge lamps, 305
    hollow cathode lamps, 304
    lasers, 306
    metal vapor lamps
      Osram, 304
      Philips, 304
  history of, 9

SUBJECT INDEX

Atomic fluorescence (*cont'd*)
  instrumentation for, 303
  monochromators for, 308
  sample cell for,
    laminar flow burners, 307-308
    nonflame cells, 308
    total consumption burners, 307
  theoretical basis of, 293-303
  treatment of analytical data, 294-297
  working curve, 294
Atomic spectra, origin of, 5, 11-15

Balmer equation, 5, 14-15
Basic definitions and units, 319-320
Bohr atom, 5, 16-21
Boling burner, 263
Boltzmann constant, 34
Buffer, spectroscopic, 174-175
Burners
  laminar flow, 217-219, 262-264, 307-308
  total consumption, 216, 307

Cauchy constants, 52
Chemiluminescence, 226-227
Comparators, 118-121

Densitometers, 118-121
Detection limits of elements, table
  atomic absorption, 361-363
  flame emission, 361-363
Dispersion, reciprocal linear, 63

Einstein coefficient, 35-36
Electrodeless discharge lamps, 9, 254-257, 305
  microwave-excited, 9, 255
Electron spin, 29
Electrodes, 122-123
  carbon, 122
  graphite, 122
  self, 122
  shapes and sizes, 122-123
Elliptical orbits, 21
Energy level diagrams,
  Grotrian, 21-23, 27, 30
Excitation sources, 104-114
  arc
    alternating current, 105-106
    direct current, 104-105
      rectifier circuit for, 105
    plasma, 108
      dc-excited, 108-110
      radiofrequency excited, 110-111

Excitation sources (*cont'd*)
  flame, 212-219
  laser, 112-113
  multisource units, 113-114
  spark, 106-108

Filters, light, 47-52
  absorption, 47
  circular variable, 51
  interference, 49
  interference wedge, 50
Flame emission
  aspirators for, 216-218
  background radiation, 233
  burners for, 214-218
    laminar flow, 217-218
    total consumption, 216
  effect of temperature on excitation, 220
  excitation process, 219
  fuel-oxidant control in, 218-219
  historical development, 6
  instrumentation requirements, 212
  multielement analysis in, 235-236
  nebulization of sample, 230
  surface tension, effects of, 231
  treatment of analytical data, 236-242
    background correction, 238
    sample bracketing, 239-240
    standard addition, 241
    working curves, 237
  viscosity, effects of, 231
Flames
  oxidizing region, 215
  reducing region, 215
  temperatures of, 214-215
Fourier transform spectroscopy, 139
Fraunhofer lines, sun's spectrum, 3

Gratings, diffraction
  blaze of, 65
  concave, 66
  dispersion of, 63
  echelle, 67-69
  ghosts in, 65
  Lyman, 65
  Rowland, 65
  groove spacing, 65-66
    interferometric control of, 66
  holographic, 67
  invention of, 3
  plane, 60-63
  replicas of, 66

SUBJECT INDEX

Gratings, diffraction (cont'd)
  resolving power of, 63-64
  theory of, 61-62

Hadamard transform spectroscopy, 139
Hartmann diaphragm, 101
Hartmann formula, 53

Interferences in
  arc–spark emission,
    control of, 170-176
      buffers, 172
      internal standard, 170-172
      moving plate technique, 159
      matrix effects, 163
      spectral band, 157
      spectral line, 157
  atomic absorption
    chemical, 287
    control of, 289-294
      background, 292-293
        blank solutions, 292
        deuterium lamp, 293
        two-line compensation, 293
      chemical, 290-293
        chelating agents, 291
        chemical separations, 292
        flame temperature, 290
        fuel-to-oxidant ratio, 291
        releasing agents, 291
      ionization, 290
      spectral, 289
    flameless sampling, 288
    ionization, 286
    spectral, 285
  atomic fluorescence, 309-311
    chemical, 310
    physical, 310
    spectral, 309
  flame emission, 231-235
    cation–anion, 227
    cation–cation, 228
    chemiluminescence, 228-229
    control of, 231-235
      background radiation, 233
      ionization, 233
      oxide, 234-235
      physical, 235
      spectral, 231-232
    oxide formation, 228
    physical, 229-231

Interferometers, 59-60
  measurement of wavelength of light, 5

Kirchhoff's law, 7

Lasers, 39-41
  dye laser, 39
  gas laser, 41
  mechanism of action, 39
  tunable, 41
Lenses, 70-74
  chromatic aberration, 72
  collimating, 70-71
  coma in, 72-73
  condensing, 70-71
  image size, 70-72
  spherical aberration in, 72-73
Letzen linien principle, 169
Light, *see* Radiation, electromagnetic
Logarithm table, 357-359

Metastable atomic energy states, 39

Orbits, atomic
  elliptical, 6
  energy of, 19
Order sorters, 118

Periodic table of the elements, 365
Petry arc–spark stand, 117
Photographic emulsions, 126
  calibration curves, 185-186
    H and D curve, 185
    percent transmittance, 189
    Seidel function, 185-186
  calibration methods, 180-182
    iron lines, 182
    relative line intensities, 181
    step sector, 181
    stepped neutral filter, 181
    two-line method, 182-183
  contrast, 128
  Eberhard effect, 130-131
  gamma, 127, 128
  graininess, 131-132
  granularity, 132-133
  intermittency effect, 130
  reciprocity law, 129
  resolving power, 132-133
  solarization, 127
  spectral sensitivity, 133-135
Photographic plates and films, 135-139
  developing process, 135

Phototubes, light-sensitive
  characteristics of, 142-143
    dark current, 142
    fatigue, 143
    linearity of response, 143
    noise in, 143
    transit time, 142
  solar blind, 143-144
  spectral response of, 140-141
Planck equation, 16
Prisms
  Cornu, 58
  dispersion, 53
  glass, 59
  index of refraction, 52
  Littrow, 58-59
  materials for, 57
  resolving power, 55
    Rayleigh criteria for, 56

Quantum numbers, selection rules, 24

Radiation, electromagnetic
  amplitude, 12
  frequency, 12
  interference phenomena, 12
  phase, 12
  polarization of, 13
  reflections of, 12
  resonance detection of, 144-145
  refraction of, 2, 12, 52
  wavelength of, 3
Raies ultimes principle, 169
Reduced mass, 20
Ritz combination principle, 18
Rotating disk electrode, 117
Rydberg constant, 20
Rydberg corrections, 28

Saha equation, 225
Schrödinger equation, 25
Sectors, rotating
  logarithmic, 103-104
  step, 103
Seidel function, numerical values of, 355
Snell's law, 2, 52-53
Spectra, molecular, 41-44
  Doppler broadening, 44
  collisional broadening, 44
Spectral lines
  charts of, 347-352

Spectral lines (*cont'd*)
  fluorescence, types of
    direct line, 38
    resonance, 38
    sensitized, 38
    stepwise, 38
  gf values of, 36
  intensity of
    Boltzmann factor, 34
    statistical weight, 33
  multiplicity of, 29
  oscillator strengths of, 35
  series (Balmer, Brackett, Lyman, Pfund, Ritz–Paschen), 15
  tables of,
    by element, 337-345
    by wavelength, 321-335
  transition probability, 35
  widths
    Doppler effect, 37
    natural width, 37
    pressure effect, 37
    Stark effect, 37
Spectrochemical analysis
  applications of
    alloys, 206
    animals, 209
    geology, 207
    men, 209
    metals, 206
    oil, 207
    plants, 208
    soils, 208
    water, 207
  qualitative, 147-159
    excitation of sample, 148
    interferences to
      by continuum, 158
      by spectral bands, 157-158
    line identification, 149-157
      by comparison spectra, 151-152
      by spectral charts, 155-156
      by wavelength measurements, 149-150
      by wavelength tables, 156-157
  quantitative, 169-210
    background correction in, 193
    buffers for, 172-175
    calculating board for, 192
    carrier distillation techniques, 199-200
    cathode layer excitation, 201

SUBJECT INDEX 375

Spectrochemical analysis *(cont'd)*
  quantitative *(cont'd)*
    comparison standards for, 177-178
      purity of chemicals, 177-178
    controlled atmospheres for, 200
    direct read-out techniques, 194-195
    excitation of the sample, 175-176
    fractional distillation of solid samples, 198-199
    internal standards for, 170-172
    laser volatilization of samples, 200-201
    matrix effects in, 174-175
    sample preparation, 179-181, 205-206
      liquid samples, 179
      powder samples, 179-180
      self-electrodes, 179
    selection of spectral lines for, 177
    types of samples, 180, 195-198, 202-203
      biological, 180
      gases, 202
      liquid, 180, 195
      metallic, 196-197
      nonconducting, 180
      organic, 180, 197
      powder, 197
      radioactive, 202-203
  semiquantitative, 159-166
    determination of a concentration level, 160
    Harvey method, 160-165
      detection limits, 162
      electrodes for, 161
      matrix effects in, 164-165
    Wang method, 165-166
      buffer for, 165
      standards for, 165
    microsamples, 166
Spectrometers
  adjustment of, 94
    entrance slit, 94-95

Spectrometers *(cont'd)*
  adjustment of *(cont'd)*
    tilt, 59
  care of, 94-96
  concave grating instruments, 83
    direct reading, 92-93
    Eagle, 86
    grazing incidence, 89, 96
    Paschen–Runge, 85
    Rowland, 84
    Seya–Namioka, 90-91
    Wadsworth, 88-89
  plane grating instruments
    crossed-beam, 82
    Czerny–Turner, 81
    double-grating, 82-83
    Ebert, 78
    Ebert–Fastie, 79
  prism instruments
    Cornu, 76
    Littrow, 76, 77
  selection of, 93-94
  slit, 99-100
  vacuum instruments, 90-92
Spectroscopic term symbols, 30
Spectroscopy, history of, 1-10
Stallwood jet, 116
Stark effect, 26, 31, 33,
Step filter, 101-102

Time-resolved spectroscopy, 203-205
  analytical applications, 205
  characteristics of, 204-205
  components for, 203-204

Vidicon detectors of radiation, 145-146

Wang burner head, 264
Wavelength of solar spectrum, 4
Wavenumber, definition, 20

Zeeman effect, 26, 31, 33